Nutrition and Care of Livestock During Natural Disaster

THIS BOOK IS DEDICATED TO

CAPT SASWATA DAS

(1983-2009)

He laid down his life in the service of the country in the best tradition of the Army.

- Commanding Officer, 28 Inf. Div. Sig. Regt.

Nutrition and Care of Livestock During Natural Disaster

Editor

N. Das

Plant Animal Relationship Division
Indian Grassland and Fodder Research Institute,
Jhansi - 284 003 (Uttar Pradesh)

2011

Studium Press (India) Pvt. Ltd.

Nutrition and Care of Livestock During Natural Disaster

ISBN: 978-93-80012-25-4

Published by:

Studium Press (India) Pvt. Ltd.
4735/22, 2nd Floor, Prakash Deep Building
(Near Delhi Medical Association),
Ansari Road, Darya Ganj, New Delhi-110 002
Tel.: + 91-11-43240200-15 (15 lines); Fax: 91-11-43240200
jngovil@gmail.com; jngovil@hotmail.com

Printed at:
Salasar Imaging Systems
Delhi-110035 (India)

LIST OF CONTRIBUTORS

A.B. Mandal: Central Avian Research Institute, Bareilly, Izatnagar - 243 122 (U.P), E-mail: abmcari@rediffmail.com

A.K. Misra: *Plant Animal Relationship Division*, Indian Grassland and Fodder Research Institute, Jhansi - 284 003 (U.P.), E-mail: asimkmisra@gmail.com

Anil Kumar: *Plant Animal Relationship Division*, Indian Grassland and Fodder Research Institute, Jhansi - 284 003 (U.P.), E-mail: anil.igfri@gmail.com

B.P. Kushwaha: *Plant Animal Relationship Division*, Indian Grassland and Fodder Research Institute, Jhansi - 284 003 (U.P.), E-mail: bp_kush@yahoo.com

Bhupinder Singh: *Dairy Cattle Nutrition Division*, National Dairy Research Institute, Karnal - 132 001 (Haryana).

Bindu Madhuri: *Deptt. of Poultry Science*, College of Veterinary Science, Sree Venkteswar Veterinary University, Rajendra Nagar, Hyderabad (A.P.), E-mail: bindu_madhuri@yahoo.com

K.K. Singh: *Plant Animal Relationship Division*, Indian Grassland and Fodder Research Institute, Jhansi - 284003, (U.P.), E-mail: krisksingh@gmail.com

K.S. Das: Central Institute for Research on Buffaloes, Nava (Punjab), E-mail: kalyanlpm@gmail.com

K.K. Singhal: *Dairy Cattle Nutrition Division*, National Dairy Research Institute, Karnal - 132 001 (Haryana), E-mail: k_ksinghal@yahoo.co.in

M.M. Das: *Plant Animal Relationship Division*, Indian Grassland and Fodder Research Institute, Jhansi - 284 003 (U.P.), E-mail: mmdas1964@rediffmail.com

M.M. Roy: Central Arid Zone Research Institute, Jodhpur (Rajasthan), E-mail: mmroyster@gmail.com

N. Das: *Plant Animal Relationship Division*, Indian Grassland and Fodder Research Institute, Jhansi - 284 003 (U.P.), E-mail: ndas31@gmail.com

P. Saxena: *Crop Improvement Division*, Indian Grassland and Fodder Research Institute, Jhansi - 284 003 (U.P.), E-mail: pradeepsax@yahoo.com

P. Das: Indian Veterinary Research Institute, Bareilly, Izatnagar - 243 122 (U.P.), E-mail: pdas@ivri.up.nic.in

R.P. Singh: Central Avian Research Institute, Bareilly, Izatnagar - 243 122 (U.P.).

S.K. Nag: *Plant Animal Relationship Division*, Indian Grassland and Fodder Research Institute, Jhansi - 284 003 (U.P.), E-mail: nagsk_67@rediffmail.com

S.K. Das: *Crop Production Division*, Indian Grassland and Fodder Research Institute, Jhansi - 284 003 (U.P.), E-mail: skdas1964@gmail.com

S.B. Maity: *Plant Animal Relationship Division*, Indian Grassland and Fodder Research Institute, Jhansi - 284 003 (U.P.), E-mail: sb_maity@yahoo.in

S.K. Mahanta: *Plant Animal Relationship Division*, Indian Grassland and Fodder Research Institute, Jhansi - 284 003 (U.P.), E-mail: mahantask@rediffmail.com

S.K. Tomar: *Dairy Cattle Nutrition Division*, National Dairy Research Institute, Karnal - 132 001 (Haryana), E-mail: santoshktomar@gmail.com

S.M.K. Naqvi: Central Sheep and Wool Research Institute, Avikanagar - 304 501 (Rajasthan), E-mail: naqvismk@yahoo.co.in

Sharmila Roy: *Crop Improvement Division*, Indian Grassland and Fodder Research Institute, Jhansi - 284 003 (U.P.), E-mail: roysharmilaigfri@gmail.com

Sultan Singh: *Plant Animal Relationship Division*, Indian Grassland and Fodder Research Institute, Jhansi - 284 003 (U.P.), E-mail: singh.sultan@rediffmail.com

Sunil Kumar: National Research Center for Agro-forestry, Jhansi - 284 003 (U.P.), E-mail: sktiwari98@gmail.com

V.P. Maurya: Central Sheep and Wool Research Institute, Avikanagar - 304 501 (Rajasthan), E-mail: vijaipmaurya@yahoo.co.in

V. Sejian: Central Sheep and Wool Research Institute, Avikanagar - 304 501 (Rajasthan), E-mail: drsejian@gmail.com

PREFACE

"Building culture of prevention is not easy. While the cost of prevention had to be paid in the present, its benefits lie in the distant future. More over, the benefits are not tangible; they are disasters that did not happen." Kofi Annan, UN Secretary General, 1999.

A disaster (Greek meaning "Bad Star") is a natural or man-made event that negatively affects life, property, livelihood or industry often resulting in permanent changes to human societies, ecosystems and environment. Except volcanic activity, all types of disasters are common in India. Floods, droughts, cyclones, tsunami, earthquakes and landslide have been a recurrent phenomenon. Although there is no clear-cut definition of disaster which is acceptable to all, but the main ideas are that it is an event concentrated in time and space, which contributed a social, economical, cultural and political devastation and affects people and communities. Almost all the definitions of disaster are centered on the human suffering. Rarely there is any mention about the loss or injury or suffering of livestock although these animals contribute significantly to the livelihood of rural farmers. During natural calamities, attention usually goes primarily towards human welfare, however, welfare of animal is also of paramount importance considering their causalities from drought and flood prone diseases, epidemics and different feed poisoning.

Although the damage to livestock during these disasters may not be as big as it is for loss of other property and human lives but it should be noted that livestock in India are kept by small scale farmers and landless labourers whose livelihood are mainly dependent on the earning from livestock rearing. The impact of disasters on animals is also higher and long lasting in developing countries than in developed countries. For a countries like India which has substantial animal population and where the livelihood of majority of rural poor are dependent on the income from livestock rearing operations, disaster preparedness, management and recovery plans that include a livestock component are important and essential. A view that disaster plan especially for livestock is unaffordable in developing countries like India is unjustified and can be harmful in the long run not only to animals, but also to humans. A state of preparedness and ability to respond quickly to natural disasters will definitely mitigate the suffering and loss of life to a great extent.

The book "Nutrition and Care of Livestock during Natural Disaster" is a collection of 20 chapters comprising the various issues related to the livestock rearing during natural disasters. The first chapter deals with the definition of disaster and its various classifications as described by different agencies. All the relevant information related to various categories of livestock resources (cattle, buffalo, sheep, goat, equines, camel, avian etc) maintained in the natural calamities prone districts of India and their economic importance are documented in the subsequent two chapters. The fourth chapter deals with the various effects of earthquake on animal's life, their dwellings and other consequences. It covers both direct and indirect post-earthquake effect and management of animals required immediately after the disaster.

Statistics on feed and fodder availability (roughages, concentrates and green fodder) for feeding of livestock in disaster prone areas are mentioned in the chapter 5. Feeding strategies for different categories of livestock during natural calamities are discussed in this chapter. The role of various types of droughts on different types of damages to the plants as well as the physiological adaptations in plants to conserve water are described. The Chapter 7 has highlighted the available options to produce quality fodder from arable lands and grasslands, management of grasslands, browse and top feed resources and to suggest long as well as short term strategies to maintain forage supplies for livestock in the affected areas. The book also provides guidelines on the preparedness and strategies to counteract the feed scarcity during calamities by adopting suitable technologies like complete feed-block feed and fodder bank etc. Apart from bringing out how water is a vital but often overlooked nutrient for livestock, the chapter 9 draws upon its importance under different calamities like drought, earthquake etc when there is acute shortage of drinking water.

The animal care varies from disaster to disaster; in cyclones, high winds and hurricanes physical injuries are predominate whereas in drought and snow, starvation and dehydration may be problems. Details of the five primary emergency actions (evacuation, feeding/watering, identification, treatment and prophylactic measures), that must be taken post-disasters to prevent injuries and death of animal are described. The risk of diseases increases after disasters because of exposure to many kinds of disease causing bacteria, viruses, fungi, parasites spread by flood waters and debris. Environmental stress during drought also reduces the immunity of under-fed animals. The details of the various diseases that are commonly found after disaster along with the treatment and prevention measures are mentioned. The possibility of expanding the use of agro-industrial by-products and non-conventional feeds including plant secondary metabolites as an alternative source of animal feeding has found a place in the book.

Unique characteristics and physiological adaptation of different livestock breeds evolved in the flood and drought prone areas are discussed in the book. Continued advances in these breeds of livestock are needed but quite often the feed intake of these improved breeds becomes low because of environmental stress. An attempt has also been made to present direct and indirect adverse impacts of flood on poultry, particularly chickens and the need for flood related pro-active mitigation efforts, preparedness plan and response approach to minimize loss of poultry as well as short- and long- term strategies for rehabilitation of affected poultry farmers in the aftermath of flood. The last chapter portrays the strategies for disaster mitigation and management. The approach to disaster management in both the Pre-National Disaster Management Act (2005) Era in India and Post-National Disaster Management Act Era has been reviewed. Details of the four phases of disaster management *viz.* mitigation, preparedness, response and recovery are described. People Centric Early Warning along with early-warning process (forecast, warning and reaction) is depicted. Safety Codes for buildings and constructions are given for ready reference. Selected reference are also provided in each chapter to direct the readers into greater depth and breadth as they become intrigued with certain topics.

The editor is grateful to the authors of various chapters who have made a good effort in brining out issues related to post-disaster livestock rearing. We appreciate the commitment of our authors in timely submitting the chapters. The encouragement and guidance of Dr. K.A. Singh, Director, IGFRI, Jhansi are gratefully acknowledged. We would like to take this opportunity to extend our heartfull thanks to Shray Jain who have ably shepherded us through the dealing with the publisher 'Studium Press'. Our thanks are also go to Dr. J. N. Govil and his team for guiding us through the final steps of printing and publication process.

N. Das

ABOUT THE EDITOR

Dr. N. Das is currently working as Principal Scientist and Head, Plant Animal Relationship Division at Indian Grassland and Fodder Research Institute, Jhansi (UP). He has 25 years of research experience in the field of Livestock Production and management. As deputy advisor Planning Commission, Government of India, he was associated with the identification of policies and strategies for development of animal husbandry sector in India. He was also associated with the Mid-term appraisal of the Ninth (1997-2002) and formulation of the Tenth Five Year Plan (2002-2007) programme in the livestock sector. He has guided many postgraduate and PhD students at Indian Veterinary Research Institute, Izatnagar. Dr. Das has 50-research publication of national and international repute. He is one of the pioneer worker in the field of animal behaviour. He has also edited another book 'Forage for Sustainable Livestock Production'.

CONTENTS

1

NATURAL DISASTER AND ITS IMPACT ON LIVESTOCK IN INDIA

N. Das

ABSTRACT

A disaster is a natural or man-made event that negatively affects life, property, livelihood or industry often resulting in permanent changes to human societies, ecosystems and environment. Different organizations have defined disaster in their own ways. Disasters can be classified according to their genesis. Natural Disasters (like drought, flood, tsunami, cyclone, earthquake, landslides, fire and epidemic and exotic diseases) are the results of fury of the nature, whereas induced disaster (like industrial and chemical pollutions, war, terrorism, economic recession, global warming and over exploitation of natural resources) are man made. India is one of the top four most disaster prone zones in the world. Except volcanic activity, all types of disasters are common in India. The country can be divided into five categories based on disaster affinity viz. i) Northern mountain region (snow storms, landslides, avalanches and earthquakes), ii) Indo-Gangetic-Brahmaputra Plains (flood), iii) Decan Plateau (drought and earthquake), iv) Western Desert (drought) and v) Coastal areas (cyclones). The impact of disaster may be primary (loss of live, property and physical infrastructure etc.), secondary (decline in production and income, unemployment etc.) and tertiary (migration of people and livestock, spreading of diseases, hampering development process etc.). Disaster affects the animals in the same way as humans. The affected animals add burden to the owner who is already in need of emergency assistance. A disaster may cause severe business disruptions by adversely affecting the well-being of animals that ultimately, have an impact on trade in livestock and their produces. The prices of livestock go down drastically due to increased cost of rearing. Losses also include replacement value of animals and loss of production. In addition

Plant Animal Relationship Division, Indian Grassland and Fodder Research Institute, Jhansi (UP), India *reduced fertility prolong the recovery period. As the farmers are compelled to slaughter animals at a younger age (as a source of nutrition and of sale for cash), it takes several years to return to the original size of flock.*

1.0 DISASTER

A disaster (Greek meaning "Bad Star") is a natural or man-made event that negatively affects life, property, livelihood or industry often resulting in permanent changes to human societies, ecosystems and environment. There are many definition of disaster. Different organizations have defined disaster in their own ways.

- **The World Health Organization (WHO)** defines disaster as 'any occurrence that causes damage, economic destruction, loss of life and destruction in health and health services on a scale sufficient to warrant an extraordinary response from outside the affected community or area'.
- **UNDRO** "Disaster Management Training Manual defines" disaster as a serious disruption of the functioning of a society, causing widespread human, material, or environmental losses which exceed the ability of the affected society to cope using only its own resources.
- **The United Nations** define "A disaster is an event that is concentrated in space and time and that subject a society to severe danger and such serious losses of human life or such major material damage that the local social structure breaks down and the society is unable to perform any or some of its key functions".
- **The High Powered Committee of the Government of India**, in its October 2001 report defined disaster as "an occurrence of a severity and magnitude that normally results in deaths, injuries and property damage and that cannot be managed through the routine procedures and resources of government. It usually develops suddenly and unexpectedly and requires immediate, coordinated and effective response by multiple government and private sector organizations to meet human needs and speedy recovery".
- **The National Disaster Management Act 2005** defines disaster as "a catastrophe, mishap, calamity or grave occurrence affecting any area, arising from natural or manmade causes, or by accident or negligence which results in substantial loss of life or human suffering or damage to, and destruction of, property, or damage to, or degradation of, environment, and is of such a nature or magnitude as beyond the coping capacity of the community of the affected area".
- The definition of disaster provided by the **Centre for Research on the Epidemiology of Disasters (CRED)** is relatively simpler. CRED defines a disaster as a situation or event which overwhelms

local capacity, necessitating a request to national or international level for external assistance; an unforeseen and often sudden event that causes great damage, destruction and human suffering. Then it goes on to add that for a disaster to be entered into the database, at least one of the following criteria must be fulfilled: (1) ten or more people recorded killed (2)100 people reported affected (3) declaration of State of Emergency and (4) call for international assistance.

Although there is no clear-cut definition of disaster which is acceptable to all, but the main ideas are that it is an event concentrated in time and space which contributed a social, economical, cultural and political devastation and affects people and communities. Disaster events are sudden, highly disruptive and time neutral (short or long lasting). No disaster is exactly identical and its impact and consequences vary from region to region and from community to community. But it needs to mention that loss or injury or suffering of livestock and pets are not mentioned in any of the above definitions although these animals contribute significantly to the livelihood of rural landless and marginal farmers.

1.1 Classification of Disasters

Disasters can be classified according to their genesis:

1.1.1 *Natural Disasters*

They are the results of fury of the nature like

- Drought
- Famine induced migration of people and livestock
- Flood, Tsunami, Katrina etc.
- Tornado, Cyclone
- Earthquake, Volcanic eruptions, Landslides
- Fire
- Epidemic and exotic diseases

1.1.2 *Induced (Man-made) Disasters*

- Industrial and chemical pollutions
- War and territorial disputes
- Impulsive market crash
- Global warming
- Over exploitation of natural resources

1.1.3 *Meteorological Disasters*

Meteorological disasters are drought, flood, hurricane, severe storms, snow and ice, tornado and severe winds, wildfires etc.

1.1.4 *Geological Disasters*

Geological disasters like avalanche, landslides, earthquakes, tsunami, volcanic eruptions etc.

1.1.5 *Based on Onset and Duration of Disaster*

- *Cataclysmic Disaster* (earthquakes, cyclones and tsunamis): The onset of such type of disaster is very rapid. In this type of disaster, events that occur in seconds, cause most of the damage and destructions resulting into tremendous amount of sufferings and losses.
- *Continuing Disaster* (droughts, famines, epidemics, civil war etc.): These disasters are of longer duration and situation deteriorates with time. Affected area and population may be extremely large.

1.1.6 *As per HPC Report*

The High Powered Committee (HPC) of Government of India, in its report submitted to the Government of India in October 2001, classified the disaster in five categories as follows:

1.1.6.1 *Water and Climate Related Disasters*

- Over exploitation of natural resources
- Floods and drainage management
- Cyclones
- Tornadoes and hurricanes
- Hailstorm
- Cloud burst
- Heat wave and cold wave
- Snow avalanches
- Droughts
- Sea erosion
- Thunder and lighting

1.1.6.2 *Geologically Related Disasters*

- Landslides and mudflows
- Earthquakes
- Dam failures/dam bursts
- Mine fires

1.1.6.3 *Chemical, Industrial and Nuclear Related Disasters*

- Chemical and industrial disasters
- Nuclear disasters

1.1.6.4 *Accident Related Disasters*

- Forest fires
- Urban fires
- Mine flooding
- Oil spill
- Major building collapse
- Serial bomb blasts
- Festival disasters and fires
- Electrical disasters and fires
- Air, road and rail accidents
- Boat capsizing
- Village fire

1.1.6.5 *Biologically Related Disasters*

- Biological disaster and epidemics
- Pest attacks
- Cattle epidemics
- Food poisoning

1.2 Impact of Disasters

Drought, earthquakes, floods, cyclones and tsunami affect the livestock most. The impact of disaster may be primary (direct), secondary (indirect) and tertiary.

1.2.1 *Direct Impact*

- Loss of human, livestock and other animals
- Damage of private and public property
- Damage of infrastructure (power, telecommunication, bridges etc.)
- Health hazards

1.2.2 *Indirect Impact*

- Decline agricultural and livestock production
- Decline in industry and service output
- Fall in earning potential (due to disability and trauma etc.)
- Unemployment
- Negative impact on export/import

1.2.3 *Tertiary Impact*

- Migration of people and livestock
- Spread of diseases
- Fund reallocation

- Overall investment climate
- Long term development

2.0 INDIA AND DISASTER

India is one of the top four most disaster prone zones in the world. Except volcanic activity, all types of disasters are common in India. Floods, droughts, cyclones, earthquakes and landslide have been a recurrent phenomenon. The country can be divided into five categories based on disaster affinity.

A. **Northern Mountain Region (Including Foothills)**: Snow storms, landslides, snow avalanches and cold waves are very common in this zone. Land and soil degradation also occur due to heavy rainfall. Earthquakes also occur in this area.

B. **Indo-Gangetic-Brahmaputra Plains**: Here floods are annual phenomenon.

C. **The Deccan Plateau**: It is a drought prone area due to erratic rainfall. Earthquakes also happen.

D. **The Western Desert**: Drought is very frequent in this area due to scanty rainfall.

E. **Coastal Areas**: Every year five to seven cyclones come from Bay of Bengal and Arabian Sea and they cause extensive damage of life (both human and livestock) and property.

3.0 DISASTER AND LIVESTOCK

Disaster affects the animals in the same way as humans. The affected animals add burden to the owner who is already in need of emergency assistance. Animals which are left as stray in large scale disaster also compete with human for shelters and roads. Death of milking and drought animals result economic losses to the farmers and the local community. A disaster affects the livestock producers and may cause a severe business disruption by adversely affecting the well-being of animals that ultimately, have an impact on trade in livestock and their produce. Disasters cut down important source of income in rural areas and also hinder the distribution of good and foods.

3.1 Floods and Cyclone

According to one estimate, 12% of India's land area is prone to flood, about 8% is cyclone prone and a long coastline is exposed to tsunamis and storm surges. Floods affected 37% of India's land area in 1998. Of the total of 62 major rivers, eighteen are flood prone and drain an area of 150 Mha.

Table 1 Summary of districts having substantial multi-hazards prone areas in India

State	Name of Districts Having Substantial Multi-Hazard Prone Areas			
	Earthquake and flood	**Cyclone and flood**	**Earthquake, cyclone and flood**	**Earthquake and cyclone**
Andhra Pradesh	Adilabad, Karim Nagar, Khammam	Krishna, Nellore, Srikakulam, Vishkapatnam, Vizianagram	East Godavari, Guntur, Prakasam, West Godavari	–
Assam	All 22 districts listed in Table 2	No cyclone, but speed can be 50 m/s in districts of Table 2 causing local damage except Dhubri	-	–
Bihar	All 25 districts listed in Table 2	–	–	–
Goa	–	–	North and South Goa	–
Gujarat	Banaskantha, Danthe GS, Gandhinagar, Kheda, Mahesana, Panchmahals, Vadodara	–	Ahmedabad, Bharuch, Surat, Valsad	Amreli, Bhavnagar, Jamnagar, Janagad, Kachchh
Haryana	All 8 districts listed in Table 2	–	–	–
Kerala	Idduki, Kottayam, Palakad, Pathnamthitta	–	Alappuzha, Ernakkulum, Kannur, Kasargode, Kollam, Kozhikode, Malappuram, Thiruvananthapuram, Trissur	–
Maharashtra	–	–	–	Bombay, Rayagad, Ratnagiri, Sindhudurg, Thane
Orissa	Ganjam	Baleshwar, Cuttack, Puri	Dhenkanal	–
Punjab	All 12 districts	–	–	–
Uttar Pradesh	All 50 districts	–	–	–
West Bengal	Birbhum, Darjeeling, Jalpaiguri, Kooch Bihar, Malda, Murchidabad, West Dinajpur	–	Bardhaman, Calcutta, Hugli, Howra, Mednipur, Nadia, North and South 24 Parganas	Bankura
Union Territories		Delhi	–	Yanam (Py) Diu
Total	**139 districts**	**6 districts**	**29 districts**	**16 districts**

Source: http://www.vitcdmm.org/journel_on_disasters.pdf

Table 2 Multi-hazard prone districts in selected states

Assam	Barbeta, Bongaigaon, Cachar, Darrang, Dibrugarh, Dhemaji, Dhuburi, Golaghat, Gailaknadi, Goalpara, Jorhat, Karbi Anglong, Kamrup, Karimganj, Kokrajhar, Lakhimpur, Morigaon, Nagaon, Nalbari Sibsagar, Sonitpur, Tinsukia.
Bihar	Banka, Begusarai, Bhagalpur, Bhojpur, Darbhanga, Gopalganj, Jamui, Khagaria, Kishanganj, Madhepura, Madhubani, Mazaffarpur, Munger, Nalanda, Nawada, Pashchim Champaran, Patna, Purba Champaran, Purnia, Samstipur, Saran, Saharsa, Sitamarhi, Supaul, Vishali
Haryana	Ambala, Bhiwani, Faridabad, Fandabad, Gurgaon, Hisar, Jind, Kurukshetra
Punjab	Amritsar, Bathinda, Faridkot, Firozpur, Gurdaspur, Hoshiarpur, Jalandhar, Kapurthala, Ludhiana, Patiala, Rup Nagar, Sangrur
Uttar Pradesh	Agra, Aligarh, Allahabad, Azamgarh, Bahraich, Ballis, Barabanki, Bareilly, Basti, Bijnor, Budaun, Bulandshahr, Deoria, Etah, Etawah, Faizabad, Farrukhabad, Fatehpur, Firozabad, Ghaziabad, Ghazipur, Gonda, Gorakhpur, Hardoi, Haridwar, Jaunpur, Kanpur (Dehat), Kanpur (Nagar), Kheri, Lucknow, Maharajganj, Mainpuri, Mathura, Mau, Meerut, Mirzapur, Moradabad, Muzaffarnagar, Nainital, Pilibhit, Partapgarh, Raebareli, Rampur, Saharanpur, Shahjahanpur, Siddharth Nagar, Sitapur, Sultanpur, Unnao, Varanasi

Floods are mainly due to heavy rainfall in association with depressions and cyclones originated in the seas. Since 1980 our country has experienced more than 65 floods either because of monsoon heavy rains or cyclones. Floods in the Indo-Gangetic-Brahmaputra plains are an annual feature. Intense storms originating from sea are known in different parts of the world by different names like typhoons in Pacific ocean, hurricanes in north Atlantic sea and cyclone in the Indian ocean. Among various natural calamities, tropical cyclones are known to claim a higher share of death and destruction world over. Cyclones which occur between April and May are called pre-monsoon cyclonic storms whereas post-monsoon cyclonic storms occur between October and December. Coastal India frequently suffers from tropical cyclones, storm surges, high winds, flooding and coastal erosion. India has a very long coastal line of 5700 km. This is made of 2700 km of the east coast bordering Bay of Bengal and 3000 km of the west coast bordering the Arabian Sea. The Bay of Bengal and the Arabian Sea together generate about 5–6 tropical cyclones annually of which 2–3 could be quite severe. The experience has shown that nearly four times more cyclones occur in the Bay of Bengal than in the Arabian sea. In the north Indian ocean, cyclone seasons are May-June and mid-September to mid-December. Months of May, June, October and November are known for severe storms. The states most exposed to cyclone related hazards, including strong winds, floods and storm surges are coastal states like West Bengal, Orissa, Andhra Pradesh, Gujarat and Maharashtra.

Animal death is very common during floods and cyclones when they are washed off from the farms. Animals who are alive also suffer from physical stress, they have to stand in contaminated flood waters for long and thus become susceptible to infection of hooves and skin. In addition, the cuts acquired from disaster debris make animals more vulnerable to tetanus and toxins contaminated in flood water. Common illnesses caused by sewage-contaminated water in case of animals are tetanus, hepatitis, dysentery and food poisoning. Besides, other pathogenic virus, bacteria and parasites present in floodwater can enter the body of animals. Major livestock loss that was observed in the Orissa cyclone in the year 1999 is given in Table 3. Orissa also experience a cyclone in the year 1982 when about 12000 livestock died. In Andhra cyclone (1977) about 5.75 lakh animal died.

Table 3 Loss of livestock during 'Super Cyclone' of Orissa in 1999

Species	Total livestock population before cyclone	Population loss	Percentage loss of livestock	Estimated loss (Rs. Million)
Cow	1,818,277	346,200	19.04	187.71
Bullocks	1,904,209	52,937	2.78	132.43
Calves	2,219,726	90,343	4.07	36.09
Buffaloes	330,287	13,476	4.08	20.20
Sheep	812,210	103,151	12.70	25.78
Goats	2,268,055	196,187	8.65	49.05
Pigs	43,840	2,812	6.43	0.45
Poultry	4,735,273	1,151,145	24.31	28.78

Source: Bandhyopadhyaya, 1999

3.2 Earthquake and Tsunami

About 60% of the landmass is prone to earthquakes of various intensities. 55% of total areas is in seismic zone III –IV and vulnerable to earthquakes. Earthquakes of magnitude 6+ on the Richter scale have frequented India every 2 years. The major earthquake prone areas in India belong to the Himalayan belt in the northern region from Kashmir to Assam. In the east, the earthquake prone areas are Nagaland, Manipur, Mizoram and the Kutch region in the west. Earlier the peninsular was considered safe, but after the devastating earthquake of Latur, this part of the country is now equally vulnerable. Seismic activity is also noticed along Narmada, Son and Godavari rivers and sometimes over the Eastern and the Western Ghats. Most of the Himalayan earthquakes tend to be shallow earthquakes. Earthquake induces ground shaking, ruptures, faults, generating landslides, differential ground movements, liquefaction, tsunamis etc., causing staggering losses of both human and livestock lives and damages to infrastructure and environment. Of the numerous earthquakes, those of the Rann of Kuchh (1918), Shillong (1987), Kangra (1905), Bihar - Nepal

(1934, 1988) and Assam (1950) did cause extensive damage to the life and property. The death and destruction brought about by the Uttarkashi, Latur, Chamoli and Jabalpur earthquakes are known to everybody. Table 4 presents the livestock losses that occurred in the recent earthquakes in India.

Table 4 Effect of earthquakes in India on the livestock

Date	Place	Magnitude (Richter scale)	Livestock losses (number)
26 January 2001	Gujarat	7.9	20,000
30 September 1993	Latur & Osmanabad (Maharashtra)	6.4	15,800
20 October 1991	Uttaranchal	6.1	307,000

Source: Sen and Chander, 2003

The actual movement of the ground is rarely the direct cause of death or injury to human and animals. Partial or total building collapse, flying glass from broken windows, overturned large furniture falling on people and animals, fires resulting from broken gas lines, electrocution from fallen power lines and exertion and heat leading to heart failure are more often the actual causes of injury or mortality during an earthquake (Sen and Chander, 2002). Besides immediate casualities of livestock in earthquake, the mortalities due to malnutrition because of insufficient fodder also take place. After earthquake, the supply of electricity and water gets interrupted; as a result the whole rearing system is disturbed. Water supplies can become contaminated be seepage around broken water mains. Damage to roadways and other means of transportation result in shortage of food and other resources for livestock. Aftershocks may cause additional damage to the building and other properties. Animal may become nervous and apprehensive and may bite, kick or scratch during earthquake.

Tsunamis are long high sea wave caused by underwater earthquake or volcanic eruptions. A series of waves with 15 to 30 minutes between waves, strike the sea coast. Tsunamis can occur in all oceans, but they are most common in the pacific. The coastal region in Tamil Nadu, Andhra Pradesh, Kerala and Pondicherry are more prone to tsunami. Because of tsunami (December, 2004) in Tamil Nadu, a total of 18,448 animals were reported to have died in the disaster. The production loss of milk, meat and draught work due to this death toll was estimated at Rs. 37 crore. In the same tsunami, the livestock loss in Pondicherry, Andhra Pradesh and Kerala was Rs. 3.80, 0.24 and 5.08 crore respectively (Sharma, 2006).

3.3 Drought

Drought may be defined as a severe shortage of food and/or water usually during periods of low rainfall. The state revenue department declares an area drought affected, when more than 50% standing crop are damaged in that area (Bhasin, 2002). About 19% of India's total land area, with 12% of its populations is drought prone. The most remarkable fact is that about 68% of the total cultivated area is susceptible to drought; out of this 33% of the area receives less than 750 mm and is chronically drought prone; the remaining 35% of the area with 750–1125 mm rainfall is also subject to erratic distribution of rainfall leading to damage of crops and grasses. Of the 35 states and union territories, as many as 27 are prone to disaster. The eleven agriculturally and drought vulnerable states are Andhra Pradesh, Bihar, Gujarat, Haryana, Karnataka, Maharashtra, Madhya Pradesh, Orissa, Rajasthan, Tamil Nadu and Uttar Pradesh. It can starts any time, last indefinitely and attain many degree of severity. Due to erratic behaviour of monsoon and uneven and untimely distribution of rainfall across the country, drought has become a frequent phenomenon in India. During past 200 years (1800–2000 A.D.) about 24 events of droughts are reported (Table 5). The three major droughts in the last century were observed during 1904–05, 1965–66 and 1986–87. The Bengal famine of 1943 was perhaps the worst ever off in this subcontinent in the last Century. According to one report, nearly 3 million people perished in one year. The drought during 1987 showed a tremendous effect on the life of people and livestock. Out of 35 meteorological subdivisions of the country, only 14 received normal rainfall, 18 were deficient and 3 scanty. Agriculture was badly affected in 43% of total cropped area. Rajasthan and Gujarat became the worst affected state (Negi, 2002). Recently (2006–2007) Bundelkhand area (in U.P and M.P) experienced drought like situation for consecutive three years. In certain parts of India drought is so common and frequent that it is often termed as 'normal' to that area. In arid and semi-arid region of country, 2–3 years of drought in every 5 years is very common.

Table 5 Reported events of drought in India over the past 200 years

Period	Years
1801–1850	1801, 1804, 1806, 1812, 1819, 1825, 1832, 1833, 1837
1851–1900	1853, 1860, 1862, 1866, 1868, 1873, 1877, 1883, 1891, 1897, 1899
1901–1950	1901, 1904, 1905, 1907, 1991, 1918, 1920, 1939, 1941
1951–2007	1951, 1965, 1966, 1971, 1974, 1977, 1978, 1979, 1982, 1983, 1985, 1987, 1988

Source: Mandal *et al.*, 2006

The effect of drought is observed in long term and depends on its duration, time of the year, extent of damage of pasture and grazing pressure. Since it impacts very large areas for months and years, its affects the economy seriously apart from destruction of ecological resources, food shortage and causing starvation of million of animals. Losses due to drought during the period 1999–2001 are presented in Table 6.

Table 6 Losses due to droughts in India (1999–2001)

Year	District	Village affected (no)	Population affected (lakh)	Damage to crop areas (lakh ha)	Cattle population affected (in lakh)
1999	125	–	369.88	134.22	345.60
2000	110	54883	378.88	367.00	541.67
2001	103	22255	88.19	67.44	34.28
	338	77138	836.21	568.66	921.55

Source: Annual Reports, National Disaster Management Division, Ministry of Agriculture

The immediate impact of drought is seen on agricultural production (grain, feeds and fodders). As animal consume mostly the bye-products of crop, the impact of drought is felt at the later stage. Moreover, mild meteorological or agricultural drought may not necessarily affect the livestock. Prolongation of drought situation leads to less vegetative growth and less biomass production. It results scarcity/shortage of feed resources. Moreover the quality of feed is often poor during drought. The consecutive drought during 2000–2004 in Rajasthan caused heavy losses to the sheep husbandry. The sheep population declined from14.5 million to 10.0 million in 2003 with overall reduction of 31% (Singh and Shinde, 2006). The effect of drought on cattle loss in the year 1999–2000 is presented in Table 7. Little (1992) reported that about 70% of the total livestock died in Kenya during the drought of 1984. Livingstone (1986) reported similar large losses in Kenya during the drought of 1961–62.

Table 7 The Effect of drought on cattle in 1999–2000

State	District (number)	Cattle (million)
Chhattisgarh	16	3.610
Gujarat	17	7.130
Madhya Pradesh	22	3.443
Orissa	24	Not reported
Rajasthan	31	40.000

Source: Sen and Chander, 2003

The exposure of animals to high temperature reduces feed intake and in many cases, energy intake is less than energy expenditure resulting in loss of production. For food, sheep and goat are mostly dependent on the availability of biomass in the pasture. However during drought when day temperature becomes high, these animals prefer shades during major part of the day and as a result dry matter intake may reduce up to 25%. Further, nutrient digestibility may reduce to about 30% for dry matter and about 35% for protein in heat stressed animal. Shortage of feed and fodder lead to under nutrition, deterioration of body condition, lowering of immunity and disease resistance capabilities that result in reduction in production, reproduction and increase in morbidity and mortality rates. In drought areas, where food and water supply are in scarcity, animals are likely to consume unsuitable food and water, which may cause body weight loss, exacerbate diseases and even death losses. In many areas of desert, well have increased salt and other noxious solutes in them, cause health hazard to animals.

Environmental stressors and poor nutrition contributes to starvation death in young and weak animal during drought. The large surface area of young animals readily absorbs heat and insufficient thermoregulatory mechanism cause dehydration and hyperthermia. Monthly distribution of mortality rate of lambs in desert area has shown that maximum number of mortality occur in the month of April-June. In male animals, intensive summer heat causes seminal degeneration, resulting in poor quality semen. Spermatogenesis is decreased and production of abnormal spermatozoa is increased. Heat stress also reduces sexual drives and breeding activity of breeding males. Estrous and conception rates are declined in heat stressed females. The reduction of estrous behaviour and disturbance in secretion of hormones controlling follicular development and ovulation hamper seriously (Mittal, 2002). The quality and availability of feed is often poor during hot dry season in arid and semi-arid region of the country. Poor nutrition of ewes maintained on the range is also responsible for fetal dwarfing in heat stressed ewes. In heat stressed pregnant animals, more blood flow to body surface, lungs and respiratory tract in an adaptive response to facilitate body heat loss through evaporative cooling and as a result blood flow in gravid uterus is reduced and this suppress the growth and development of placenta and foetus (Dubey and Shinde, 2002). Inadequate placental function leads to fetal hypoglycemia and hypoxemia with reduced glycogen and fat reserve and these in turn make the fetus susceptible to environmental stressor. Low birth weight is common in sheep and goats exposed to intense heat. Dehydration and hyperthermia also reduced resistance to diseases. Recognizing the signs of heat stress will help to extend the amelioration measures (like access to shade, providing plenty of water, spraying water on animals etc.) in time and this will save the life. Personal water conservation measures should be practiced to avoid depletion of water supplies both before and extended periods of drought. Creation of artificial sheds in the pasture and avoidance of long distance grazing will keep animal cool.

Scarcity of feed and water force farmers to migrate along with their animals to the non-drought affected areas. Livestock migration not only causes hardship to both farmers and livestock but also causes social chaos. Migrating livestock causes damage of standing crops and compete with available feed resources. Hence, local people of non-drought affected areas do not want to accept the migrating livestock in their areas and this initiate clashes between two groups. Such clashed do not yield any fruitful solution; rather aggravate the situation of poor farmer. Further, overcrowding and lack of proper feed and water during the migration cause stress, heat stroke, dehydration, trauma and transit metabolic diseases in animals.

During disaster hundreds of animals are died which are immediate losses to the farmers. The prices of livestock go down drastically due to increased cost of rearing. Losses should include replacement value of animals and loss of production. sA reduced production and further management cost directly influence the trade of livestock products. In addition, reduced fertility prolong the recovery period. As the farmers are compelled to slaughter animals at a younger age (as a source of nutrition and of sale for cash), it takes several years to return to the original size of flock. Impact of disaster takes many years to recover.

4.0 REFERENCES

Bandhyopadhyaya, S.K. 1999. FAO Mission Report on Agricultural Rehabilitation off Cyclone Affected Areas of Orissa. FAO, Rome.

Dubey, S.C. and Shinde, A.K. 2002. Disposal of animal carcass during natural disaster. National Seminar on 'Biological Threats and Disaster Management – Role of Veterinarians'. 9–11 August, 2002, Punjab Agriculture University, Ludhiana.

Little, Peter D. 1992. The Elusive Granary: Herder, Farmer and State in Northern Kenya, Cambridge University Press: Cambridge. 1992.

Livingstone, Ian 1986. The common property problem and pastoralists economics behaviour. *Journal of Development Studies* **23**: 5–19.

Mandal, D.K., Tyagi, S. and Kumar, Mahesh 2006. Measures for saving livestock from disaster of drought. National Symposium on Technological Interventions for Livestock Improvement and Production. 17–19 February, 2006. New Delhi.

Mittal, J.P. 2002. Management of sheep during drought conditions in arid and semi-arid belt of the country. National Seminar on 'Biological Threats and Disaster Management – Role of Veterinarians'. 9–11 August, 2002, Punjab Agriculture University, Ludhiana.

Negi, D.S. 2002. Drought and livestock farming. *Indian Dairyman* **54(8)**: 33–35.

Sen, A. and Chander, M. 2003. Disaster management in India: the case of livestock and poultry. *Rev. Sci. Tech. Off. Int. Epiz.* **22(3)**: 915–930.

Sen, A. and Chander, M. 2002. Disaster management in India: The case of livestock. National Seminar on 'Biological Threats and Disaster Management – Role of Veterinarians'. 9–11 August, 2002, Punjab Agriculture University, Ludhiana.

Sharma, N. 2006. Market Instability of Livestock & Poultry Products During Disaster. National Symposium on Technological Intervention for Livestock Improvement & Production Thrust Area: Disaster Management. pp. 121–128.

2

LIVESTOCK IN DROUGHT AND FLOOD PRONE AREAS IN INDIA

ANIL KUMAR AND K.K. SINGH

ABSTRACT

Under the situation of any natural disaster or calamities, contingent plans mainly address the savings of human life and no attention is paid to livestock at the time it is needed most. The most obvious reason is the first priority to human beings. Other possible factor may be the lack of any contingent plans for the livestock that include number of livestock being affected in the disaster prone regions and type of inputs needed for their survivability. The list comprising 139 drought prone districts in India are given for ready reference. Rajasthan has the highest drought prone area (19.5%) followed by Gujarat (13.8%) Karnataka (13.0%), Maharashtra and A.P. (11.4% each). There are 54.9 million cattle, 29.15 million buffaloes, 37.67 million sheep, 43.84 million goats and 2.36 million pigs in the drought prone districts in India. Flood prone zones are endowed with the large livestock population. It accounts for about 70.93 million cattle, 53.38 million buffaloes, 15.59 million sheep, 51.48 million goats, 6.39 million pigs and 0.61 million other livestock species including equines and camels. Of the 33.5 m ha flood prone area, 7.35 m ha area get affected every year, which is 21.9% of the flood prone area. Taking into account this figure, it has been estimated that annually, 0.43 m cattle, 0.23 m buffalo, 1.4 m sheep and 2.9 m goats are affected by flood for which contingency plan need to be made.

1.0 INTRODUCTION

The climate plays an important role in determining the destiny of a region. Major activities of the human beings are governed by the climatic

---

Plant Animal Relationship Division, Indian Grassland and Fodder Research Institute, Jhansi (UP), India phenomenon. A favorable climate brings prosperity to a region by allowing bumper crops and permitting favorable conditions for the other activities. Livestock production is also governed to a large extent by the soil condition and climatic phenomenon. In periods of adversity, the agricultural activity is severely affected. Several contingency plans are initiated for agricultural sector. But a long term perspective for providing succour to the livestock is hardly taken into account. However, it is the livestock sector, which can sustain the livelihood and income of farmers when crop fails. Besides, they can also provide better nutrition to the people during such trying times.

The disaster preparedness for livestock sector primarily encompasses the drought and flood, which causes havoc to the livestock sector not only in terms of starvation and deaths, but also because of migration, underfeeding which severely impacts their production. In the past, no attempt has been made to map the livestock falling in the zone of vulnerability and to assess their requirements in terms of feed requirement, production performance and the strategies to minimize the impact of drought and flood on the poor livestock. This chapter attempts to identify the drought and flood prone regions of the country and the livestock, which live in these regions.

2.0 DROUGHT PRONE AREA

Drought is a temporary reduction in water or moisture availability significantly below the normal or expected amount for a specific period. This condition occurs either due to inadequacy of rainfall, or lack or irrigation facilities, under-exploitation or deficient availability for meeting the normal crop requirements in the context of the agro-climatic conditions prevailing in any particular area. The drought prone areas were defined by the Central Water Commission (CWC) in the past which have been taken up by various agencies for study and programme implementation. The list as mentioned by the CWC has 99 districts covering an area of 1.08 million

Fig. 1 Drought prone districts (updated) in India

sq km spread over 14 states in India. However, over a period of time there have been several additions by carving out new districts from the existing ones. But they may not find mention in the list being used by various agencies (Agricultural Research Data Book, 2005) presently. Therefore an exercise was undertaken to identify the newly carved out districts from the list of 99 drought prone districts identified by CWC. The total number of drought prone districts as of now is 139 (Table 1).

Table 1 List of drought prone districts (139) in India

States	Drought prone districts
Andhra Pradesh	Anantpur, Chittoor, Hyderabad, Cuddapah, Kurnool, Mahbubnagar, Nalgonda, Prakasam, Rangareddy
Bihar	Aurangabad, Begusarai, Bhojpur, Buxar, Gaya, Jamui, Jehanabad, Kaimur, Lakhisarai, Munger, Nawada, Rohtas, Sheikhpur
Gujarat	Ahmadabad, Amreli, Anand, Banaskant, Bharuch, Bhavnagar, Dahod, Jamnagar, Kheda, Kutch, Mehsana, Panchmaha, Rajkot, Sabarkantha, Surendranagar
Haryana	Bhiwani, Faridabad, Gurgaon, Jhajjar, Mahendrag, Rewari, Rohtak
Jammu & Kashmir	Doda, Udhampur
Jharkhand	Garhwa, Palamu
Karnataka	Bagalkot, Bangalore, Belgaum, Bellary, Bijapur, Chamarajn, Chikmagalur, Chitradurga, Dharwad, Gadag, Gulbarga, Hassan, Haveri, Kolar, Mandya, Mysore, Raichur, Tumkur
Madhya Pradesh	Anuppur, Betul, Datia, Dewas, Dhar, Jhabua, Khandwa, Khargone, Shahdol, Shajapur, Sidhi, Ujjain
Maharashtra	Ahmednagar, Aurangabad, Beed, Jalna, Latur, Nasik, Osmanabad, Pune, Sangli, Satara, Solapur
Orissa	Boudh, Kalahandi, Kendujhar, Nawarangpur
Rajasthan	Ajmer, Banswara, Barmer, Bikaner, Churu, Dungapur, Jaisalmer, Jalore, Jodhpur, Jhunjhunu, Nagaur, Pali, Rajsamand, Udaipur
Tamilnadu	Coimbatore, Dharmapuri, Dindigul, Erode, Kanyakuma, Karur, Madurai, Namakkal, Perambalur, Ramanatha, Salem, Sivaganga, Theni, Thoothukudi, Tiruchira, Tirunelveli, Virudhunagar
Uttar Pradesh	Allahabad, Banda, Chandauli, Chitrakut, Hamirpur, Jalaun, Mahoba, Mirzapur, Sant Ravidas Nagar, Sonbhadra, Varanasi
West Bengal	Bankura, Midnapore, Purulia

2.1 Area and Human Population in Drought Prone Districts

Of the 3.28 million sq km of the country, 1.10 million sq km in 14 states covering 139 districts are drought prone. About 34.78 percent geographical area of the country is prone to drought. In the drought prone areas 289.9 million people reside (28.2% of India), of which 83.7 million are in urban area and 206.2 million in rural area (Table 2). Rajasthan has the highest drought prone area (19.5%) followed by Gujarat (13.8%) Karnataka (13.0%), Maharashtra and AP (11.4% each) besides others.

Table 2 Area and human population in drought prone districts of India

Sl no	State/ District	Area (sq km)	No of persons		
			Total	**Urban**	**Rural**
	Andhra Pradesh	**125306**	**30743800**	**10174801**	**20568999**
		(11.38)	*(10.60)*	*(12.15)*	*(9.97)*
1	Anantapur	19130	3640478	919563	2720915
2	Chittoor	15151	3745875	811030	2934845
3	Hyderabad	217	3829753	3829753	0
4	Cuddapah	15359	2601797	587753	2014044
5	Kurnool	17658	3529494	817464	2712030
6	Mahbubnagar	18432	3513934	371355	3142579
7	Nalgonda	14240	3247982	432678	2815304
8	Prakasam	17626	3059423	467368	2592055
9	Rangareddi	7493	3575064	1937837	1637227
	Bihar	**32007**	**22409542**	**2433797**	**19975745**
		(2.91)	*(7.73)*	*(2.91)*	*(9.69)*
10	Aurangabad	3305	2013055	170057	1842998
11	Begusarai	1918	2349366	107623	2241743
12	Bhojpur	2395	2243144	312414	1930730
13	Buxar	1703	1402396	128974	1273422
14	Gaya	4976	3473428	475949	2997479
15	Jamui	3098	1398796	103244	1295552
16	Jehanabad	1569	1514315	111612	1402703
17	Kaimur	3362	1289074	41775	1247299
18	Lakhisarai	1228	802225	117740	684485
19	Munger	1419	1137797	317847	819950
20	Nawada	2494	1809696	138443	1671253
21	Rohtas	3851	2450748	326806	2123942
22	Sheikhpura	689	525502	81313	444189
	Gujarat	**152004**	**33190720**	**11840000**	**21350720**
		(13.80)	*(11.45)*	*(14.14)*	*(10.35)*
23	Ahmadabad	8087	5816519	4663533	1152986
24	Amreli	7397	1393918	312958	1080960
25	Anand	2941	1856872	507971	1348901
26	Banas Kantha	10757	2504244	275501	2228743
27	Bharuch	6527	1370656	352560	1018096
28	Bhavnagar	9981	2469630	935038	1534592
29	Dahod	3655	1636433	156323	1480110
30	Jamnagar	14125	1904278	836256	1068022
31	Kachchh	45652	1583225	474892	1108333
32	Kheda	4219	2024216	406450	1617766
33	Mahesana	4371	1837892	411717	1426175
34	Panch Mahals	5210	2025277	253362	1771915
35	Rajkot	11203	3169881	1625862	1544019
36	Sabar Kantha	7390	2082531	225129	1857402
37	Surendranagar	10489	1515148	402448	1112700

(Contd...)

(Contd...)

Sl no	State/ District	Area (sq km)	No of persons Total	Urban	Rural
	Haryana	**16715**	**8677969**	**2631252**	**6046717**
		(1.52)	*(2.99)*	*(3.14)*	*(2.93)*
38	Bhiwani	4778	1425022	270393	1154629
39	Faridabad	2151	2194586	1221344	973242
40	Gurgaon	2714	1660289	369004	1291285
41	Jhajjar	1834	880072	195097	684975
42	Mahendragarh	1899	812521	109636	702885
43	Rewari	1594	765351	136174	629177
44	Rohtak	1745	940128	329604	610524
	Jammu & Kashmir	**16241**	**1435438**	**164322**	**1271116**
		(1.47)	*(0.50)*	*(0.20)*	*(0.62)*
45	Doda	11691	691929	47695	644234
46	Udhampur	4550	743509	116627	626882
	Jharkhand	**12749**	**3133823**	**167732**	**2966091**
		(1.16)	*(1.08)*	*(0.20)*	*(1.44)*
47	Garhwa	4044	1035464	42639	992825
48	Palamu	8705	2098359	125093	1973266
	Karnataka	**143534**	**41806425**	**14905404**	**26901021**
		(13.03)	*(14.42)*	*(17.80)*	*(13.05)*
49	Baglakot	6575	1651892	478520	1173372
50	Bangalore Rural	5815	6537124	5759987	777137
51	Bangalore	2190	1881514	407396	1474118
52	Belgaum	13415	4214505	1012691	3201814
53	Bellary	8450	2027140	706850	1320290
54	Bijapur	10494	1806918	396089	1410829
55	Chamarajanagar	5101	965462	148090	817372
56	Chikmagalur	7201	1140905	222724	918181
57	Chitradurga	8440	1517896	274238	1243658
58	Dharwad	4260	1604253	881917	722336
59	Gadag	4656	971835	342183	629652
60	Gulbarga	16224	3130922	852621	2278301
61	Hassan	6814	1721669	304673	1416996
62	Haveri	4823	1439116	299020	1140096
63	Kolar	8223	2536069	625523	1910546
64	Mandya	4961	1763705	282715	1480990
65	Mysore	6854	2641027	982128	1658899
66	Raichur	8441	1669762	420837	1248925
67	Tumkur	10597	2584711	507202	2077509
	Madhya Pradesh	**86255**	**16117346**	**3420812**	**12696534**
		(7.83)	*(5.56)*	*(4.08)*	*(6.16)*
68	Anuppur	3701	667155	168622	498533

(Contd...)

(*Contd...*)

Sl no	State/ District	Area (sq km)	No of persons		
			Total	Urban	Rural
69	Betul	10043	1395175	259119	1136056
70	Datia	2691	628240	137549	490691
71	Dewas	7020	1308223	358347	949876
72	Dhar	8153	1740329	288184	1452145
73	Jhabua	6778	1394561	121031	1273530
74	East Nimar	10776	1713134	459921	1253213
75	West Nimar	8030	1529562	235459	1294103
76	Shahdol	6251	908148	229532	678616
77	Shajapur	6195	1290685	239230	1051455
78	Sidhi	10526	1831152	261031	1570121
79	Ujjain	6091	1710982	662787	1048195
	Maharashtra	**125412**	**35747168**	**11704315**	**24042853**
		(11.39)	*(12.33)*	*(13.98)*	*(11.66)*
80	Ahmednagar	17048	4040642	803697	3236945
81	Aurangabad	10107	2897013	1087150	1809863
82	Bid	10693	2161250	387070	1774180
83	Jalna	7718	1612980	307865	1305115
84	Latur	7157	2080285	490261	1590024
85	Nasik	15530	4993796	1937556	3056240
86	Osmanabad	7569	1486586	233256	1253330
87	Pune	15643	7232555	4200837	3031718
88	Sangli	8572	2583524	633218	1950306
89	Satara	10480	2808994	398121	2410873
90	Solapur	14895	3849543	1225284	2624259
	Orissa	**24330**	**3382833**	**221608**	**3161225**
		(2.21)	*(1.17)*	*(0.26)*	*(1.53)*
91	Boudh	3098	373372	18025	355347
92	Kalahandi	7920	1335494	100219	1235275
93	Phulabani	8021	648201	44094	604107
94	Nawarangpur	5291	1025766	59270	966496
	Rajasthan	**214952**	**25326912**	**5385381**	**19941531**
		(19.52)	*(8.73)*	*(6.43)*	*(9.67)*
95	Ajmer	8481	2181670	874676	1306994
96	Banswara	5037	1501589	107363	1394226
97	Barmer	28387	1964835	145404	1819431
98	Bikaner	27244	1674271	595036	1079235
99	Churu	16830	1923878	536196	1387682
100	Dungarpur	3770	1107643	80856	1026787
101	Jaisalmer	38401	508247	76394	431853
102	Jalor	10640	1448940	109994	1338946
103	Jhunjhunun	5928	1913689	395116	1518573
104	Jodhpur	22850	2886505	977082	1909423
105	Nagaur	17718	2775058	477337	2297721
106	Pali	12387	1820251	390887	1429364

(*Contd...*)

(Contd...)

Sl no	State/ District	Area (sq km)	No of persons		
			Total	Urban	Rural
107	Rajsamand	3860	987024	128723	858301
108	Udaipur	13419	2633312	490317	2142995
	Tamil Nadu	**83394**	**34424407**	**15196189**	**19228218**
		(7.57)	*(11.87)*	*(18.15)*	*(9.32)*
109	Coimbatore	7469	4271856	2820203	1451653
110	Dharmapuri	9622	2856300	455946	2400354
111	Dindigul	6058	1923014	673252	1249762
112	Erode	8209	2581500	1193963	1387537
113	Kanniyakumari	1684	1676034	1093927	582107
114	Karur	2901	935686	311256	624430
115	Madurai	3696	2578201	1444176	1134025
116	Namakkal	3404	1493462	545232	948230
117	Perambadur	3694	1189170	158205	1030965
118	Ramanathapuram	4175	1187604	302394	885210
119	Salem	5245	3016346	1390184	1626162
120	Sivgangai	4143	1155356	326084	829272
121	Theni	2869	1093950	591841	502109
122	Thoothkudi	4621	1572273	664773	907500
123	Tirunelveli	6810	2723988	1308246	1415742
124	Tiruchchirappalli	4511	2418366	1139162	1279204
125	Virudhunagar	4283	1751301	777345	973956
	Uttar Pradesh	**41237**	**18217656**	**4024051**	**14193605**
		(3.74)	*(6.28)*	*(4.81)*	*(6.88)*
126	Allahabad	5482	2992286	864694	2127592
127	Banda	4460	1537334	244018	1293316
128	Chandauli	2541	1643251	173558	1469693
129	Chitrakut	3164	766225	76560	689665
130	Hamirpur	4282	1043724	173808	869916
131	Jalaun	4565	1454452	340526	1113926
132	Mahoba	2884	708447	154895	553552
133	Mirzapur	4521	2116042	286506	1829536
134	Sonbhadra	6788	1463519	275430	1188089
135	Sant Ravidas Nagar	1015	1353705	173485	1180220
136	Varanasi	1535	3138671	1260571	1878100
	West Bengal	**27222**	**15339999**	**1474579**	**13865420**
		(2.47)	*(5.29)*	*(1.76)*	*(6.72)*
137	Bankura	6882	3192695	235248	2957447
138	Medinipur	14081	9610788	983905	8626883
139	Puruliya	6259	2536516	255426	2281090
	Grand Total	**1101358**	**289954038**	**83744243**	**206209795**
		(100)	***(100)***	***(100)***	***(100)***

Figures in parentheses indicate the percent of the total area/population

2.2 Livestock in Drought Prone Area

There are 54.9 million cattle, 29.15 million buffalo, 37.67 million sheep, 43.84 million goats and 2.36 million pigs and 198.8 million total livestock

Table 3 Livestock in drought prone districts in different states of India

Sl. no.	State/ District	Cattle	Buffalo	Sheep	Goat	Pig	Total livestock
	Andhra Pradesh	**3812198**	**3739772**	**11993629**	**3047665**	**193491**	**22817900**
		(6.9)	*(12.8)*	*(31.8)*	*(7.0)*	*(8.2)*	*(13.5)*
1	Anantapur	695384	410604	1905972	503275	38691	3570328
2	Chittoor	835510	141448	980837	247413	11155	2217141
3	Hyderabad	4357	31400	2057	11841	413	51843
4	Cuddapah	159290	447169	973122	371208	12489	1965208
5	Kurnool	429309	458913	1149281	386974	13061	2444490
6	Mahbubnagar	730914	356269	3356769	509341	43717	4998481
7	Nalgonda	533505	592271	1821782	404137	40390	3392464
8	Prakasam	119337	1029356	1156889	294830	19189	2620325
9	Rangareddi	304592	272342	646920	318646	14386	1557620
	Bihar	**3307710**	**1936864**	**224041**	**2341602**	**341556**	**8202535**
		(6.0)	*(6.6)*	*(0.6)*	*(5.3)*	*(14.5)*	*(4.9)*
10	Aurangabad	405769	169187	47575	227503	20657	871585
11	Begusarai	279385	100613	1251	147906	3791	536924
12	Bhojpur	221491	230796	43698	134142	17097	656984
13	Buxar	138215	161808	20012	70757	8462	405266
14	Gaya	639713	278060	18042	442026	122914	1511043
15	Jamui	384321	64611	16535	295883	50362	813085
16	Jehanabad	132625	144326	9001	102728	20692	411199
17	Kaimur	212994	191130	29608	104455	15677	559119
18	Lakhisarai	114499	53185	104	92906	7501	270923
19	Munger	143180	53176	712	141082	6393	347562
20	Nawada	321337	121643	8046	326846	47328	825748
21	Rohtas	255396	321916	28789	195110	12527	818036
22	Sheikhpura	58785	46413	668	60258	8155	175061
	Gujarat	**5111949**	**5144219**	**1889713**	**3356012**	**228908**	**15841370**
		(9.3)	*(17.6)*	*(5.0)*	*(7.7)*	*(9.7)*	*(9.4)*
23	Ahmadabad	225893	276655	20862	119940	20534	671279
24	Amreli	255287	176843	141016	141336	10561	726634
25	Anand	131565	291016	15506	66867	8028	518699
26	Banas Kantha	494841	715314	179775	330969	4530	1743801
27	Bharuch	142385	132901	7178	135821	15771	437473
28	Bhavnagar	336163	306832	290425	195992	43755	1180537
29	Dahod	515074	213771	5239	383050	1281	1120498
30	Jamnagar	348931	209616	232530	173022	16043	985390
31	Kachchh	334987	178033	495253	459442	31788	1516960
32	Kheda	200911	488435	38019	104524	10459	848708
33	Mahesana	160023	478161	15281	81871	3223	750399
34	Panch Mahals	644507	516791	4148	450364	7600	1627914

(Contd...)

(*Contd...*)

Sl. no.	State/ District	Cattle	Buffalo	Sheep	Goat	Pig	Total livestock
35	Rajkot	439208	273953	274546	218139	23044	1231554
36	Sabar Kantha	588416	682959	69346	315027	9343	1678086
37	Surendranagar	293758	202939	100589	179648	22948	803438
	Haryana	**361353**	**2197268**	**271669**	**303344**	**42299**	**3221957**
		(0.7)	*(7.5)*	*(0.7)*	*(0.7)*	*(1.8)*	*(1.9)*
38	Bhiwani	93510	441647	88108	74397	10135	727205
39	Faridabad	52947	412611	19823	24801	5580	517361
40	Gurgaon	63388	350550	26028	77495	9141	530450
41	Jhajjar	37252	256386	26112	10617	5604	339070
42	Mahendragarh	30785	226759	38740	69628	2228	376982
43	Rewari	21924	205199	26126	34657	733	294678
44	Rohtak	61547	304116	46732	11749	8878	436211
	Jammu & Kashmir	**861195**	**326304**	**1254717**	**606411**	**351**	**3122038**
		(1.6)	*(1.1)*	*(3.3)*	*(1.4)*	*(0.0)*	*(1.8)*
45	Doda	483948	100905	677834	195612	0	1505271
46	Udhampur	377247	225399	576883	410799	351	1616767
	Jharkhand	**1109073**	**192227**	**85858**	**541209**	**127591**	**2057590**
		(2.0)	*(0.7)*	*(0.2)*	*(1.2)*	*(5.4)*	*(1.2)*
47	Garhwa	356797	58860	36272	139305	27481	619623
48	Palamu	752276	133367	49586	401904	100110	1437967
	Karnataka	**7029727**	**3095034**	**6702671**	**4009428**	**243995**	**21112865**
		(12.8)	*(10.6)*	*(17.8)*	*(9.1)*	*(10.3)*	*(12.5)*
49	Baglakot	238338	200808	409559	266913	35782	1152304
50	Bangalore Rural	412678	75635	263054	169326	4523	925890
51	Bangalore	158217	14945	72516	31114	7464	285272
52	Belgaum	507583	701196	902555	508776	27983	2657926
53	Bellary	368723	187438	396575	221861	15368	1190371
54	Bijapur	232510	146099	270290	322400	28460	1000981
55	Chamarajanagar	268132	37451	116517	105833	998	529044
56	Chikmagalur	365174	106077	62969	47096	5093	586491
57	Chitradurga	313717	178505	715592	220272	7832	1436144
58	Dharwad	193476	89882	33583	60946	5388	383599
59	Gadag	138272	60805	208666	100677	4315	513032
60	Gulbarga	819620	224109	459515	562104	40044	2106666
61	Hassan	541241	194373	154360	99604	3804	993641
62	Haveri	304092	122924	197981	127678	4925	758252
63	Kolar	467316	122969	633201	203363	15375	1449343
64	Mandya	317323	169932	339533	204886	9633	1041642
65	Mysore	521550	69608	198784	160661	3184	960596
66	Raichur	360120	176492	382513	273933	10733	1203903
67	Tumkur	501645	215786	884908	321985	13091	1937768
	Madhya Pradesh	**5551759**	**2021052**	**115043**	**2674075**	**58335**	**10443512**
		(10.1)	*(6.9)*	*(0.3)*	*(6.1)*	*(2.5)*	*(6.2)*

(*Contd...*)

(Contd...)

Sl. no.	State/ District	Cattle	Buffalo	Sheep	Goat	Pig	Total livestock
68	Anuppur	260707	62462	462	57523	4087	386795
69	Betul	569483	150158	3070	183863	8370	916970
70	Datia	139671	142467	14656	113056	3557	414185
71	Dewas	385822	151937	99	149432	4672	693123
72	Dhar	539286	401161	6793	373637	3245	1326761
73	Jhabua	629968	124029	14694	488286	727	1260306
74	East Nimar	460390	141153	17713	171463	4939	798239
75	West Nimar	425503	97955	3034	185566	2045	717895
76	Shahdol	421758	92739	5883	93748	5722	619923
77	Shajapur	388755	235225	19890	193084	4881	844525
78	Sidhi	970726	194035	18153	474997	12296	1670520
79	Ujjain	359690	227731	10596	189420	3794	794270
	Maharashtra	**6795469**	**2873523**	**2253862**	**5226398**	**185734**	**17377211**
		(12.4)	*(9.9)*	*(6.0)*	*(11.9)*	*(7.9)*	*(10.3)*
80	Ahmednagar	1257220	192196	363999	960735	30169	2809603
81	Aurangabad	522743	83415	81980	381606	23053	1097719
82	Bid	595049	208707	126634	444116	11066	1387039
83	Jalna	629500	225923	48298	204414	23259	1134125
84	Latur	386487	245660	45867	176986	9031	865982
85	Nashik	998461	232287	324277	701601	30627	2294768
86	Osmanabad	373490	166118	40611	236171	5137	822916
87	Pune	717994	244966	410967	508763	18895	1906810
88	Sangli	271018	527134	217386	389468	9708	1417907
89	Satara	373385	358171	340138	388270	8188	1471300
90	Solapur	670122	388946	253705	834268	16601	2169042
	Orissa	**1451081**	**261967**	**236684**	**623129**	**94780**	**2667734**
		(2.6)	*(0.9)*	*(0.6)*	*(1.4)*	*(4.0)*	*(1.6)*
91	Boudh	236212	41755	76001	111996	1209	467173
92	Kalahandi	437562	81335	84632	217058	5905	826516
93	Phulabani	336862	68277	5401	212383	45280	668205
94	Nawarangpur	440445	70600	70650	81692	42386	705840
	Rajasthan	**5819322**	**3564047**	**7381984**	**1E+07**	**102055**	**27425657**
		(10.6)	*(12.2)*	*(19.6)*	*(23.1)*	*(4.3)*	*(16.2)*
95	Ajmer	307863	275673	392945	602604	26951	1611401
96	Banswara	658666	259946	22705	451193	1036	1400817
97	Barmer	537242	130863	1067210	1460772	8003	3303788
98	Bikaner	608597	132732	928892	686507	4551	2435908
99	Churu	215234	194524	381005	595899	4324	1444833
100	Dungarpur	417378	215991	124158	375028	2789	1141709
101	Jaisalmer	243250	2205	890181	588000	1427	1773313
102	Jalor	246936	356496	563130	451248	7067	1642947
103	Jhunjhunun	122858	378942	162537	490318	5286	1188109
104	Jodhpur	519972	180087	884191	1036696	3087	2660530
105	Nagaur	363013	420007	747003	1082967	9591	2645044
106	Pali	288187	285992	892895	632287	12981	2129256
107	Rajsamand	251863	200184	120641	499334	4097	1082979

(Contd...)

(Contd...)

Sl. no.	State/ District	Cattle	Buffalo	Sheep	Goat	Pig	Total livestock
108	Udaipur	1038263	530405	204491	1164316	10865	2965023
	Tamil Nadu	**4859445**	**1143323**	**4335566**	**5356241**	**173346**	**15904488**
		(8.9)	*(3.9)*	*(11.5)*	*(12.2)*	*(7.4)*	*(9.4)*
109	Coimbatore	362570	40912	206835	286499	12665	915202
110	Dharmapuri	597308	119009	560950	427055	6323	1712390
111	Dindigul	245116	68112	214143	351211	6443	889915
112	Erode	398572	230004	506015	562270	7288	1708053
113	Kanniyakumari	101712	6077	1143	100698	1266	210911
114	Karur	117781	65486	218514	173591	10205	586132
115	Madurai	226507	12380	216416	238588	2650	697027
116	Namakkal	222381	182202	146217	388832	21824	963287
117	Perambadur	284208	22341	96175	372142	2022	778079
118	Ramanatha-puram	128566	3877	245304	234727	33270	646623
119	Salem	593847	176521	371026	497814	5164	1645629
120	Sivgangai	280572	15183	227672	234746	4781	764763
121	Theni	101683	11955	52247	83454	6056	257024
122	Thoothkudi	163343	20933	202419	342180	12752	743599
123	Tirunelveli	418694	78777	487273	390570	14164	1395133
124	Tiruchchira-ppalli	354301	49577	257271	366753	12355	1042372
125	Virudhunagar	262284	39977	325946	305111	14118	948349
	Uttar Pradesh	**3115865**	**2297181**	**308226**	**1546003**	**194312**	**7493950**
		(5.7)	*(7.9)*	*(0.8)*	*(3.5)*	*(8.2)*	*(4.4)*
126	Allahabad	143750	716174	9587	145471	23527	1050775
127	Banda	386718	232638	12491	151273	17887	803746
128	Chandauli	173580	156042	23723	86744	7499	450439
129	Chitrakut	420334	138847	18547	95788	10267	684939
130	Hamirpur	266208	133391	25872	128723	25076	580326
131	Jalaun	237087	239162	30048	257389	26522	792572
132	Mahoba	200091	102969	23772	154531	25765	508468
133	Mirzapur	454720	195126	83329	145181	23149	906275
134	Sonbhadra	553126	148027	31371	242407	18777	994648
135	Sant Ravidas Nagar	104260	91216	32980	33584	8464	271877
136	Varanasi	175991	143589	16506	104912	7379	449885
	West Bengal	**5715769**	**356396**	**618901**	**4096031**	**371483**	**11159026**
		(10.4)	*(1.2)*	*(1.6)*	*(9.3)*	*(15.8)*	*(6.6)*
137	Bankura	1334543	141835	151190	1052730	135922	2816315
138	Medinipur	3484055	57226	118378	2026262	126566	5812673
139	Puruliya	897171	157335	349333	1017039	108995	2530038
	Grand Total	**54901915**	**29149177**	**37672564**	**43844717**	**2358236**	**168847833**
		(100)	***(100)***	***(100)***	***(100)***	***(100)***	***(100)***

Figure in parentheses indicate percent of the total population in the drought prone area.

in the drought prone districts in India (Table 3) which constitute 29.6% of cattle population of India, 29.8% of buffalo, 61.3% of sheep, 35.3% goat, 17.4% pig and 34.8% of the total livestock.

Livestock play an important role in the livelihood support system under difficult environment. This is reflected in the number of livestock per thousand human beings (PTH). Classification of the 139 drought prone districts according to dependence of livestock (Table 4) have revealed that there are 23 districts where there are over 900 livestock PTH, 13 districts have between 700 – 900, 53 districts between 500 – 700, 31 districts between 300 – 500 and 19 districts with less than 300 number of livestock PTH. The dependence of human beings on different livestock species in the drought prone districts have been shown in Figs. 2 to 7.

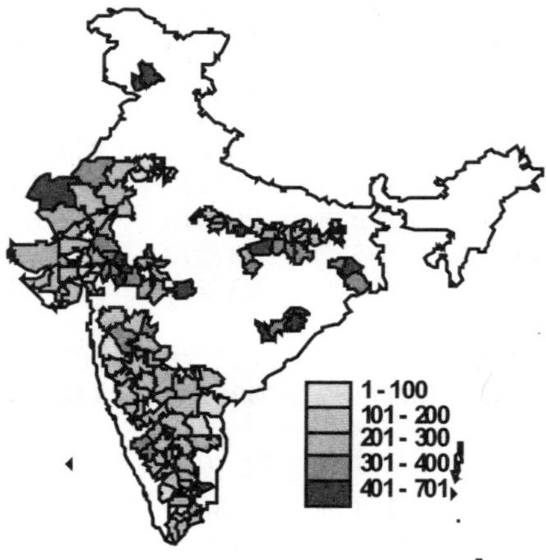

Fig. 2 No. of cattle per 1000 human beings in drought prone districts

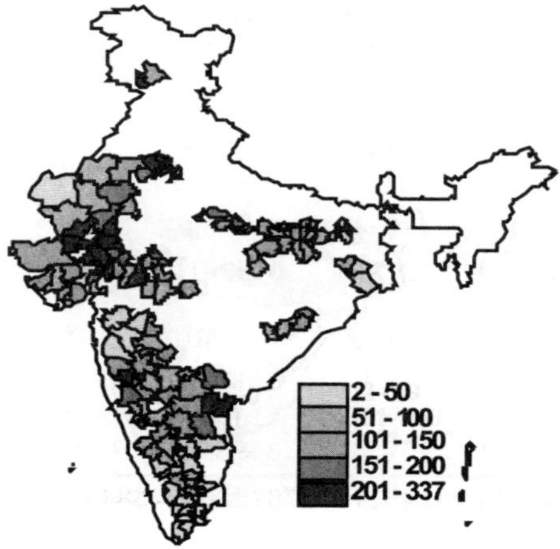

Fig. 3 No. of buffalo per 1000 human beings in drought prone districts

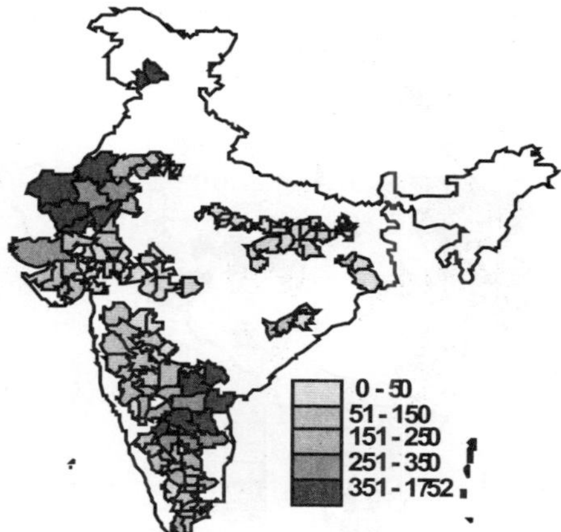

Fig. 4 No. of sheep per 1000 human beings in drought prone districts

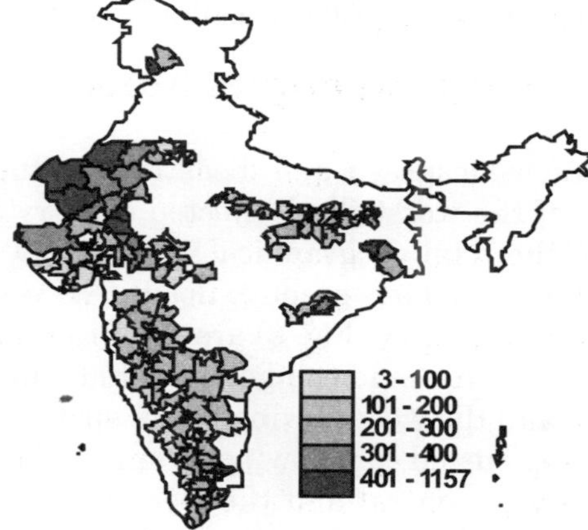

Fig. 5 No. of goat per 1000 human beings in drought prone districts

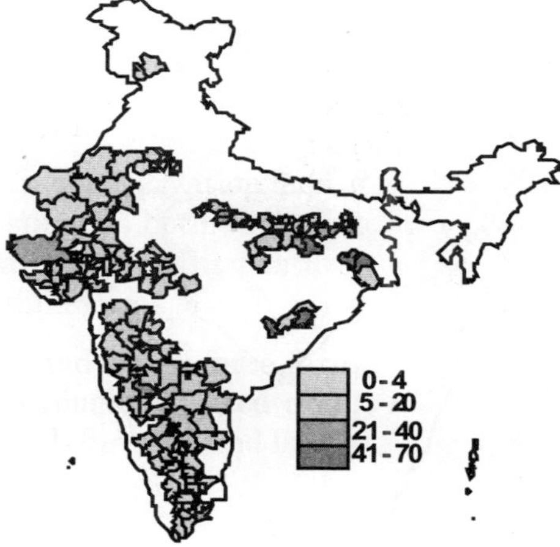

Fig. 6 No. of pig per 1000 human beings in drought prone districts

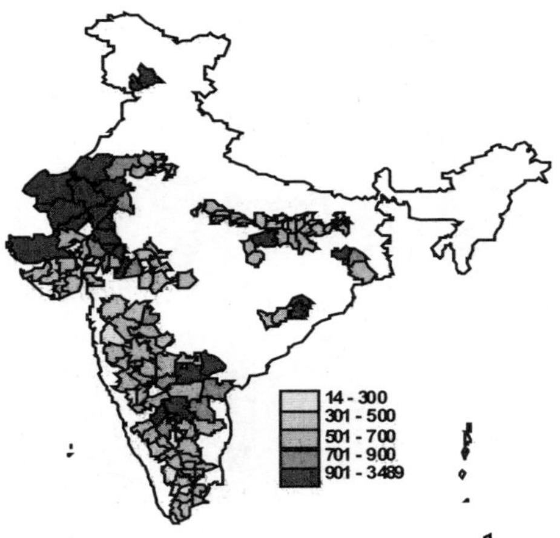

Fig. 7 No. of total livestock per 1000 human beings in drought prone districts

3.0 LIVESTOCK IN FLOOD PRONE AREA

Floods are recurrent, widespread and disastrous natural hazards of the world. India is one of the worst flood affected country in the world after Bangladesh. Out of the total geographical area of 329 million hectare in the country, more than 40 m ha is prone to floods (NDMA, 2008). The major flood-prone areas in the country (Fig. 8) are the Brahmaputra, Ganga and Meghna river basins in the Indo-Gangetic-Brahmaputra plains in north and northeast India and the west flowing rivers such as the Narmada and Tapti in north-west region and east flowing rivers like Mahanadi, Godavari, Krishna and Cauvery in central and the Deccan region. On an average floods affect 75 lakh ha of land and result in a loss of more than 1600 lives

Fig. 8 Major flood prone areas in the country

Table 4 Distribution of drought prone districts according to dependence on livestock

State/No of districts	Number of livestock per 1000 human beings									
	Over 900		700–900		500–700		300–500		<300	
	23		13		53		31		19	
	Districts	Livestock PTH	Districts	Livestock PTH	Districts	Livestock PTH	Districts	Livestock PTH	Districts	Livestock PTH
Andhara	Anantapur	981	Cuddapah	755	Chittoor	592	Rangareddi	436	Hyderabad	14
Pradesh	Mahbubnagar	1422	Prakasam	856	Kurnool	693				
	Nalgonda	1044	Panch Mahals	804						
Bihar					Jamui	581	Aurangabad	433	Begusarai	229
							Gaya	435	Bhojpur	293
							Kaimur	434	Buxar	289
							Lakhisarai	338	Jehanabad	272
							Munger	305		
							Nawada	456		
							Rohtas	334		
							Sheikhpura	333		
Gujarat	Kachchh	958	Sabar Kantha	806	Amreli	521	Bharuch	319	Ahmedabad	115
					Banas Kantha	696	Bhavnagar	478	Anand	279
					Dahod	685	Kheda	419		
					Jamnagar	517	Mahesana	408		
					Surendranagar	530	Rajkot	389		
Haryana					Bhiwani	510	Gurgaon	319	Faridabad	236
							Jhajjar	385		
							Mahendragarh	464		
							Rewari	385		
							Rohtak	464		
J & K	Doda	2175								
	Udhampur	2175								
Jharkhand					Garhwa	598				
					Palamu	685				

(Contd...)

(*Contd...*)

State/No of districts	Number of livestock per 1000 human beings									
	Over 900		700–900		500–700		300–500		<300	
	23		13		53		31		19	
	Districts	Livestock PTH	Districts	Livestock PTH	Districts	Livestock PTH	Districts	Livestock PTH	Districts	Livestock PTH
Karnataka	Chitradurga	946	Raichur	721	Baglakot	698	Mysore	364	Bangalore	152
			Tumkur	750	Belgaum	631	East Nimar	466	Bangalore Rural	142
					Bellary	587	Ujjain	464	Dharwad	239
					Bijapur	554	West Nimar	469		
					Chamarajanagar	548				
					Chikmagalur	514				
					Gadag	528				
					Gulbarga	673				
					Hassan	577				
					Haveri	527				
					Kolar	571				
					Mandya	591				
Madhya Pradesh	Jhabua	904	Dhar	762	Anuppur	580				
	Sidhi	912	Jalna	703	Betul	657				
					Datia	659				
					Dewas	530				
					Shahdol	683				
					Shajapur	654				
Maharashtra					Ahmadnagar	695	Aurangabad	379	Pune	264
					Bid	642	Latur	416		
					Osmanabad	554	Nashik	460		
					Sangli	549				
					Satara	524				
					Solapur	563				
Orissa	Boudh	1251			Kalahandi	619				
	Phulabani	1031			Nawarangpur	688				

(*Contd...*)

(Contd...)

State/No of districts	Number of livestock per 1000 human beings									
	Over 900		700–900		500–700		300–500		<300	
	23		13		53		31		19	
	Districts	Livestock PTH	Districts	Livestock PTH	Districts	Livestock PTH	Districts	Livestock PTH	Districts	Livestock PTH
Rajasthan	Banswara	933	Ajmer	739	Jhunjhunun	621				
	Barmer	1681	Churu	751						
	Bikaner	1455								
	Dungarpur	1031								
	Jaisalmer	3489								
	Jalor	1134								
	Jodhpur	922								
	Nagaur	953								
	Pali	1170								
	Rajsamand	1097								
	Udaipur	1126								
Tamilnadu					Dharmapuri	600	Dindigul	463	Coimbatore	214
					Erode	662	Thoothkudi	473	Kanniyakumari	126
					Karur	626	Tiruchchirappalli	431	Madurai	270
					Namakkal	645			Theni	235
					Perambadur	654				
					Ramanathapuram	544				
					Salem	546				
					Sivgangai	662				
					Tirunelveli	512				
					Virudhunagar	542				
UP			Chitrakut	894	Banda	523	Allahabad	351	Chandauli	274
			Mahoba	718	Hamirpur	556	Mirzapur	428	Sant Ravidas Nagar	201
					Jalaun	545			Varanasi	143
West Bengal	Puruliya	997	Bankura	882	Sonbhadra	680				
					Medinipur	605				

annually. Average annual damage to crops, houses and public utilities (Table 5) is more than 1800 crores. In the last 10 years, the flood damage has shown an increasing trend with an average loss of Rs. 4745 crores (1996-2005) as against an average of Rs. 1805 crores during last 53 years.

Table 5 Average annual loss due to Floods

S. No.	Items	Loss
1.	Area affected	7.351 million hectare
2.	Population affected	40.967 million
3.	Human lives lost	1793 number
4.	Cattle lost	85599 number
5.	Houses damaged	1452904 number
6.	Houses damaged	370.607 crore
7.	Crop area damaged	3.725 million hectare
8.	Crop damaged	1095.132 crore
9.	Public utilities damaged	1186.456 crore
10.	Total losses	2706.243 crore

Source: Central Water Commission, Ministry of Water Resources, Government of India

The flood affected area in the country has over the years shown an increasing trend. It has increased from an annual average of 6.48 m ha in the 1950s to over 9 m ha in the 1970s and 1980s. This increase is definitely

Table 6 Flood prone area in different states in India

S. No.	State	Area liable to Floods (million ha)
1.	Andhra Pradesh	1.390
2.	Assam	3.150
3.	Bihar	4.260
4.	Gujarat	1.390
5.	Haryana	2.350
6.	Himachal Pradesh	0.230
7.	Jammu & Kashmir	0.080
8.	Karnataka	0.020
9.	Kerala	0.870
10.	Madhya Pradesh	0.260
11.	Maharashtra	0.230
12.	Manipur	0.080
13.	Meghalaya	0.020
14.	Orissa	1.400
15.	Punjab	3.700
16.	Rajasthan	3.260
17.	Tamil Nadu	0.450
18.	Tripura	0.330
19.	Uttar Pradesh	7.336
20.	West Bengal	2.650
21.	Delhi	0.050
22.	Pondicherry	0.010
	Total	**33.516**

Source: http://mowr.gov.in/index3.asp?sslid=355&subsublinkid=359&langid=1

an indication of the country's growing flood proneness. The flood prone areas in different districts are mentioned in Table 6.

Floods, which are often described as natural disasters, turn out to be social disasters as well, tending to choose their victims by class. It is usually the poor who are severely affected because they live on the periphery of the human habitat. One disaster makes the poor more vulnerable to the next and, in this way, converts a disaster into a disaster process.

3.1 Livestock Population in Flood Prone Areas

Livestock play an important role in the livelihood of people in flood prone areas because most of these areas are under developed and agrarian in character. Livestock rearing is an important vocation for many people. If these areas are well connected near the roads, the milk produced in these hinterlands is transported outside; otherwise they are made into ghee to be sold later on.

Most of the studies on flood related problems have overlooked the miseries of the marooned livestock in the flood affected area. In the pandemonium that follows flood, the poor livestock gets the least attention. People would rush away with one or two goats that they own leaving behind many. The chain of Food Corporation of India has helped in minimizing the loss of human life due to starvation but the poor livestock remain neglected. A fair assessment has not yet been made to assess the actual number of animals lost during flood and the resultant loss in productivity. With the increasing dependency of population on livestock for their income and employment, people rue the misery their livestock are subjected during floods.

Restoring livelihood of people after a bout of natural calamity is the quickest way to rehabilitate them. Initiating agricultural activities may take some time, but livestock can provide succour right in the midst of such misfortune, provided prior planning and coordination of various stakeholders are in place. Ahmed (1999) noted that there prevails a serious concern throughout flood period about lack of work for the many able bodied people. However, if the livestock sectors are taken well care of they have the potential to provide income and employment right in the midst of such misfortune provided the back up support system is in place beforehand.

Flood prone zones are endowed with the large livestock population. It accounts for about 70.93 million cattle, 53.38 million buffaloes, 15.59 million sheep, 51.48 million goats, 6.39 million pigs and 0.61 million other livestock species including equines and camels. Of the 33.5 m ha flood prone area, 7.35 m ha area get affected every year, which is 21.9% of the flood prone area. Taking into account this figure, it has been estimated that annually, 0.43 m cattle, 0.23 m buffalo, 1.4 m sheep and 2.9 m goats are affected by flood for which contingency plan need to be made.

4.0 REFERENCES

Ahmed, I. 1998. Governance of floods. Seminar. June 1998. pp. 34–41.

Centre for Science and Environment. 1991. Floods, flood plains and environmental myths. Centre for Science and Environment. New Delhi.

National Remote Sensing Agency. 2007. Flood Inundation Mapping. Hyderabad.

Roy, D. 1998. Drowning people, killing hope. Seminar. June 1998. pp. 42–45.

Sen, B., Hashemi, S.M. and Ahmed, I. 1998. Impact of flood on rural economy: Results from a large scale rapid survey. National seminar on 'The Deluges of 1998' organized by the Nagrik Durjog Mokabila Udyog. 31 December 1998.

Singh, J.P. 1997. Flood hazard in the Rapti, Burhi-Rapti region: Problem, adjustment and management. Indian Grassland and Fodder Research Institute, Jhansi.

3

ECONOMIC IMPORTANCE OF LIVESTOCK IN VULNERABLE ZONES

A.K. Misra, K.K. Singh and M.M. Das

ABSTRACT

Animal husbandry plays an important role in livelihood security and economic sustenance of people in the arid and semiarid drought prone zones of country because crop farming is a big gamble due to uncertainties of normal rainfall. The contribution of cattle, buffalo, sheep and goat in the arid and semiarid states of the country as per the recent census of 2003 is 63.5 million tons of milk, 0.755 million tons of meat and 38.7 million kg of wool. Out of total production of the country, share of arid and semiarid states is 72% to milk, 50.76% to meat and 79.83% to wool. Most of the herds or flocks are reared on grazing land in arid and semi-arid zones. The rangeland supports full spectrum of production systems from sedentary to traditional nomadic and semi-nomadic systems. The main constraints to animal production on rangelands in arid and semi-arid zones are progressive shrinkage of grazing lands due to intensive land reclamation for raising crops, simultaneous increase in livestock population resulting in over stocking, over grazing by all kind of livestock leading to soil erosion and ecological degradation and the long dry spell (8-9 months) which does not allow proper rejuvenation of vegetation. Ruminant livestock play an important role in rice wheat system prevailing in the flood prone regions. Buffalo and pigs have unique importance in the flood-affected region as both have occupied about half of the total population of the country. Of the total buffalo and pigs population about 55 and 47 percent are found in this region. The population of livestock is still increasing rapidly in this zone. The chapter also mentions the details of the animal production systems prevalent in vulnerable zones of the country (viz. landless animal production system, crop based animal production system and rangeland based animal production system).

Indian Grassland and Fodder Research Institute, Jhansi (UP), India

1.0 INTRODUCTION

Livestock are promising sub-sectors in the agricultural sector which have substantial growth prospects and presently account for 24.7 percent of agricultural GDP and 4.4 percent of total GDP of India. During natural disaster, the agricultural activity is severely affected. Several contingency plans are initiated for agricultural sector. But a long term perspective for providing help to the livestock is hardly taken into account. However, it is the livestock sector, which can sustain the livelihood and income of farmers when crop fails. Besides, they can also provide better nutrition to the people during such trying times. Vulnerability and impact of disaster are high, on the poorer sections of the community. Many of the poor are landless or small holders. This section largely depends on animals, for livelihood.

There are 54.9 million cattle, 29.15 million buffalo, 37.67 million sheep, 43.84 million goats and 2.36 million pigs and 198.8 million total livestock in the drought prone districts in India which constitute 29.6% of cattle population of India, 29.8% of buffalo, 61.3% of sheep, 35.3% goat, 17.4% pig and 34.8% of the total livestock.

Flood prone zones are endowed with the large livestock population. It accounts for about 70.93 million cattle, 53.38 million buffaloes, 15.59 million sheep, 51.48 million goats, 6.39 million pigs and 0.61 million other livestock species including equines, avian and rabbit. Of the 33.5 m ha flood prone area, 7.35 m ha area get affected every year, which is 21.9% of the flood prone area. It has been estimated that annually, 0.43 m cattle, 0.23 m buffalo, 1.4 m sheep and 2.9 m goats are affected by flood for which contingency plan need to be made. Buffalo and pigs have unique importance in the flood-affected region as both have occupied about half of the total population of the country. Of the total buffalo and pigs population about 55 and 47 percent are found in this region. The population of livestock is still increasing rapidly in this zone.

In present chapter attempts have been made to focus the economic importance of livestock and the production system of livestock that exists in different drought or flood prone and vulnerable areas of the country.

2.0 ECONOMIC IMPORTANCE OF LIVESTOCK IN VULNERABLE AREA

2.1 Arid and Semi-Arid Drought Prone Regions

The drought prone regions of the country cover the 138 districts spread over 14 states. Animal husbandry plays an important role in livelihood security and economic sustenance of people in the arid and semiarid drought prone zones of country because crop farming is a big gamble due to uncertainties of normal rainfall. The contribution of cattle, buffalo, sheep and goat in the arid and semiarid states of the country as per the recent census of 2003 is 63.5 million tons of milk, 0.755 million tons of meat and 38.7 million kg of wool. Out of total production of the country, share of arid and semiarid states is 72% to milk, 50.76% to meat and 79.83% to wool. The increase of animal production in the arid and semi-arid states of the country during the 1997–2003 was 7.77% milk, 17.44% meat and 3.63% wool, the rise of animal production during the same period in the country was 22.12% milk, 8.41% meat and 6.54% wool. Milk production in all the states except Rajasthan registered an increasing trend during 1997 and 2003. Wool production in the Rajasthan is almost same during 1997 and 2003. The wool production in the Punjab has declined sharply from 1202 thousand kg in 1997 to 554 thousand kg in 2003 with concomitant decrease of sheep population from 4.36 lakh heads in 1997 to 2.20 lakh in 2003. Meat production has risen in these states during 1997 and 2003. Drought was prevalent during the period in the most of the states of arid and semiarid region and people were compelled to sale their animals specially sheep and goats due to lack of feed and fodder, this could be perhaps one of the reasons for rise of meat production. Narayan and Rajora (2004) reported that during drought years, income from agriculture decreased to 10%, whereas the income from livestock always remains more than 50% of the normal rainfall year, but during drought year of 2002 milk and meat production rather increased. This shows livestock in drought prone areas are dependable source of income for rural people during drought situation. In semiarid zones, nearly 15–20% of land is available as rangeland beside community land, fallow land and wasteland. In arid zones, rangeland occupied 45% (Shankar *et al.,* 1988) and in extreme arid districts of Rajasthan nearly 95% of land is available as grazing land after harvesting of kharif crops. The livestock population in arid region far exceeded the human population. Most of the herds or flocks are reared on grazing land in arid zones. Animal production from these lands is important both in terms of the product produced and number of people dependent on these lands for their livelihood. The rangeland supports full spectrum of production systems from sedentary to traditional nomadic and semi-nomadic systems. It is widely opined that main constraints to animal production on rangelands in arid and semi-arid zones are progressive shrinkage in grazing lands due to intensive land reclamation for raising crops, simultaneous increase in livestock population resulting in over stocking, over grazing by all kind of livestock depleting the grazing resources leading to soil erosion and ecological degradation, the long dry spell (8–9 months) does not allow proper rejuvenation of vegetation as a result livestock face serious feed scarcity during lean periods, additionally water scarcity during summer months is common.

The arid zone is gifted with best hardy, dairy and draught breeds of cattle, meat and milk breeds of goat and meat and wool breeds of sheep and well-known draught breeds of camel. Many well known breeds of livestock in the region are declining at a fast rate due to uncontrolled breeding and cross breeding of native germplasm by exotic one. An immediate attention for their conservation and propagation is needed. There is an urgent need for rehabilitation of native ranges by perennial grasses to ensure adequate supply of forage to grazing animals in the years to come (Singh, 1995). The animal diseases of various natures in the regions also cause economic losses. Systematic disease control program in the most intensive production areas of arid and semiarid regions can bring down the incidence of diseases to well controlled status. Numerous geophysical disasters are affecting the livestock-agriculture causing considerable loss of animal life and production in the arid and semiarid zones. The economic consequences of disaster involving livestock are great because many livestock are important source of employment and revenue for rural peoples (Dubey *et al.,* 2002). The programs for mitigating the effect of disasters on livestock need attention in the Indian perspective during the new millennium. In future, prime objective should be to sustain the animal productivity in the arid and semiarid regions with eco-conservation and production of safe animal produces and efforts are required for strengthening the sanitary and phyto-sanitary measures for meeting the domestic as well as global demand of safe and quality animal produce.

2.2 Flood Prone Regions

Flood is a common feature occurring every year in many states of the country leaving thousands dead, hundreds of thousands displaced and agriculture lands that many rely on for their livelihoods in ruins. In Asian scenario, heavy monsoon season starting in mid June swelled rivers across the region and sent floodwaters and mudslides crashing into cities and villages, destroying homes and businesses and affecting an estimated 35 million people. India, Bangladesh and Nepal are the countries getting affected every year. Widespread devastation also was reported in Pakistan, China, North Korea and Indonesia. No information available on the number of livestock affected during the flood but it can be presumed that all the livestock of the region become vulnerable. No contingency plan for the relief of livestock so far been observed under operation in any flood affected region. Majority of the livestock become the victims of flood and remain at risk for starvation. Calving and lambing season begins soon, and if livestock do not receive fodder, the animals and people who depend on them will experience a domino effect. Consequences extend beyond animal starvation—whole villages can be affected. Lands cannot be tilled; crops, for human consumption and livestock fodder, will not grow; animals will be unable to produce, dairy may stop, leaving animals unable to feed newborn calves; and lost milk production. Where do hungry, homeless animals graze when land is submerged? There becomes intense pressure on the strips of land that are available and even the land above water is simply not enough to accommodate so many hungry animals. Restoring the lands will not be an easy task. Because of sand and silt, regaining fertile ground and grazing areas could take as long as a year, increasing pressure on the herds.

Ruminant livestock play an important role in rice wheat system prevailing in the flood prone regions. Rice, wheat, maize are the main crops in this region. Pulses, oil seed, sugarcane, vegetables, fruit crops are also widely grown. The farmer's preference of the livestock is in the order of buffalo followed by cattle and goat. Goat rearing is not well accepted by all classes of people. It is mostly confined to backward classes and landless laborers who are unable to rear large animals. As a result, goat rearing has not been developed as an industry. Holding size is generally 1–3 goats per family depending on availability of surplus labor in the family. Village goat is mostly of Bengal and other local breeds. However, crosses with other breeds like Jamunapari, Barbari, Sirohi and Jakharana are also available. Since goat is confined to only downtrodden people no scientific feeding practices is followed. Female or child of the family mainly rears goat. Feed requirement is met up mostly by grazing for 6–8 hours a day on common property resources on the bunds of canals, rivers, and ponds or nearby forest. Goat is also kept on grazing after the crop is harvested. During other seasons it is stall fed with local grasses collected from bunds or grazing area or fodder trees available in the villages. Very little concentrate like broken rice or maize or wheat or pulse chunies is given along with rice bran and mustard cake. Some farmers who have land either own or taken on lease, occasionally offer green fodder like Sudan or Berseem. The feeding of large ruminants is mainly from crop residues grown by the farmers. The straw of wheat/paddy constitute the basal dry roughage of the animals. The minor source of dry roughage includes pulse straw. The advanced farmer cultivate green fodder like maize, sorghum, berseem, hybrid napier etc. The use of readymade concentrate pellets is common among the farmers who are the members of any milk producer cooperative societies. These societies provide readymade mix or pellets on farmer's door with reasonable price tag. The balanced concentrate mixture available in the market is mostly used by the livestock owners of peri-urban/urban areas. In rural areas home made concentrate mixture consisting of oil cakes along with wheat bran/wheat flour are usually offered to animals. The concentrate mixture is mixed with wheat straw and water to make "*Sani*" and offered to animal at the time of milking. The green fodder, if available, is also added in the "*Sani*" to fortify the ration. The crop husbandry is a major source of income for the farmers. Sale of milk also sizably contributes in daily income pool of the medium and small farmers.

2.3 Animal Production Systems in Vulnerable Zones

Livestock production involves both ruminants and non-ruminants and a variety of systems integrated with crops. The system varies as a function of agro-climatic zone and intensity of farming operations. The prevailing animal production systems in vulnerable zones fall into one of the three categories and it is appropriate to discuss these here briefly:

1. Landless animal production system
2. Crop based animal production system
3. Rangeland based animal production system

2.3.1 *Landless Animal Production System*

These are of two categories, i) Urban and peri-urban industrial landless system, and ii) Rural landless animal production system.

2.3.1.1 *Urban and Peri-Urban Industrial Landless System*

These landless systems are normally large, mainly industrial, very intensive and vertically integrated pig and poultry enterprises whose economic outputs are higher than the ruminant enterprises. The system involves the use of largely imported production inputs at high cost *e.g.*, germplasm, feed, supplements and medicine and technologies. This system is very efficient with production cycles of four to five crops of broilers of eight week each and an average efficiency of feed conversion of about 1.8–2.5.

Intensive production is the key feature of this system, in which the use of grain based feed stuffs, is the norm and maize represents a major item of cost of production. There are two concerns associated with this production system. One is the rising cost of maize and protein supplements. Secondly, in many areas there are serious problems of pollution, surface water contamination and human health hazards *e.g.*, bird flu.

2.3.1.2 *Rural Landless Animal Production System*

Rural landless animal production system refers mainly to ruminants, and is prevalent in the whole India, particularly in the arid and semi-arid zone of north India. These involve zero grazing practices and extensive system, which is, associated with the resource poor families with seasonal migration for search of feed for livestock. This system is very common in arid and semi arid regions of India.

2.3.1.3 *Rangeland Based Animal Production System*

This system is mainly found in arid and semi-arid regions of India. Sparse vegetation, containing mainly native grasses and shrubs are characteristics of range based production system.

2.3.2 *Crop Based Animal Production System*

The existence of interactive mixed crop-livestock systems is a typical characteristic of the Indian agrarian economy. This system form the back bone of Indian agriculture, and is especially important in terms of land area involved, extent of poverty, integrated natural resource management, food security and opportunity for increased food grain production. Mixed crop-livestock systems are partially closed and thus are environmentally the most benign. In this system, the waste products of one enterprise (crop production) are used by another enterprise (animal production), which in tern returns its own waste (manure) back to the first enterprise. In Indian subcontinent, small scale mixed crop and livestock farming is the lifeline of most of the small and marginal farms. They accounts for only 34% of arable land but contain 59% of bovine, 64% of ovine, 70% of the pig and 75% of the poultry (Rao *et al.*, 2004).

In the mixed farming systems, the animal component is varied and integrated with cropping in various ways. Both ruminants and non-ruminants are involved and the choice of one or more species is dependent on overriding influence of preference, market demand, income generation potential contribution to crop cultivation and livelihood. In India, mixed farming provides 92% of milk, 65% of ruminant meat, 55 % of pork and poultry and 20% of eggs. This production system is unlikely to change in near future, however, there will be increasing intensification and a shift especially from extensive to system combining arable cropping. It indicates that ruminant production in mixed farming system will continue to be important in future. The principle aim should therefore be to improve feeding and nutrition, and maximum use of the available feed resources, notably crop residues and leguminous forages as feed.

There are significant regional variations in the composition of livestock population and feed availability. For example, buffaloes are often found in more well-endowed and irrigated areas, while cattle, sheep and goat are predominant in rain fed areas. The suitability of livestock in different agro climatic regions not only serves the purpose of augmenting income, employment and food security but also acts as a storehouse for capital and an insurance against crop failure.

3.0 EXISTING FEEDING SYSTEMS IN THE VULNERABLE STATES

3.1 Andhra Pradesh, Karnataka and Tamil Nadu

Animal husbandry activity is one of the major components in the state of A.P., Karnataka and Tamil Nadu. This zone may be classified as moderately drought prone area. The main dry crop residues that are commonly fed to livestock are ragi straw, sorghum stover, maize kadbi, paddy straw, sugarcane tops, legume straw from pulses and groundnut haulms. The green forages are from fallow land, forest and cultivated fodder are mainly hybrid napier, pearl millet, guinea grass, leacauna, foxtail millet, sorghum, lablab. Primary grazing areas consist of permanent pasture and tree crops (like Glyricidia, Neem, Drum stick, Banana, Mulberry, Subabool etc.). The oil cakes (from Seasum, mustard, linseed, niger, safflower, sunflower, cottonseed), brans and chunis constitute the major portion while grain constitutes a small fraction of concentrate mixture. No specific grazing/ forage feeding is followed. Controlled grazing from June to March while open grazing in the forest area is common in the scanty rainfall zone of Rayalseema. In Nilgiri hills, the buffaloes are solely maintained on grazing. No supplemental feeding of concentrate is done.

3.2 Assam and Arunachal Pradesh

In Assam and Arunachal Pradesh, livestock is considered as one of the important homestead activities in the farm households while fishery enhances family income. Pig farming is most developed; goat is coming the next for meat production. Co-operatives have been initiated among pig farmers in the few far-east states. Crossbred cows are used mostly as milch animals. In Brahamaputra valley of Eastern Himalayas, swamp buffaloes (*Kaziranga*) are reared by the local inhabitants for both draught power as well as milk production. Mithun and yak are predominantly reared through grazing on pasture. In the rural areas, cattle, buffalo, sheep and goat are reared under traditional system of management *i.e.* the animals are let loose throughout the day and in the evening they are tied in animal shed. In urban areas, the animals are stall fed with cut grasses and tree leaves along with feeding roughage and concentrates. The cultivated forage resources are maize, cowpea, rice bean, Guinea grass, Dinanath grass, oat etc. Community grazing system is rarely seen even in rural areas.

3.3 Bihar

The advanced farmer in Bihar cultivate green fodder like maize, sorghum, berseem, hybrid napier etc. Fodder production practices are popular in Begusarai, Samastipur and Muzaffarpur districts because of provision of good quality seeds and extension services provided by milk cooperatives (Sohane *et al.*, 2001). The farmers usually chaffed the green fodder and offered to animal for better utilization. The use of concentrate pellets is common by the members of milk producer cooperative societies. The livestock owner of peri-urban/urban areas mostly uses the balanced concentrate mixture available in the market. In rural areas home-made concentrate mixture consisting of oil cakes along with wheat bran/wheat flour are usually offered to animals. The concentrate mixture is mixed with wheat straw and water to make "*Sani*" and offered to animal at the time of milking. The green fodder, if available, is also added in the "*Sani*" to fortify the ration.

3.4 Gujarat and Maharashtra

In this region, farmers keep milch animals only or in combination with poultry or goats. The household keeping flocks of sheep do not seem to keep any other type of livestock or poultry, however few goats are kept in sheep flocks to lead sheep flock at the time of grazing or migration. Crossbred cattle are found in the smallholder mixed farming system. Both nomads and stationary farmers rear cattle and sheep. Buffalo keeping is important activity among the local inhabitants. The dry roughage resources are comprised of wheat straw, groundnut haulms, bajra straw, maize straw, sorghum stover etc (Garg *et al.*, 2003). During rainy season, cows and buffaloes are fed less dry fodder but rest of the year the dry fodder is the major component of animal feeding. Plain zone has more area under forage while western zone has maximum pasture lands where free grazing is practiced. Cut and carry methods are used for grasses grown on farmer's land. Green fodder commonly grown in this area is bajra, cowpea, guar. hybrid napier, local grass, maize, lucerne, moth, sorghum. Green feeds in Surat consist mainly of sugarcane tops and in other places it is sorghum. Creeper of sweet potato is also fed to animals as green supplement. Although quality of the roughages is a major limiting factor, farmers having crossbreds tried to adjust to the needs of the cows by feeding concentrates. The major concentrate commonly used for feeding includes oil cakes (cottonseed, coconut, sunflower, mustard), grains (bajra, guar, maize, barley), by-products (wheat bran, gram chuni). Compounded feed manufacturer for pelleting and urea supplementation utilizes molasses available from sugarcane industries.

3.5 Jammu and Kashmir

The Jammu and Kashmir region is abundance in high yielding milch cattle and buffaloes. In the mid hills region, cattle are reared for milk as well as draught power, buffaloes and goats are maintained for milk production only. In the high altitude of western Himalayas are predominated by sheep and goat due to abundant availability of alpine pasture. Sheep and goats are normally maintained on grazing. Grazing and harvested grass constitute the main fodder for donkeys, ponies, mules etc.

3.6 Madhya Pradesh

In MP, small ruminant in particular plays an important role in ensuring rural livelihood against drought. The predominant farming is mixed crop livestock farming. The livestock sector in this region is mainly comprised of buffalo, cattle, sheep and goat. The ravines are predominated by the local buffalo breed (*Bhadawari*), local goat and *Jaluni* sheep. The local breeds of buffalo are being replaced by crossing with high yielding *Murrah* buffalo. Due to abundance of shrubs (*Jharberi, Pilua* and *Deshi Karaunda*) and trees (*Ruani, Chhonkra, Vilayati Babul*) in the huge tracts of this region, sheep and goat are concentrated and contributed significantly to the income of small and landless farmers. A system of animal grazing known as "*Chutta Pashu*" grazing system or "*Annapratha*" or "*Stray Animal Grazing*" is very common in Bundelkhand region of this zone (Shukla *et al.,* 2000). Grazing is done mostly on degraded lands. Forage resources come from natural grasslands, forest area and wastelands. The high yielding cattle and buffaloes are raised under stall-feeding system on balanced ration comprising of roughage and branded concentrate mixture. The basal roughage like wheat straw, groundnut haulms, pulses straw are mixed with home made concentrate mixture and water to make the "*Sani*".

3.7 Rajasthan

Livestock play an important role in the rural economy of Rajasthan. Small ruminants dominate the livestock population and their numbers are increasing fast, while the large ruminant population is static. The majority of the tribal families maintain cattle, goats and backyard poultry. Very few tribal families keep buffalo. In central Rajasthan, the main crops are maize, wheat or pearl millet, along with pulses. The dominant livestock species are cattle and goats. Lucerne is the crop of choice for fodder production and farmers are not interested in producing green fodder during rainy season. Among tribal families, milk is not a major part of their diet and the main objective of keeping cattle are for draught purposes or as source of fuel and manure. The roughage resources include pala, moth chara, groundnut haulmns, guar, phalghati. Creepers like *Tinosperma* and leaves of *Alangium* and *Moringa* species are fed to milch cows and buffaloes. The pods of *Acacia* and *Prosopis* species are mostly fed to goat and buffalo. The leaves of *Azadirachta* and *Acacia* species are fed mainly to goat. The majorities of the households do not feed mineral mixture and salt regularly. The majority of the farmers do not maintain regular feeding of milk cows and group feeding are practiced once a day. The mahuan (*Madhulica indica*) flowers are mostly fed to bullocks during the working season and if surplus quantity is available these are offered to milch animals. Farmers have been using cottonseed and cotton seed cake for a long time and always-claimed beneficial effects for the animal and for the fat content of milk. In the rainy season, green grasses like bhusal, anjan, sevan, bokoria, gokhru are available. The crops of choice are maize, sorghum, pearl millet, wheat, pulses and groundnut which are good sources of fodder and well-suited to rain fed farming. The farmers, especially the women, collect the crop residues and store them meticulously. Special care is taken in the collection and storage of crop residues from pulses and groundnut crops. During winter, it is a common practice to mix small quantities of green lucerne with cereal straws. While in summer, the cereal straws are supplemented with the leaves of phog, khess, bawali, khejri, neem etc. and pods of pulses, groundnut leaves and acacia pods are fed to animals. The common livestock feeding practices in the tribal region include grazing in the forest and the use of crop residues supplemented with the leaves, flowers and pods of forest trees. Better producing animals are given concentrates which are generally home-made using farm produces like grains, by-products of pulses, tree pods, flowers, etc.

3.8 Uttaranchal

Buffalo is the main milch animal in the Uttaranchal hills. Two main operations on pastures and grazing land are: grazing and hay-cutting. Grazing in the forest areas and sub-alpine and alpine pastures is the mainstay for the animals.

3.9 West Bengal

In West Bengal, the goats occupy a significant proportion of livestock population because of their characteristics meat quality and its ability to survive under poor management conditions. The livestock are maintained mainly on grazing and agricultural crop residues like paddy straw with little access of agricultural byproducts. Agricultural lands are invariably the only grazing sites for rural livestock after harvesting the crops. Animals are tethered also in the river sores, edge of the ponds, canals, and sides of the road. In addition to grazing mostly on bunds and roadside grasses, the animal receives 4–6 kg paddy straw as bulk roughage (Das *et al.*, 2003). Besides paddy straw, cattle and pigs are also fed husk and kitchen waste. Other crop residues which are used for feeding the animals are black grams, pigeon peas etc. Ragi straw is generally not collected from the field but it is left there as stable for open grazing. Areas close to forest, cattle, sheep and goats are taken for grazing and they were set free in the forest while in the evening they are taken back to the village. The feeding of large ruminants is mainly from crop residues grown by the farmers. The straws of wheat/paddy constitute the basal dry roughage of the animals. The minor source of dry roughage includes pulse straw.

3.10 Western Uttar Pradesh and Haryana

In this region, crossbred cattle as well as high-yielding buffalo breeds are concentrated, due to a higher emphasis of crossbreeding programmes and commercial orientation of dairying. *Jamunapuri* goat is famous in this region for its high milk yield. Livestock rearing is an important subsidiary occupation in the rural areas of Uttar Pradesh. Buffalo enterprise plays a crucial role in providing income to the marginal and small farmers. The '*Bhadawari*' breed of buffalo is crossed with *Murrah* for enhancing the milk productivity. The buffaloes are maintained on green and dry fodder produced in own farms. The animals are fed with concentrate like cottonseed cake, mustard cake, barley husk and wheat flour, rice bran, wheat bran, gram churi etc. depending upon their availability. The concentrate is either home grown or procured from the market. Dry roughage includes wheat straw, sorghum kadbi, pearl millet kadbi etc. A number of livestock owner also allocate their substantial lands for cultivation of green fodder. The commonly grown fodder is berseem, oat, maize, bathua etc. Most of the farmers fed crop residues and concentrate mixture in the form of '*Sani*' (Thakur and Tomar, 2001).

4.0 REFERENCES

Das, A., Ghosh, T.K. and Haldar, S. 2003. Mineral distribution in soil, feeds and grazing cattle of different physiological stages in the red laterite and new alluvial agro climatic zones of West Bengal. *Indian Journal of Animal Sciences*, **73(4)**: 448–454.

Dubey, S.C., Shinde, A.K. and Bhati, P. 2002. Strategies and challenges during drought to prevent animal life and production losses. Vth National Seminar on strength, Challenges and Opportunities in Small ruminant diseases in new millennium, Jaipur, December, 30–31, 2002.

Garg, M.R, Bhandari, B.M. and Sherasia, P.L. 2003. Macro mineral status of feeds and fodder in Kutch district of Gujarat. *Animal Nutrition and Feed Technology*, **3**: 27–30.

Narayan, Pratap and Rajora, M.P. 2004. Strategies for enhancing biomass production of rangelands. Proceeding of National Seminar on Opportunities and Challenges in feeding of sheep, goat and rabbit for sustainable production. CSWRI, Avikanagar, Feb. 10–12, 2004.

Rao, P., Parthsarathy, Birthal, P.S., Kar Dharmendra, Wickramaratne, S.H.G. and Shreshta, H.R. 2004. Increasing livestock productivity in mixed crop-livestock systems in south Asia. Report of a project, National Cenre for Agricultural Economics and Policy Research, New Delhi, ICRISAT, Patancheru, A.P., India. pp. 168.

Singh, Punjab 1995. Rangelands and their improvement in India: A review. Annals of Arid Zone, **34(3)**: 157–162.

Shukla, S.K., Solanki, K.R. and Dwivedi, R.P. 2000. Plant multipurpose trees and shrubs for live fence in Bundelkhand region. *Indian Farming* **1**: 11–15.

Sohane, R.K., Jha, P.B., Chaudhary, M.K. and Kumari, Asha 2001. Data base information on feeds, fodders and livestock in North Bihar. Proceedings of 10th Animal Nutrition Workers Conference held at NDRI, Karnal from November 9–11, 2001. Abstract No. 097.

Thakur, S.S. and Tomar, S.K. 2001. A study on the feeding practices being followed by dairy farmers of Karnal area. Proceedings of 10th Animal Nutrition Workers Conference held at NDRI, Karnal from November, 9–11, 2001, Abstract No. 088

4

EARTHQUAKE AND ITS IMPACT ON ANIMALS

S.K. NAG AND N. DAS

ABSTRACT

Earthquake is one of the many natural disasters which can cause a great damage, loss or destruction of human and animal lives apart from other effects. It occurs without warning and involves violent shaking of the ground and everything over it. On an average about 1,000 earthquakes with intensities of 5.0 or greater are recorded each year in different parts of the world. During the last 15 years India has experienced 10 major earthquakes. The impacts of earthquake normally encountered due to shaking and ground rupture. After a devastating earthquake, there is damage to building, animal sheds, roadways, fire and electrocution from fallen power line. Apart from shaking and ground rupture there may be some other effects like landslides and avalanches, fires, soil liquefaction, tsunami, floods etc. The leading causes of death and injury of large animals during earthquake are collapsed barns, dehydration, electrocution and accidents resulting from fencing failure. The necessary actions to be taken for the management of animals after earthquake are evacuation, providing temporary shelter, proper veterinary care and hospitalization, arrangement of feed, water and medicines, handling equipment and tools. In earthquake prone area animal sheds must be designed to resist the stresses caused by the tremors. A disaster plan should be made to protect properties, facilities and animals. Farm disaster kit should be made ready so that necessary equipments or tools are readily available in the event of an earthquake. To cater the needs of animals and their owners, a committee of individuals and groups who are legalized and / or qualified to provide resources and expertise for the care of animals, should be formed.

Indian Grassland and Fodder Research Institute, Jhansi (UP), India

1.0 INTRODUCTION

Earthquake is one of the many natural disasters which can cause a great damage, loss or destruction of human and animal lives apart from other effects. It occurs without warning and involves violent shaking of the ground and everything over it. It results from the release of accumulated stress of the moving lithospheric or crustal plates. The earth's crust is divided into seven major plates, some 50 miles thick, which move slowly and continuously over the earth's interior and several minor plates.

Earthquakes are tectonic in origin; that is the moving plates are responsible of the occurrence of the violent shaking.

On average about 1,000 earthquakes with intensities of 5.0 or greater are recorded each year in different parts of the world including many which occur under the ocean. Notable earthquakes have occurred at different time in different places like Lisbon, Portugal (1755); New Madrid, Mo. (1811 and 1812); Charleston, S.C. (1886); Assam, India (1897 and 1950); San Francisco (1906); Messina, Italy (1908); Gansu, China (1920); Tokyo, Japan (1923); Chile (1960); Iran (1962); Managua, Nicaragua (1972); Guatemala (1976); Hebei, China (1976); Mexico (1985); Armenia (1988); Luzon, Philippines (1990); N Japan (1993); Kobe, Japan (1995); Izmit, Turkey (1999); Central Taiwan (1999); Oaxaca state, Mexico (1999); Bam, Iran (2003); NW Sumatra, Indonesia (2004) and Pakistan (2005). Some of the earthquakes like those of Lisbon, Chilean, and Sumatran's were accompanied by tsunamis.

1.1 Earthquake Risk in India

During the last 15 years, the country has experienced 10 major earthquakes that have resulted in over 20,000 human deaths. As per seismic zone map of the country (IS 1893: 2002), over 59 percent of India's land area is under threat of moderate to severe seismic hazard. In fact, the entire Himalayan belt is considered prone to great earthquakes of magnitude exceeding 8.0, and in a relatively short span of about 50 years, four such earthquakes have occurred: 1897 Shillong (M8.7), 1905 Kangra (M8.0), 1934 Bihar-Nepal (M8.3), and 1950 Assam-Tibet (M8.6). Scientific publications have warned of the likelihood of the occurrence of very severe earthquakes in the Himalayan region, which could adversely affect the lives of several million people in India. The north-eastern part of the country continues to experience moderate to large earthquakes at frequent intervals including the two great earthquakes mentioned above. Since 1950, the region has experienced several moderate earthquakes. On an average, the region experiences an earthquake with magnitude greater than 6.0 every year. The Andaman and Nicobar Islands are also situated on an inter-plate boundary and frequently experience damaging earthquakes. Some of the recent and devastating earthquakes that happened in India were at Uttarkashi (Magnitude 6.5) in 1991, Latur (Killari, Maharashtra) in 1993, Jabalpur in 1997 and Gujarat in 2001. Table 1 gives some of the major earthquakes in India.

Table 1 Some major earthquakes in India

Date	Earthquake	Time	Magnitude	Max. intensity	Human deaths in India
16 June 1819	Great Kachchh	11:00	8.3	IX	1,500
12 June 1897	Great Shillong	17:11	8.7	XII	1,500
08 February 1900	Coimbatore	03:11	6.0	VII	Not Known
04 April 1905	Great Kangra	06:20	8.0	X	19,000
15 January 1934	Great Bihar-Nepal	14:13	8.3	X	11,000
26 June 1941	Great Andaman	??	8.1	X	Thousands
15 August 1950	Great Assam	19:31	8.6	XII	1,530
21 July 1956	Anjar	21:02	6.1	IX	115
10 October 1956	Bulandshahar	??	6.7	VIII	Many
28 December 1958	Kapkote	??	6.3	VIII	Many
02 September 1963	Badgam	07:04	5.5	VII	Hundreds
10 December 1967	Koyna	04:30	6.5	VIII	200
23 March 1970	Bharuch	20:56	5.2	VII	30
19 January 1975	Kinnaur	??	6.5	VIII	Not Known
29 July 1980	Pithoragarh	18:28	6.8	??	>150
31 December 1984	Silchar	04:53	5.6	??	20
26 April 1986	Dharamshala	13:05	5.5	??	6
21 August 1988	Bihar-Nepal	04:39	6.6	IX	1,004
20 October 1991	Uttarkashi	02:53	6.4	IX	768
30 September 1993	Killari (Latur)	03:53	6.2	VIII	7,928
22 May 1997	Jabalpur	04:22	6.0	VIII	38
29 March 1999	Chamoli	00:35	6.6	VIII	63
26 January 2001	Bhuj	08:46	7.7	X	13,805
14 September 2002	Diglipur	03:58	6.0	VII	-
26 December 2004	Great Sumatra	06:28	9.3	XII	10,749
08 October 2005	Kashmir	09:20	7.4	X	1,308
14 February 2006	Sikkim	06:25	5.7	VII	2

Source: http://ndma.gov.in/ndma/

2.0 IMPACT OF EARTHQUAKE

2.1 Direct Effect

The effects or impacts of the earthquakes normally depend on the intensity at which they appear. The magnitude of earthquake is expressed in Richter scale. The earthquakes of magnitude of 3 or lower being mostly imperceptible while those of magnitude 6.5 or more causing serious damage over large areas. The impacts normally encountered due to shaking and ground rupture following a devastating earthquake are as follows:

- Damage to building, utility lines, bridges or dams
- Damage to roadways and other means of transportation causing food and other resource shortages for both man and animals
- Partial or total building collapse, flying glass from broken windows, overturn of large furniture falling on people and animals are the major causes for mortality and injury.

Fire and electrocution from fallen power line are also responsible for animal mortality and injury.

Apart from shaking and ground rupture there may be some other effects like

- Landslides and Avalanches
- Fires
- Soil liquefaction
- Tsunami
- Floods

The 1991 Uttarkashi earthquake which took place in the Tehri region of Garhwal Himalaya, in proximity of the Tehri dam, was very damaging which killed many people and also animals and damaged completely or partially about 48000 houses.

Among the recent Indian earthquakes the Latur earthquake was the most devastating stable continental region (SCR) earthquake in the world which occurred in the typical rural setting of India on September 30, 1993 having an intensity of 6.4. There was extensive damage to the life and property and the severity of the destruction was compounded by the nature of village settlements. While most of the engineered structures survived the earthquakes, the non engineered ones were totally damaged. Overall 8000 human lives and 16000 livestock's were killed apart from injury caused to 16000 people. About 52 villages were razed to the ground, 27000 houses with all amenities and infrastructures damaged, 30000 houses collapsed, 211000 houses in 13 districts suffered damages of varying degree. The disposal of dead bodies of human and cattle was a stupendous task.

The earthquake of magnitude 6.0 occurred in Jabalpur on May 22, 1997 which caused widespread devastation in and around Jabalpur and left 48 people dead, many injured and homeless.

The 2001 Gujarat earthquake occurred on January 26, 2001 having epicenter at Bhuj and intensity of 8.1 on the Richter scale. It was one of the two most deadly earthquakes to strike India in the recorded history. The death toll was about 20000 while the number of injured was 167000. More than 20000 cattle were reported to be killed due to fallen debris and the earthquake destroyed near a million homes. After the earthquake the animals were helter skelter in pain, dropping dead in odd places. Even army personnel found it difficult to locate and burn the carcasses.

2.1.1 *Damage of Animal Sheds During Earthquake*

During earthquake, the most critical and weak element of the farmhouse is its wall, especially if it is made of loosely cemented stones or bricks. Foundation of the building is not damaged generally. Roofs fall in only when the walls and wooden frames collapse. If the wooden frames remained standing, even in a fitted condition the roof remains intact and the people/livestock survive. However, due to shaking the large mass of the roof may cause damage to the walls which may crumble down if it were not built properly. When walls are not interconnected, they may separate at the corners and junction (with roof); this leads to the destruction of "integral box action" of the enclosure leaving the walls to fend for themselves individually and these overturned as a whole or in part. With the fall or damage of walls, the frame of the roof moves laterally and eventually collapses. The roof comes down with the frames and crushes everything underneath. In cases where frames were held by the fallen debris of the walls, the roof does not collapse.

(i) Most injuries and death of livestock in earthquake result from building collapse and falling objects. Secondary injuries result from panic. As a result of severe earthquake there might be death of large number of animals in the affected household or in the streets; (ii) With fences and sheds destroyed by the impact of earth quake, livestock move in the open, often with injuries; (iii) Many animals are trapped inside the debris and they need immediate evacuation; (iv) After an earthquake approximately 60% of the displaced households will have companion animals requiring medical care and sheltering; (v) Animal may become nervous and apprehensive and may bite, kick or scratch during post earthquake period; (vi) Dead bodies of animals are lying down openly, which becomes a serious public health concern.

2.2 Indirect Effect

(i) After a strong earthquake, apart from loss of precious human lives, thousands of animals are either killed or injured, leaving countless rural communities, which had been dependent on livestock for their livelihood and food, destitute and hungry; (ii) There is damage to barns, fences and other enclosures that may require immediate repair and additional expenditure by the livestock owners; (iii) Loss of pet animals causes significant impact on public mental health due to emotional involvement of the owners and feeling of guilt and anger due to loss, injury or death of the pet animals.

3.0 MANAGEMENT OF ANIMALS AFTER EARTHQUAKE

Aftershocks are very common in earthquakes. Sometimes earthquakes of small magnitudes precede the strong one. The incidence of earthquake may also repeat number of times in quick succession. Thus, whenever, an earthquake happens, following measures need to be taken to prevent further damages or loss of animals.

A **Temporary animal shelter:** An animal sheltering/stabling plan should be developed based on the operational area concept *to include identifying transportation requirements*. The activities of animal sheltering facilities should be monitored including the tracking of animals. Animals may not be allowed into Red Cross shelters for health and space reasons. Prepare an emergency plan and know where you will take all of your animals in the event of a disaster. Temporary animal shelters may be set-up; however, these will fill rapidly. An ideal situation during disasters is to have a pre-determined place that will provide a living space for your animals. Survey the area for the best location for shelter. If the pasture area meets the following criteria, then the large animals may be better off in the pasture than being evacuated. Please ensure that the shelter place does not have

- non native trees
- overhead power lines
- debris or sources of flowing debris
- barbed wire fencing
- less than one acre in size

B Store adequate feed and fodder for at least 2-3 days.

C All the family members and animals should take tetanus vaccination.

D If the persons/pet animals are in indoors during earthquake, remain indoors and take cover under a sturdy piece of furniture (such as heavy desk, table or bed) and hold out one of the legs of pet animals.

E During earthquake, tools, equipment and other objects in the farmhouse are likely to fall and cause injury to man and animals.

F Stay away from glass windows, mirrors or sky lights that can be shattered during earthquakes.

G Include a pair of bolt cutter in disaster kit to open gate when they become damaged or unable to be opened.

H Purchase earthquake insurance for farmhouse and machineries.

I Keep contact number for medical/veterinary care.

J Most of the livestock houses in villages are built on weak foundation. Since safe building construction practices are not followed the animals need to be kept in open. Wherever housing facilities are good, all almirah and cabinets should be firmly secured to the walls to prevent them from falling. All such cupboards and cabinets should be locked to prevent them from opening in the event of an earthquake.

K If animals are required to be kept inside, either keep them loose or *tie them closely so that they have the freedom to change their standing places*.

L Evacuation of enlarged (trapped) or injured animals.

M The behavior of animals may change dramatically after an earthquake. Normally quiet and friendly animals may become furious. Monitor animals closely and only release them in safe and secure enclosures.

N The leading causes of death and injury of large animals in natural calamities including earthquake are collapsed barns, dehydration, electrocution and accidents resulting from fencing failure. Precautions should be taken to protect the farm animals from these hazards, no matter what the disaster potential of the area is. Following points need to be kept in mind, while planning preparedness for animal evacuation.

- Evacuate the animals as soon as possible. Evacuation team should be ready to leave once the evacuation is ordered.
- Work within the community to establish safe shelters for farm animals. Potential facilities include fairgrounds, other farms, racetracks, humane societies and convention centers. Survey the community and potential host communities along with the planned evacuation route.
- Contact the local emergency management authority and become familiar with at least two evacuation possible routes well in advance.
- Set up safe transportation including trucks and trailers suitable for livestock and appropriate for each type of animal, along with experienced handlers and drivers.
- Make sure that all the disaster supplies are available at the evacuation site. These include feed, water, veterinary supplies, handling equipment, tools etc.

O If evacuation is not possible, during or after the disaster, then it must be decided whether to confine large animals to available shelter on your farm or leave them loose in pastures. While it may seem that animals will be safer inside the barns, in many circumstances, confinement can reduce their ability to protect themselves.

P Veterinary medical care and carcass disposal: The medical care for all the injured animals should be arranged immediately following an earthquake or any other natural disaster. This should be done in a very coordinated manner. Veterinary doctors should be called from different area and teams should be formed for providing veterinary assistance to treat the animals. Facilities should be identified in operational area for providing emergency medical care, hospitalization, safe storage of feed and medicines. Earthquake-proof, semi-permanent veterinary clinics and hospitals should be established in the earthquake prone area to provide ongoing treatment. Vaccination programme should also be initiated among the injured animals to prevent the spread of diseases.

Q Listing of resources available in the area should be developed to conduct search, capture and treat the companion animals and livestock. After earthquake large numbers of animal casualties are noticed. Therefore, a proper carcass disposal plan should be in place in each operational area and appropriate agencies may be entrusted with the job keeping in view the animal disease and public health concerns.

R Reconstruction of damaged building: Repair and seismic retrofitting should be the key priority while undertaking repair works. As foundations are not generally damaged, walls can be raised on old foundation after incorporating earthquake-resisting measures such as reinforced concrete ring beams. The seismic retrofitting scheme consists of the following elements.

- Install angle iron nailed to wooden posts and beams.
- Reduce the roof thickness
- Construct a seismic band of iron-cement around each room, just above the door lintel level.
- If the wall is not damaged much, it can be strengthened by installing "through" or bond cements in the wall. The through hole is filled with cement concrete and reinforcing iron bars of 8mm diameter. These are constructed '*in situ*' at vertical and horizontal spacing of iron.
- Water proofing the parapets with concrete cement and sand plaster embedded in chicken wire mesh. This helps maintain the dry strength of the mud mortar in the walls.

The methodology of retrofitting can also be adopted with minor modification in the existing building at a cost of 15-20% of building replacement cost.

4.0 PREPAREDNESS

It is vital that animal facilities (which include animal shelters, kennels, veterinary clinics, hospitals, pet shops, horse stables and other facilities where animals are housed) prepare for disasters, both to protect the animals housed there and to safeguard the facilities. A lack of planning can increase the magnitude of the disaster for humans and animals alike, while a proactive approach to planning and preparation for potential emergencies allows an organization to mitigate damage and downtime. The mission for response to earthquake with regard to animal lives is to save surviving animals stranded or at risk of starvation or disease, to coordinate evacuation, veterinary medical care, sheltering and carcass disposal following the disaster.

4.1 Building (Shed)

In areas where earthquakes occur frequently buildings must be designed to resist the stresses caused by the tremors. While the intensity of tremors can be much greater in loosely compacted soil than in firm soil or solid bedrock, one- and two-story buildings are at greater risk on the firm ground or bedrock because of the shorter resonance periods. Casualties are most likely to be caused by the collapse of walls causing the roof to fall, and the failure of projecting elements such as parapets, water tanks, non-monolithic chimneys and loose roof coverings. The outbreak of fire caused by the fracture of a chimney or a break in the mains supply line presents an additional hazard.

While small buildings, having timber frame walls or a wooden ring beam supported by the posts of a mud and pole wall, can resist quite violent earthquakes, the following measures will increase the resistance of a large building to collapse due to earth tremors:

- Use a round or rectangular shape for the building. Other shapes such as "L" "T" or "U" should be divided into separate units. To be effective, this separation must be carried down through to the foundation.
- Avoid large spans, greatly elongated walls, vault and dome construction and wall openings in excess of one third of the total wall area.
- Construct a continuously reinforced footing that rests on uniform soil at a uniform depth - even on sloping ground.
- Securely fix the roof either to a continuously reinforced ring beam on top of the walls, or to independent supports, which will not fail even if the walls collapse.
- Avoid projecting elements, brittle materials and heavy materials on weak supports.
- Avoid combustible materials near chimneys and power lines.

Ductile structures have many joints that can move slightly without failing, *e.g.*, bolted trusses. Such structures have a greater capacity to absorb the energy of earthquake waves. A symmetrical, uniformly distributed ductile framework and with the walls securely fixed to the frame, is suitable for large buildings. Masonry walls are sensitive to earthquake loads and tend to crack through the joints. It is therefore important to use a good mortar and occasionally reinforcing will be required.

4.2 Farm Disaster Kit

Make a disaster kit so that supplies remain on hand in the event of a disaster. Place the kit in a central location and let every one know where it is. Check the contents regularly to ensure fresh and complete supplies. The kit includes the following items:

- Current list of all animals, including their locations and records of feeding, vaccinations and tests.
- Supplies for temporary identification of animals such as plastic neckbands, permanent markers to label them with the name of the owner/farm, address and telephone numbers.
- Basic first aid kit.
- Handling equipment like halters, cages and appropriate tools for each kind of animals.
- Water, feed and buckets.
- Tools and supplies needed for sanitation.

4.3 General Precautions

- Make a disaster plan to protect property, facilities and animals. Create an emergency telephone number list of neighbours, veterinarians, poison control office, agricultural colleges and universities etc.
- Make sure every animal has durable and visible identification.
- Ensure that poultry have access to high areas in which to perch, as well as, to food and clean water.
- Perform regular safety checks on all utilities, building and facilities.
- Use only native and deep rooted plants and trees to prevent storm damage.
- Install a hand pump and obtain enough large containers to water the animals for at least a week in the event of any problem in water due to natural disaster.
- Identify alternate water and power sources for the crisis period.
- Review and update disaster plans, supplies and information regularly

4.4 Plan for Animal Identification and Care

To cater the needs of animals and their owners in small and large emergencies, disasters and evacuations, a committee of individuals and groups, who are legalized and/or qualified to provide resources and expertise for the care of animals in disasters, should be formed. The goal of the committee is to develop a programme for providing assistance and relief during emergency situations. The responsibilities of the committee are:

- To be the principal agency for management of animals during emergency situations like earthquakes, floods, droughts etc.
- To familiarize the members with emergency operations, procedures.
- To become trained according to requirement of emergency management agency.
- To identify qualified group of persons who will care for animals in emergency/disasters.
- To identify resources that would be required to respond to the need of animals in distress.
- To maintain current list of committee members and community resources which support planning and operations for the care of animals in emergencies.
- To maintain emergency management methods for liability protection and resources funding.

The animal care committee should seek funding support for their efforts. They should also identify suppliers of cages, food and accessories for care of animals in emergencies, disasters and evacuations. The main responsibility of the animal care professionals which include veterinarians, veterinary technicians, breeders etc is to provide care for injured animals.

5.0 REFERENCES

Disaster planning for Animal facilities, The Humane Society of the United States.

Disaster preparedness for livestock.

File://F:\Farm structures-animal housing.htm

Low cost constructions resistant to earthquake and hurricanes, United Nations Department of Economic and Social Affairs, New York, United Nations.

http://www.animaldisasters.com/planning guidelines.htm

http://www.animalcare.lacounty.gov

http://www.arkanimals.com

http://en–wikipedia.org/wiki/2001_Gujarat_earthquake

http://latur.nic.in/html/earthquake.htm

File://E:\Animals%20and%20Earthquake-2.htm

File://E:\Animals%20and%20Earthquake-3.htm

5

FEED AND FODDER REQUIREMENT AND AVAILABILITY IN THE VULNERABLE ZONES OF INDIA

K.K. Singh, Anil Kumar and A.K. Misra

ABSTRACT

Livestock play a key role in the vulnerable zones, particularly in the drought and flood prone zones by providing livelihood to about 50 percent human population of the country. In addition, this zone is also providing shelter to about 76 percent livestock population. On an average there is a shortage of 34 and 37 percent dry matter in the drought and flood prone zones. However, this situation is further exaggerated depending upon the intensity of drought or flood. Ensuring availability of adequate feed resource both in quantitative and qualitative terms would be the major concern in the years to come for increasing the livestock productivity. Feeding alone accounts for about 60-70 percent of the total cost of livestock production. As such there is a shortage of about 34 percent of dry matter, 43 percent crude protein and 45 percent total digestible nutrients in the drought-prone zone and about 37 percent dry matter, 46 percent crude protein and 42 percent total digestible nutrients in flood prone zones. Therefore, the judicious utilization of existing feed resources would be the important issue for making the livestock more remunerative. The chapter also contains the major feed resources available in this zones/states. Attempts have been made to generate information regarding feed-fodder requirement at maintenance level during drought and flood period assuming 4 and 2 months duration, respectively.

1.0 INTRODUCTION

India possesses about 3.28 million square km of geographical area. Out of which about 1.33 million square km *i.e.*, about 40.55 percent of the total

Indian Grassland and Fodder Research Institute, Jhansi (UP), India

geographical area of the country is under drought and flood prone zone. These zones provide livelihood to about 49.79 percent of human population and shelter to about 75.54 percent of total livestock population. Out of 584 districts in India, 357 districts come under drought and flood prone zone. These zones (drought and flood) provides asylum to about 68 percent cattle, 84 percent buffalo, 86 percent sheep, 77 percent goats, 65 percent pig, 73 percent camel and 33 percent other livestock (Table 1).

Table 1 Area and population (million) in the drought and flood prone zones of India

Particulars	India	Drought prone area	Flood prone area	% of India Drought	% of India Flood
Area (million km^2)	3.28	1.10	0.23	34.27	7.01
District (Nos.)	584.00	139.00	218.00	23.80	37.33
Human population	1026.57	291.27	511.16	28.37	49.79
Cattle population	185.23	54.90	70.93	29.64	38.29
Buffalo population	97.94	29.15	53.38	29.76	54.50
Sheep population	61.47	37.67	15.59	61.28	25.36
Goat population	124.29	43.84	51.48	35.27	41.42
Pig population	13.48	2.36	6.39	17.51	47.40
Camel population	0.63	0.41	0.05	65.08	7.94
Others population	2.99	0.39	0.61	13.04	20.40

1.1 Drought Prone Zones

The drought prone area spread over in 14 states, which include Andhra Pradesh, Bihar, Gujarat, Haryana, Jammu & Kashmir, Jharkhand, Karnataka, Madhya Pradesh, Maharashtra, Orissa, Rajasthan, Tamil Nadu, Uttar Pradesh and West Bengal. Out of 584 districts in India, 139

Table 2 Area, human and livestock population in drought prone states

States	No. of districts	Area (km^2)	Human population	Livestock population	Livestock density per 1000 human
Andhra Pradesh	9	125348	30453649	22809820	749.00
Bihar	13	32005	22363260	8196135	366.50
Gujarat	15	152018	33118148	15823950	477.80
Haryana	7	16715	8679676	3218919	370.86
Jammu & Kashmir	2	16241	1429439	3103608	2171.21
Jharkhand	2	12851	3126155	2057584	658.18
Karnataka	19	141920	41705282	21106792	506.09
Madhya Pradesh	12	89988	16768777	10438005	622.47
Maharashtra	11	125377	35777587	17370583	485.52
Orissa	4	24330	3373493	2667734	790.79
Rajasthan	14	214952	25310146	27384995	1081.98
Tamil Nadu	17	83394	33627719	15901304	472.86
Uttar Pradesh	11	41178	20167499	7487123	371.25
West Bengal	3	27222	15365528	11159026	726.24
Total	**139**	**1103539**	**291266358**	**168725578**	**579.28**

districts come under drought prone. Among various drought-prone states, Bihar has the highest human population density of 698.74 per km^2 followed by West Bengal (564.45 per km^2). According to the livestock census 2003, drought prone states of India had 54.90 million cattle, 29.15 million buffaloes, 37.67 million sheep, 43.84 million goats, 0.41 million camel, 2.36 million pigs and 0.39 million other livestock.

There is a wide variation in the density of livestock in the drought-prone states, when represented per thousand human populations (Table 2). The Jammu & Kashmir possess highest total livestock density of 2171 per thousand human population followed by Rajasthan 1081. The Jammu & Kashmir possess the highest cattle, sheep and goat density of 602, 877 and 424, respectively and Haryana possess the highest buffalo density (253) while Jharkhand and Rajasthan have highest pig (41) and camel (13) density per thousand human populations.

1.2 Flood Prone States

The flood-prone states are also spread over 14 states that include Andhra Pradesh, Arunachal Pradesh, Assam, Bihar, Delhi, Gujarat, Haryana, Kerala, Manipur, Orissa, Punjab, Uttar Pradesh, Uttaranchal and West Bengal (Table 3). Amongst these states, total 218 districts have been identified as flood prone. Among flood prone states, Delhi has the highest human population density of 9293.98 per km^2 followed by Bihar (909.06), West Bengal (903.88), Kerala (845.37) and Uttar Pradesh (746.15). The flood-prone states of India provides livelihood to about 511.16 million human

Table 3 Area, human and livestock population in flood prone states

States	No. of districts	Area (km^2)	Affected area (km^2)	Human population	Livestock population	Livestock density per 1000 human
Andhra Pradesh	11	154321	19098	37771725	24445802	647.20
Arunachal Pradesh	1	8325	406	103575	102234	987.05
Assam	14	53161	8995	18158350	9363786	515.67
Bihar	34	85379	32645	77614950	24952021	321.48
Delhi	1	1483	116	13782976	372363	27.02
Gujarat	13	73204	14010	30565861	11474505	375.40
Haryana	18	42313	6810	20270967	8461222	417.41
Kerala	13	36732	6582	31051992	3328136	107.18
Manipur	6	13366	1091	2048434	722061	352.49
Orissa	8	29236	5604	12737396	7218528	566.72
Punjab	17	50362	27150	24289296	8602597	354.17
Uttar Pradesh	60	211513	69929	157820788	56164538	355.88
Uttaranchal	4	11925	674	4720756	1567884	332.13
West Bengal	18	88752	34804	80221171	41611603	518.71
Total	**218**	**860072**	**227914**	**511158237**	**198387280**	**388.11**

populations and shelter to about 198.39 million livestock population. This zone has 70.93 million cattle, 53.38 million buffaloes, 15.59 million sheep, 51.48 million goats, 0.05 million camel, 6.39 million pigs and 0.61 million other livestock. Amongst flood-prone states, the highest livestock density per thousand human populations was found in Arunachal Pradesh (987) followed by Andhra Pradesh (647) and Orissa (566), while lowest in Delhi.

In the flood prone states, the highest cattle density (per thousand human population) was found in Orissa (371) followed by Assam (311) while lowest in Delhi (7). The flood-prone part of Haryana posses the highest buffalo density and Andhra Pradesh possess highest sheep density (280) while West Bengal possess highest density of goat (234) and Arunachal Pradesh possess the highest density of pig (344).

2.0 FEED-FODDER AVAILABILITY AND THEIR REQUIREMENTS

In the drought and flood prone zones, the livestock are mainly fed on crop residues in addition to the grazing of the available rangelands with meager amount of concentrate supplementation. The crop residue and their by-products constitute the main feed resource accounting for 70-80 percent of the total dry matter availability for various species of livestock. The green fodder and concentrate contribute about 10-20 and 7-10 percent of the total dry matter requirement.

2.1 Drought Prone Zones

The availability of dry fodder, concentrates and green fodder as fresh basis in the drought-prone states are to the tune of 104.94, 11.63 and

Table 4 Feed and fodder availability (000, tons) in drought-prone states

States	Roughage	Concentrates	Green fodder
Andhra Pradesh	10132.25	1778.04	13772.87
Bihar	4778.45	484.74	8676.71
Gujarat	11704.03	2073.96	12237.94
Haryana	4077.90	602.69	3800.37
Jammu & Kashmir	756.26	72.47	2462.52
Jharkhand	288.17	25.25	1025.65
Karnataka	22802.57	1601.42	20030.32
Madhya Pradesh	4883.05	1496.87	18773.05
Maharashtra	21062.21	736.96	15580.04
Orissa	945.70	65.05	5900.52
Rajasthan	6099.79	937.01	13600.75
Tamil Nadu	12823.63	1050.03	9172.24
Uttar Pradesh	52.91.67	310.16	7767.21
West Bengal	4584.11	395.17	3229.30
Total	**104938.12**	**11629.82**	**136029.49**

136.03 million tons (Table 4), respectively while in terms of dry matter, the availability is 94.44, 10.47 and 27.21 million tons, respectively against the estimated requirement of 103.69 million tons of dry fodder, 41.48 million tons of concentrates and 62.22 million tons of green fodder. Therefore there is a shortage of about 9% dry fodder, 56% green fodder and 75% concentrates in this zone under normal years. Thus making an over all deficit of about 34% of dry matter in this zone (Table 5). This shortage may be inflated in drought years depending upon the intensity and duration of drought.

Table 5 Dry matter availability and requirement (000, tons) in drought-prone states

States	Dry matter availability	Dry matter requirement	Deficit (000, tons)	Deficit (%)
Andhra Pradesh	13473.84	22931.89	9458.06	41.24
Bihar	6472.21	12011.41	5539.20	46.12
Gujarat	14847.78	26403.67	11555.89	43.77
Haryana	4972.61	6762.66	1790.05	26.47
Jammu & Kashmir	1238.36	3137.47	1899.11	60.53
Jharkhand	487.21	2673.06	2185.85	81.77
Karnataka	25969.66	25380.43	+589.22	+2.32
Madhya Pradesh	9496.54	16423.91	6927.37	42.18
Maharashtra	22735.26	23833.32	1098.06	4.61
Orissa	2089.78	3660.64	1570.86	42.91
Rajasthan	9053.27	25000.92	15947.65	63.79
Tamil Nadu	14320.74	15176.75	856.00	5.64
Uttar Pradesh	6595.09	12242.93	5647.84	46.13
West Bengal	5127.21	11744.52	6617.31	56.34
Total	**136879.55**	**207383.57**	**70504.02**	**34.00**

2.2 Flood Prone Zones

There is an excess availability of about 189.94 million tons against the estimated requirement of 160.64 million tons of dry roughage. Thus there is surplus of about 29.30 million tons of dry fodder, which is about 18.24% of total available dry roughage in the flood prone zones. The availability of concentrates and green fodder is to the tune of 15.08 and 139.17 million tons, respectively (Table 6) against the estimated requirements of 64.26 and 96.38 million tons. Therefore there is a shortage of about 31.24 million tons of concentrates and about 68.55 million tons of green fodder in the flood prone zones to meet the requirements of existing livestock population.

In the flood prone zones, the availability of total dry matter is to the tune of 212.29 million tons against the requirement of 321.28 million tons (Table 7). Thus there is a shortage of about 37% of total dry matter. This situation is further exaggerated during flood.

Table 6 Feed and fodder availability (000, tons) in flood prone states

States	Dry roughage	Concentrates	Greens fodder
Andhra Pradesh	17268.94	1852.84	17818.65
Arunachal Pradesh	38.35	3.62	2057.25
Assam	4571.78	498.81	4652.45
Bihar	14782.12	1104.80	12145.12
Delhi	50.60	4.24	0.00
Gujarat	8782.38	825.91	13624.45
Haryana	17943.56	1878.61	2257.66
Kerala	899.67	364.52	2703.15
Manipur	407.06	36.50	546.47
Orissa	2351.18	136.78	12345.16
Punjab	30501.41	2617.78	10123.45
Uttar Pradesh	69997.82	3757.62	45625.12
Uttaranchal	2854.92	114.31	615.12
West Bengal	19492.85	1822.08	14652.56
Total	**189942.64**	**15018.42**	**139166.61**

Table 7 Dry matter availability and requirement (000, tons) in flood prone states

States	Dry matter availability	Dry matter requirement	Deficit (000, tons)	Deficit (%)
Andhra Pradesh	20773.33	31481.04	10707.71	34.01
Arunachal Pradesh	449.223	130.58	+318.65	+144.03
Assam	5494.021	11901.42	6407.40	53.84
Bihar	16727.25	39947.42	23220.17	58.13
Delhi	49.356	1460.24	1410.88	96.62
Gujarat	11372.35	25385.17	14012.82	55.20
Haryana	18291.49	22369.60	4078.11	18.23
Kerala	1678.401	14555.47	12877.07	88.47
Manipur	508.498	1137.83	629.34	55.31
Orissa	4708.196	11030.80	6322.60	57.32
Punjab	31831.96	32209.71	377.75	1.17
Uttar Pradesh	75504.92	97813.76	22308.84	22.81
Uttaranchal	2795.331	3903.09	1107.76	28.38
West Bengal	22113.95	45496.30	23382.35	51.39
Total	**212298.3**	**338822.42**	**126524.12**	**37.34**

2.3 Major Feed Resources

The major feed resources in the drought and flood prone zones of the country are crop residues, grasses, agro-industrial by-products, cultivated fodders, biomass from community grazing land and top feed. Crop residues in general are characterized by high fiber content, higher lingo-cellulose complex and deficient in critical nutrients, mainly protein and minerals. Crop residues are of various kinds *viz.*, coarse, fine and leguminous. The stem portion obtained from the harvesting of grain from sorghum, bajra, maize, ragi, millets and sugarcane are collectively called coarse straw while the stem portion obtained from rice, wheat and barley are called fine straw and the straw obtained from pulses are classified as leguminous straw. Leguminous straw includes black gram, green gram, horse gram, groundnut, bengal gram and red gram.

2.3.1 *Crop Residues*

Coarse cereal straw is the predominant crop residue available in drought-prone regions (51.19%) followed by fine cereal straw (30.91%) and leguminous straw (17.90%). The availability of coarse cereal straw including sugarcane top is to the tune of 56.43 million tons (Table 8) followed by fine cereal straw (34.07 million tons) and leguminous straw (19.73 million tons).

Table 8 Crop residues availability (000, tons) in drought prone states

States	Coarse straw	Fine straw	Leguminous straw
Andhra Pradesh	2290.05	3298.72	4543.480
Bihar	402.03	4263.70	112.718
Gujarat	5291.97	2304.53	4107.530
Haryana	1132.50	2640.40	305.000
Jammu & Kashmir	0.00	756.26	0.000
Jharkhand	86.06	150.87	51.240
Karnataka	17044.78	2709.94	3047.850
Madhya Pradesh	1627.50	2218.40	1037.150
Maharashtra	18163.97	1171.93	1726.310
Orissa	152.04	759.08	34.580
Rajasthan	3095.29	1562.29	1442.212
Tamil Nadu	5766.54	4663.27	2393.820
Uttar Pradesh	1223.23	3203.71	864.730
West Bengal	154.95	4370.36	58.800
Total	**56430.91**	**34073.46**	**19725.420**

Table 9 Crop residues availability (000, tons) in flood prone states

States	Coarse straw	Fine straw	Leguminous straw
Andhra Pradesh	1784.95	9918.15	5565.84
Arunachal Pradesh	10.00	130.00	0.07
Assam	15.13	4324.29	232.36
Bihar	4273.26	10263.61	245.25
Delhi	0.00	0.00	0.00
Gujarat	4302.15	3117.93	1362.30
Haryana	4703.10	12978.20	260.32
Kerala	20.45	899.67	14.15
Manipur	10.55	407.06	1.24
Orissa	29.32	2135.63	186.23
Punjab	3450.65	26959.80	90.14
Uttar Pradesh	30224.83	37504.01	2268.98
Uttaranchal	192.55	1004.84	1657.53
West Bengal	714.70	18589.20	188.95
Total	**49721.64**	**128102.39**	**12073.29**

In the flood prone zones, fine cereal straw is the predominant crop residue (67.46%) followed by coarse cereal straw (26.18%) and leguminous straw (6.36%). The availability of crop residue in the flood-prone zones is to the tune of 189.90 million tons (Table 9). Of these crop residues, fine cereal straw constitute about 128.10 million tons followed by coarse straw (49.72 million tons) and leguminous straw (12.07 million tons).

2.3.2 *Concentrates*

Concentrates comprise of cereal grains, and their by-products, brans, chunies and oilseed cakes. The total concentrate availability in the drought-affected states is 11.63 million tons. The contribution of grain, cake, bran

Table 10 Availability of concentrate ingredients (000, tons) in drought prone states

States	Grain	Cake	Bran	Chuni
Andhra Pradesh	277.55	860.34	605.99	34.15
Bihar	138.83	34.51	303.05	8.35
Gujarat	140.52	1803.35	119.23	10.86
Haryana	96.06	193.04	312.12	1.48
Jammu & Kashmir	37.33	11.29	23.62	0.23
Jharkhand	6.19	14.08	4.61	0.37
Karnataka	251.19	1108.38	193.95	47.89
Madhya Pradesh	135.91	1118.12	208.55	34.29
Maharashtra	132.46	96.80	497.28	10.42
Orissa	10.38	43.33	10.34	1.00
Rajasthan	91.46	576.97	261.29	7.29
Tamil Nadu	234.09	110.18	688.06	17.70
Uttar Pradesh	51.72	56.67	200.71	1.06
West Bengal	92.50	36.17	258.12	8.38
Total	**1854.68**	**5667.36**	**3932.66**	**175.13**

Table 11 Availability of concentrate ingredients (000, tons) in flood prone states

States	Grain	Cake	Bran	Chuni
Andhra Pradesh	277.55	860.34	605.99	34.15
Andhra Pradesh	289.23	896.53	631.49	35.59
Arunachal Pradesh	0.00	0.00	0.00	0.00
Assam	77.67	114.83	303.63	2.69
Bihar	316.41	78.66	690.70	19.04
Delhi	0.00	0.00	0.00	0.00
Gujarat	55.96	718.15	47.48	4.32
Haryana	288.24	579.24	936.57	4.44
Kerala	12.25	303.76	48.08	0.42
Manipur	7.96	0.32	28.14	0.08
Orissa	24.58	17.97	92.30	1.93
Punjab	539.52	131.57	1945.11	1.46
Uttar Pradesh	837.75	394.32	2462.19	63.36
Uttaranchal	34.39	8.18	71.24	0.51
West Bengal	303.93	332.99	1178.94	6.22
Total	**2787.90**	**3576.52**	**8435.85**	**140.06**

and *chuni* is 15.94, 48.73, 33.82 and 1.51 %, respectively of the available concentrates in the drought-prone region (Table 10).

The total concentrate availability in the flood prone zones is 14.94 million tons. The contribution of grain, cake, bran and chuni is 18.66, 23.94, 56.46 and 0.94 %, respectively of the available concentrate ingredients in the flood-prone zones (Table 11).

A considerable variation is seen in the potential availability of concentrate ingredients across the different drought prone states of the country. Although, the contribution of concentrate ingredients to the overall dry matter availability is only 7 and 10 %, respectively in drought and flood prone zones, but they have a major role in satisfying the critical nutrient *viz.*, protein, energy and mineral requirements of livestock.

2.3.3 *Green Fodder*

Shortage of green fodder is one of the major concerns in livestock production in drought as well as in flood prone zones of the country. Area under fodder crop is stagnated over the past two decades. Green fodder available from different sources has been stagnant at about 136.03 and 139.17 million tons, respectively in drought and flood prone regions (Tables 4 and 6) of the country for the last two decades. The major sources of green forages are the cultivated fodders (berseem, oats, sorghum, maize etc.), grazing and forest feed resources. Top feed is an important source of green fodder particularly for small ruminants. During drought and flood this feed resource play major role as life saving feed. Top feeds are generally rich in protein, minerals and vitamins. The some of the important top feeds are *Prosopis cinerea*, *Albizzia lebbeck*, *Acacia* spp., *Leucaena leucocephala*, *Dalbergia sissoo*, *Ailanthus excelsa*, *Azadiracta indica* etc. should be given priority in the vulnerable zones, as these have ability to survive in adverse climatic conditions and provide life saving feed supplements when other crops and by-products are not available.

2.4 Nutrient Availability

2.4.1 *Drought Prone Zones*

The current availability of crude protein and total digestible nutrients from different feed resources is 10.67 and 63.33 million tons against the estimated requirement of 18.66 and 114.06 million tons by all livestock. Thus, as a whole there is a shortage of about 8 million tons of crude protein and 50.73 million tons of total digestible nutrients in the drought prone states.

All the states are deficient in crude protein and total digestible nutrients (Table 12). The Jharkhand has the maximum shortage (80.54%) of crude protein followed in order by West Bengal (66.81%), Rajasthan

Table 12 CP and TDN requirement and availability (000, tons) in drought-prone states

States	Requirement		Availability		Deficit	
	CP	TDN	CP	TDN	CP	TDN
Andhra Pradesh	2063.87	12612.54	1178.01	6420.52	885.86	6192.02
Bihar	1081.03	6606.27	541.39	3066.83	539.64	3539.44
Gujarat	2376.33	14522.02	1255.12	6988.60	1121.21	7533.42
Haryana	608.64	3719.46	396.42	2303.78	212.22	1415.68
Jammu & Kashmir	282.37	1725.61	117.41	613.41	164.96	1112.20
Jharkhand	240.58	1470.18	46.82	242.73	193.76	1227.45
Karnataka	2284.24	13959.24	1782.12	11621.46	502.12	2337.78
Madhya Pradesh	1478.15	9033.15	1075.78	4953.69	402.37	4079.46
Maharashtra	2145.00	13108.33	1391.46	9916.29	753.54	3192.04
Orissa	329.46	2013.35	225.70	1089.5	103.76	923.85
Rajasthan	2250.08	13750.51	838.44	4418.33	1411.64	9332.18
Tamil Nadu	1365.91	8347.21	973.07	6378.69	392.84	1968.52
Uttar Pradesh	1101.86	6733.61	493.30	3032.47	608.56	3701.14
West Bengal	1057.01	6459.48	350.82	2286.75	706.19	4172.73
Total	**18664.500**	**114061.00**	**10665.86**	**63333.05**	**7998.67**	**50727.90**

(62.47%) and Jammu & Kashmir (58.42%). The shortage of digestible crude protein is also of the same magnitude. Similarly, the gap between availability and the requirements of total digestible nutrients is greatest in Jharkhand (83.49%) while Karnataka had the lowest gap (16.75%).

2.4.2 *Flood Prone Zones*

In the flood prone states, the availability of crude protein and total digestible nutrients are to the tune of 14.39 and 94.54 million tons against the estimated requirement of 26.69 and 163.11 million tons. Thus there is a

Table 13 DCP and TDN availability and requirement (000, tons) in flood prone states

States	Requirement		Availability		Deficit	
	CP	TDN	CP	TDN	CP	TDN
Andhra Pradesh	1573.13	9522.35	2768.33	16917.6	1195.2	7395.25
Arunachal Pradesh	63.91	262.96	12.61	77.04	+51.3	+185.92
Assam	416.39	2518.39	972.12	5940.75	555.73	3422.36
Bihar	1145.09	7475	3053.02	18657.36	1907.93	11182.36
Delhi	2.78	20.89	84.17	514.38	81.39	493.49
Gujarat	910.73	5316.92	1980.5	12103.07	1069.77	6786.15
Haryana	1136.39	7914.13	1723.22	10530.81	586.83	2616.68
Kerala	195.5	877.91	406.55	2484.47	211.05	1606.56
Manipur	39.26	235.11	81.67	499.1	42.41	263.99
Orissa	485.77	2414.02	771.44	4714.34	285.67	2300.32
Punjab	1990.75	13844.52	2003.4	12243	12.65	+1601.52
Uttar Pradesh	4734.14	33041.53	9028.32	55173.05	4294.18	22131.52
Uttaranchal	146.95	1173.6	279.17	1706.07	132.22	532.47
West Bengal	1551.29	9923.64	3526.36	21549.96	1975.07	11626.32
Total	**14392.08**	**94540.97**	**26690.88**	**163111**	**12298.8**	**68570.03**

shortage of about 12.30 million tons *i.e.*, about 46.08% of crude protein and 68.57 million tons *i.e.*, about 42.04% of total digestible nutrients in the flood prone zones.

Except Punjab and Arunachal Pradesh, all other states are deficient in available crude protein. The shortage of crude protein is highest in Delhi (95.94%) followed by Kerala (64.66%), Bihar (59.94%) and Assam (57.61%). The shortage of total digestible nutrients is also of same extent in the flood prone states (Table 13). The maximum shortage exists in Delhi (96.70%) while minimum in Arunachal Pradesh and Punjab.

3.0 FEED AND FODDER REQUIREMENTS DURING CALAMITIES

Often drought persists for about 4 months. Such prolonged drought put tremendous stress on natural resources and leads to severe scarcity of feed, fodder and water. Grazing herd quickly remove the grass cover along with their roots due to non-availability of feed and fodder thus aggravating the problems of desertification. Widespread crop failures lead to acute shortage of feed and fodder as a consequence livestock suffer from malnutrition and consequently become victim of diseases. While the flood period is generally shorter of about 2 months duration. During this period nothing is left for the feeding of livestock except top feeds.

3.1 During Drought

Drought prone area as a whole require a minimum of about 322.59 million tons of roughage and about 13.83 million tons of concentrates for the maintenance feeding of livestock during 4 months of drought period. Depending upon the number of livestock population in drought prone state, each state should have minimum amount of roughage and concentrate (Table 14) as life saving feed.

On an average each district should have about 2.52 million m^3 storage space to store the required feed and forages in conventional methods (*i.e.*, in loose form) while the storage space requirement could be reduced to 0.63 million m^3 if the feeds are stored in densified form. Thus the storage space requirement could be minimized if the feed is stored in densified form. Densification reduces the space requirement by about 4 times.

Livestock producers in the drought affected zones generally meet their fodder requirement through a combination of crop residues and grazing on common lands, private lands, forests, fallow agriculture lands, harvested agriculture lands and drought affected crop lands. The top feeds are also used specially for small ruminants. In case of non-availability of above feed resources, sugarcane baggase and sugarcane tops could be utilized as

Table 14 Emergency roughage and concentrate requirement during drought

States	Maintenance feed required during 4 months drought period (000, tons)		Storage space required to store the feed material (000, m^3)	
	Roughage	Concentrate	Conventional form	Densified form
Andhra Pradesh	3567.18	1528.79	38729.43	9672.36
Bihar	1868.44	800.76	20285.92	5061.48
Gujarat	4107.24	1760.24	44592.86	11138.22
Haryana	1051.97	450.84	11421.36	2845.34
Jammu & Kashmir	488.05	209.16	5298.82	1314.71
Jharkhand	415.81	178.20	4514.50	1118.63
Karnataka	3948.07	1692.03	42864.71	10706.18
Madhya Pradesh	2554.83	1094.93	27738.16	6924.54
Maharashtra	3707.41	1588.89	40251.85	10052.96
Orissa	569.43	244.04	6182.40	1535.60
Rajasthan	3889.03	1666.73	42223.78	10545.94
Tamil Nadu	2360.83	1011.78	25631.84	6397.96
Uttar Pradesh	1904.45	816.19	20676.93	5159.23
West Bengal	1826.93	782.97	19835.19	4948.80
Total	**32259.66**	**13825.57**	**350247.76**	**87421.94**

life saving feed. However, if the drought is severe, it is difficult for individual to solve the fodder shortage problem. Thus, there should be provision of fodder loan from the fodder bank and the same amount should be returned in the normal production year to the fodder bank.

3.2 During Flood

Although the duration of flood period is shorter than the drought, but it is very difficult to manage shortage of feed and fodder during calamity. Since, the flood blocks the roads, national highway and railways. During flood, only air dropping could be the way to provide feed and fodder to livestock. But the major emphasis is given to human life saving and livestock left on their own. On an average, flood prone district require about 0.11 million tons of roughage and 0.04 million tons of concentrate as minimum maintenance requirement during flood period of two months.

On an average each district should have about 1.15 million m^3 storage space to store the required feed and forages in conventional methods (*i.e.*, in loose form) while the storage space requirement could be reduced to 0.29 million m^3 if the feeds are stored in densified form (Table 15). Further, it will be easier to air drop the densified form of feed at the place where livestock have been send. When the flood is over, densified feed could effortlessly be transported to the places where livestock have been kept.

During this period nothing is available for the feeding of livestock except top feeds if available in plenty. In the flood prone zones, livestock

Table 15 Emergency roughage and concentrate requirement during flood

States	Maintenance feed required during 2 months flood period (000, tons)		Storage space required to store the feed material (000, m^3)	
	Roughage	Concentrate	Conventional form	Densified form
Andhra Pradesh	2392.39	1025.31	25974.50	6483.62
Arunachal Pradesh	10.89	4.67	118.29	19.57
Assam	840.11	360.05	9121.16	2270.29
Bihar	2638.41	1130.75	28645.64	7151.41
Delhi	72.74	31.17	789.75	187.44
Gujarat	1711.54	733.52	18582.49	4635.62
Haryana	1489.20	638.23	16168.51	4032.13
Kerala	351.34	150.57	3814.54	943.63
Manipur	70.58	30.25	766.29	181.57
Orissa	666.67	285.72	7238.17	1799.54
Punjab	1731.33	742.00	18797.33	4689.33
Uttar Pradesh	7802.25	3343.82	84710.14	21167.54
Uttaranchal	241.26	103.40	2619.41	644.85
West Bengal	3047.47	1306.06	33086.80	8261.70
Total	**23066.20**	**9885.53**	**250433.02**	**62468.25**

owner can prefer to store the available feed and forages at the upland place where floodwater cannot reach during flood.

4.0 CONCLUSIONS

Livestock play a key role in the vulnerable zones, particularly in the drought and flood prone zones by providing livelihood to about 50 percent human population of the country. In addition, this zone is also providing shelter to about 76 percent livestock population. On an average there is a shortage of 34 and 37 percent dry matter in the drought and flood prone zones. However, this situation is further exaggerated depending upon the intensity of drought or flood. Ensuring availability of adequate feed resource both in quantitative and qualitative terms would be the major concern in the years to come for increasing the livestock productivity. Feeding alone accounts for about 60-70 percent of the total cost of livestock production. As such there is a shortage of about 34 percent of dry matter, 43 percent crude protein and 45 percent total digestible nutrients in the drought-prone zone and about 37 percent dry matter, 46 percent crude protein and 42 percent total digestible nutrients in flood prone zones. Therefore, the judicious utilization of existing feed resources would be the important issue for making the livestock more remunerative. The assessment of availability and requirement of livestock feed resources is the primary requirement for developing any short or long-term policies for the improvement of livestock sector in the drought and flood prone zones of the country.

5.0 REFERENCES

DAHD. 2003. Department of Animal Husbandry and Dairying, Government of India, Livestock Census, New Delhi. India

Feedbase. 2003. National Institute of Animal Nutrition and Physiology, Bangalore, India.

Nivsarkar A.E., Vij, P.K. and Tantia, M.S. 2000. Animal genetic resources of India: Cattle and buffalo, ICAR Publication, New Delhi, India

Ramchandra, K.S, Taneja, V.K., Sampath, K.T., Anandan, S. and Angadi, U.B. 2007. Livestock Feed Resources in Different Agro-ecosystems of India: Availability, Requirement and their management. NIANP, Bangalore, India

Singh, S., Singh, K.K., Das, M.M., Kumar, A. and Misra, A.K. 2008. Livestock Rearing: An Ecological Perspective. Indian Grassland and Fodder Research Institute, Jhansi, India.

6

REVITALIZING FEED - FODDER RESOURCES DURING NATURAL CALAMITIES

Sunil Kumar and S.K. Das

ABSTRACT

This chapter starts with the present scenario of feed and fodder availability in India under different categories of greens, crop residues and concentrates. Situation of forage resources is discussed under the special condition of drought. Various types of droughts are described along with different types of damages to the plants caused by drought. The important physiological adaptations in plants to conserve water and some other also to resist drought situation are elaborated. Different long term and short term strategies for minimizing the adverse effect of drought over the community as a whole are described. In the long term strategies, emphasis is given on management of rangelands, through introduction of suitable forage species, utilization of non-competitive lands, production of feed-blocks for ease in transport to the needy areas, introduction of high quality fodder in the crop rotation under irrigated areas, use of under-utilized and non-conventional plant species as fodder, encouragement in fodder conservation, e.g., silage and also initiatives for establishing "fodder banks". Some short term or immediate measures are also suggested like use of failed crop straws, vegetable wastes, crop residues and alternate feed resources to meet the requirement in the stress situation of drought. A short but useful depiction of management of forage resources under flood situation is made in this chapter with mention of special importance of growing MPTS (multi purpose tree species) under different levels of submergence. There is an elaborate discussion on wastelands, their classification and prospects of growing suitable fodder species depending on the wasteland class to fulfill the demand during emergencies of natural calamity. The feasibility of grazing in the forest lands and its impact is also explored. The problems and management of common property land resources are also found place in the chapter.

Indian Grassland and Fodder Research Institute, Jhansi (UP), India

1.0 INTRODUCTION

Forage and feed resources are deficit in the country. Current scenario depicts an unsatisfactory situation with respect to the availability of cultivable land from where feed and fodder are grown. Indian livestock comprising 15% of world population (209.5 million heads) and 54% of world buffalo population is in a critical state in the face of mounting deficiency of feed and fodder. Apart from these, natural calamities like, drought, flood, tsunami etc also worsen this situation. Pathak and Sunil Kumar (2004) indicated that projection of demand and supplies of forages reflects the deficit of green fodder and dry matter will be more than 65 and 25 percent by 2025. To combat this problem, locally available feed and fodder should be used efficiently to optimize cattle production. However locally available feeds like crop residue, agro industrial by products, top feed etc. may be of inferior quality that could be treated and converted to moderate quality feed by using simple technologies. These technologies and optimizing the level of inclusion in the diet can remarkably reduce the feeding cost. Further formulating own feed/mineral mixture would pave way to reduce the feeding cost. In order to reduce risk in agriculture and impart greater resilience to Indian agriculture against droughts and floods, efforts will be made for achieving greater flood proofing of flood prone agriculture and drought proofing of rain-fed agriculture for protecting the farmers from vagaries of nature. For this purpose, contingency agriculture planning, development of drought and flood resistant crop varieties, watershed development programmes, drought prone areas and desert development programmes and rural infrastructure development programmes have to be taken up. In India, cattle of intensive cropped areas obtain only about 25 percent of their feed from grazing in nearby forests and other uncultivated lands; the balance comes from residues unsuitable for human consumption. As per current estimate the availability of green fodder from cultivated areas and pastures is 462.05 million tons. Similarly, in case of crop residues, total availability is 393.83 million tons which constitute coarse straw from coarse grains, sugarcane tops, fine cereal straw and leguminous straw. Concentrates availability, particularly of oil cakes, brans and grains for feeding livestock is 35.29 million tons.

2.0 DROUGHTS AND ITS EFFECT ON FORAGE RESOURCES

Large parts of the country perennially reel under recurring drought. Over 68% of India is vulnerable to drought. The drought-prone areas of the country are confined to peninsular and western India – primarily arid, semi-arid and sub-humid regions. An analysis of 100 years of rainfall data reveals that the frequency of 'below-normal rainfall' in arid, semi-arid and sub-humid regions is 54–57%, while severe and rare droughts occurred once every eight to nine years in arid and semi-arid zones. Drought is a period or condition of unusually dry weather within a geographic area where rainfall is normally present. Drought is a temporary or prolonged shortage of rainfall in a geographical region. Seasonal or annual drought spells, which are common in arid and semi-arid zones, can have serious impacts on natural rangelands, and consequently on livestock. With decreased rangeland resources, livestock is more vulnerable to drought. At the community level, drought causes loss of crops and livestock, and, in severe cases, leads to chronic food shortages and famines. Migration of rural people from drought-stricken areas to cities, in search of food and employment, add to pressure on urban resources, causing new social problems. Overgrazing over a period of less than 10 years may reduce the annual productivity of degraded rangelands from around 100–30 kg/ha of dry matter. Overgrazing not only reduces the protective ground cover but also reduces plant diversity and consequently the dry matter available from range plants.

2.1 Classification of Drought

Drought relates to moisture deficiencies caused by soil, atmosphere and/or by both which force the plants to stop physiological functions resulting into a complete failure of plants to grow, develop and mature. Thus drought is a condition of insufficient moisture supply under which plants fail to develop and mature properly. This occurs when the plants do not get water either through precipitation or through irrigation continuously for a period of 20 days or more (Singh, 1988).

According to the moisture condition the drought may be classified into following three groups:

2.1.1 *Atmospheric Drought*

This type of drought is caused by low air humidity which is frequently accompanied by hot and dry winds. It may occur even under condition of relatively high soil moisture. Mostly dry winds cause a considerable reduction in the moisture content of plants. The atmospheric drought mainly caused due to high temperature, low humidity, high wind velocity and extensive dry areas with sparse vegetation. A plant may recover from injury when if given water, although the portions which get dry do not become green again. Wilting due to atmospheric drought is always reversible and it occurs because of either an under- developed root system or their physical inability to conduct water fast enough to compensate the water losses from the leaves and tender portions of plants during the stress period.

2.1.2 *Physiological Drought*

Presence of soluble salts in soil water increases the concentration of the solution and hence availability of water to plants is reduced, although water remains in the soil. Thus it is caused by some physiological disturbances in the plants when their roots fail to absorb water and nutrients from the soil due to increased concentration of soil solution. The absorption of water is reduced in bog soils because of poor root growth of trees also, which is due to lack of oxygen and higher level of CO_2.

2.1.3 *Soil Drought*

Inadequate soil moisture results in to acute crop stress and fall in agricultural productivity. This type of drought usually occurs after a meteorological drought, but it refers to decreased soil moisture content which is needed to grow particular crops. Five distinct categories of soil drought affecting dryland crop production have been identified depending upon the time of occurrence of drought as described below:

2.1.3.1 *Early Season Drought*

This occurs with delay in commencement of sowing rains and the droughts of 1972, 1987 and 2002 are the result of this kind of drought.

2.1.3.2 *Mid Season Drought*

Mid season drought occur in association with the breaks in south west monsoon.

2.1.3.3 *Late Season Drought*

Due to early cessation of rainy season sometimes crops may encounter moisture stress during its reproductive phase which leads to forced maturity of the crop due to rise in temperature

2.1.3.4 *Permanent Drought*

Permanent drought is associated with inadequacy of soil moisture/rainfall to meet the water requirements of the crop during most part of growing season.

2.1.3.5 *Apparent Drought*

While rainfall in the region may be adequate for one crop but may not be enough to grow the other crops.

2.2 Plant Death under Drought Conditions

The effect of drought on plants is complex. During drought the plant suffers from dehydration of its cells and tissues as well as from a considerable increase in the temperature of its body, *i.e.*, from overheating. Thus the lack of water caused by drought is usually accompanied by high body temperatures.

2.2.1 *Desiccation*

Due to high temperature a relatively high rate of water loss from plants, particularly at times when the rate of absorption of water is sluggish, often leads to death of some or all the leaves or branches on a plant as a result of desiccation. In extreme cases the entire plant is killed in the process.

2.2.2 *Metabolic Disturbances*

Relatively high temperature often induces various types of metabolic disturbances which are detrimental or even fatal to the plants. Some of such metabolic disturbances are explained below:

2.2.2.1 *Respiration*

The water stress leads to many abnormalities in the plant metabolism, in which respiration is most important. Drastic reduction in respiratory activities due to water stress has been observed in many plants (Crafts, 1968; Henckel, 1964). It has been concluded from different experiments that changes in phospholipids or protein structure of the mitochondrial respiration) may cause suppression in oxidation rates of succinate, malatepyruvate and exogenous NADH in water stressed tissues.

2.2.2.2 *Protein Metabolism*

In water stress situation, ornithine amino transferase activity and ornithine concentration are increased. Proline content is also increased under drought conditions. *De novo* synthesis of drought proteins may explain the increased protein content present under drought conditions. The genotypes, which had the highest yield, also had a major concentration of polyamines under water stress. The results also suggest that arginase activity participated in amino acid synthesis such as ornithine and, consequently, in the protein concentration under water stress (Mejia, 1998). An interruption in protein synthesis and proteolysis is directly related with ribonucleic acid (RNA) levels of the plants. Thus lower protein synthesis rate results in lower nucleic acid level due to water stress (Slatyer, 1969).

2.2.2.3 *Carbohydrate Metabolism*

Rate of photosynthesis and carbohydrate reserve of the plants are governed by direct and indirect effects to drought. Insufficient CO_2 supply to plants and dehydration result in closure of stomatal openings and lower the net photosynthesis in drought affected plants is reported. The coordination of N and C metabolism is retained during drought conditions via modulation of the activities of phosphate synthase and NR commensurate with the prevailing rate of photosynthesis. A correlation between maximal extractable foliar nitrate reductase (NR) activity and the rate of CO_2 assimilation was observed. The NR activation state and maximal extractable NR activity declined rapidly in response to drought. Photosynthesis and NR activity recovered rapidly when nutrient solution was restored at this point (Foyer *et al.*, 1998)

2.2.2.4 *Direct Effect of Heat on Protoplasm*

The death point (critical temperature) of most active living plants' cell is in the approximate range of 50–60°C but the exact temperature at which death of the protoplasm will occur depends upon the length of the period which the cells are being exposed to the lethal temperature leading to the coagulation of protoplasm. Setback in photosynthesis, respiration and conversion of starch to sugar result in stunting of the plants. Further the plants become dependent on the reserved food. Maintenance of such a condition for extended or prolonged period results into the death of the plants.

2.3 Drought Resistance in Plants

Drought resistance is related to a moisture stress environment and means the ability of one genotype to be more productive with a given amount of soil moisture than another genotype (Quizenberry, 1982). Hence, drought resistance of the plant includes the ability to withstand dehydration (drought resistance in the narrow sense of the term) and secondly, the ability to withstand overheating (heat resistance). The ability of plants to face drought is a result of a multiplicity of morphological and physiological characteristics of plants which are very difficult to be separated out. It has been found from experimentations that plants are more resistant at thermo-phase (vegetative phase) than at photo-phase (reproductive phase). The resistance to drought in a plant may be attributed to higher resistance to wilting associated with rapid recovery. High heat resistance is not always linked with high drought resistance. Succulents, especially cacti have high heat resistance but are unable to endure dehydration.

Levitt (1972) made distinctions for adaptations of plants to dry habitats in two ways, *i.e.*, escaping the drought and enduring the drought. Begg and Turner (1976) described two types of mechanisms of adaptation to the water deficit such as (i) those which provide morphological adaptation (ii) those conferring physiological adaptation. Water stress avoidance of plants is the ability to maintain a favorable water balance and turgidity even when they are exposed to severe drought conditions. Thus the drought forces the plants to bring about certain changes in the physiological process which result into changes of morphological and anatomical characteristics of plants. A favorable water balance in drought affected plants can be achieved by following two mechanisms (Singh, 1988).

2.3.1 *Mechanism for Conserving Water in Plants*

There are many mechanisms that relate to adaptations of plants to water deficit situations. A single mechanism can not accord complete drought resistance to a genotype. The drought affected plants develop certain changes in the anatomical characters of plants for an effective conservation of the cell sap against an accelerated rate of transpiration due to rise in plant body temperature. These changes can be described as under:

2.3.1.1 *Stomatal Mechanism*

Within a species, the stomata number can vary to a sufficient range. In certain plant types, mostly in cereals, it has been found that they open their stomata only for few hours early in the morning when moisture is at its maximum and they close the stomata for rest of the day hours which result in an efficient photosynthesis is with least loss of water. These types of plants survive for longer period under drought than those which open their stomata late in the morning and keep them opened for rest of the day hours that favors more water loss due to transpiration. The stomatal mechanism of opening and closing differs greatly from crop to crop or even within the varieties of the same crop.

2.3.1.2 *Increased Photosynthesis Efficiency*

Under light and dark condition photosynthesis, stomatal conductance and transpiration decreased with increase in the duration of drought. Re-watering after 15 days of drought resulted in sudden revival from the stress (Cui *et al.*, 1998). Closure of stomata to reduce the transpiration loss of water result in a lower rate of photosynthesis and best possibility to overcome this problem is an increased CO_2 assimilation by plants for given stomatal opening. The C-4 species have higher growth potentials and more efficient use of water as compared with plants of Calvin cycle (C-3) under the same moisture regime. The carboxylating enzyme, which has a high potential activity and a high affinity for CO_2, helps in a high rate of CO_2 fixation and maintenance at relatively restricted stomatal opening. These physiological changes are used to inhibit the photorespiration and thereby the net photosynthesis is enhanced.

2.3.1.3 *Low Rate of Cuticular Transpiration*

Plants develop cutine, a waxy coating, over cuticles for reducing the cuticular transpiration of water. The intensity of cutine covering depends upon the severity of the drought. The dwarf wheat varieties of durum group, triticales, barley, sorghum, and castor use waxy coating to survive the drought under rain-fed conditions. Some plants have lower rate of cuticular transpiration (*viz.*, xerophytic plants such as cactus) by which they conserve water and survive in drought conditions for longer time.

2.3.1.4 *Decreasing Transpiration*

The transpiration loss of water through other plants parts is also very significant loss. Levitt (1972) reported that when soybean is exposed to moderate light especially during drought, the plants produce lipid on the surface of leaf and provide a covering through which the transpiration of water is reduced to minimum.

2.3.1.5 *Reduced Leaf Area*

As the water loss through transpiration is directly proportionate to the leaf area, plants can survive longer during drought, if they can reduce their leaf surface area. The xeromorphic plants reduce their transpiring surface and find fit for drought conditions. In other plants the droopiness or curling of leaves at noon time, especially under dry conditions, reduce the transpiration loss of water to the extent of 55 percent in arid zones and 75 percent in desert areas (Sharma and Mohan, 1985).

2.3.1.6 *Structure of Leaf Surface*

Various morphological adaptations of leaves help to reduce the rate of transpiration loss of water for a prolonged survival under drought conditions. The most pronounced changes in the leaf surface are development of thick cuticle, deposition of more lipids and waxy covering or development of spines, etc. Some plants convert their leaves into thick flattened leaves covered with spines such as cactus etc. which have very low transpiration requirement of water for normal functioning.

2.3.1.7 *Position of Stomata and its Functioning*

The stomata present on the leaf surface or on slightly elevated position come in direct contact with hot air current and transpire more water. For the protection of plant's file from desiccation, stomata are situated in cavities or in depressions and this helps to retard the transpiration loss of water from the leaf surface. The duration of the stomatal opening also governs the extent of water transpired from the leaf surface. The plants , affected by drought , use to bear less number of stomata which are located either in depressions (in cavities) and they retard the rate of transpiration of water.

2.3.1.8 *Effect of Awns*

The awns are photosynthetically active organs of the reproductive part of a cereal plant and they have low transpiration rate. In case of moisture stress, the transfer of the photosynthates from lower leaves to sink may not be a high rate while it may be at a very high rate from flag leaf and awns. Awned varieties of cereals like wheat and rice are the best adapted ones under drought conditions. Awns contribute to about 12 percent to the total dry matter production of the drought affected plants (Fisher and Maurer, 1978).

2.3.2 *Mechanism to Accelerate Water Uptake by Roots*

Well developed and extensive root system of the plant can withdraw water from relatively deeper layer of the soil but contrary to this the shallow rooted plants are often victim of drought. The mechanism of water uptake totally depends upon the root system of the plants which may be described as under:

2.3.2.1 *Root Growth*

Drought resistant varieties should have a highly branched and well developed root system than the drought susceptible varieties (Hurd, 1971). The vertical growth of roots is more helpful for an efficient adaptation of plants to dry conditions. The density of roots/unit volume of soil and number of secondary hair roots are the proper index for finding out the drought resistance of plants. Deeper roots with well developed laterals and extensive root hairs are more beneficial to fight the drought.

2.3.2.2 *Root to Top Ratio*

Deeper the roots better is the drought resistance in plants. The water movement to deeper layers in soil results in more root elongation and hence the root growth rate exceeds considerably from that of the stem. This type of growth result is better uptake of water because of deeper roots and also in reduction of water loss through transpiration as volume of the top or shoot remains lower with lesser number of stomata and cuticles per plant.

2.3.2.3 *Osmotic Potential of the Plants*

Drought affected plants can withdraw water from soil even after wilting point, whereas, plants under irrigated conditions cannot do so. The water movement in plants depends upon their osmotic pressure. Therefore, plants grown under dry conditions use to have high osmotic pressure and they survive for relatively longer period (Ramos, 1999).

2.3.2.4 *Water Economy in Plants*

In some situations, the water spenders become water savers to fight better with drought. This type of change which comes automatically has been reported for higher drought resistance in plants. Plants use to transpire water with normal rate under regular water supply but a restricted water supply to plants reduces transpiration rate.

2.4 Minimizing the Impact of Drought on Plant Growth

Under drought situations, there are two distinct phase in which application of weather and climate can minimize the impact of drought on the community. The first is the long term strategy can be devised and precautions taken to reduce impact. The second phase is the action (suitable soil and water management, and agronomic manipulation such as suitable varieties, special cultural practices plant protection measures and efficient nutrient, which is taken during the onset of the event to reduce adverse effect. Rathore (2005) has also suggested various strategies to mitigate drought effect.

2.4.1 *Improved Management of Rangelands*

(a) Natural pastures and grazing lands must be improved by including multipurpose top feed fodder tree species and leguminous components (Table 1). These trees will come to rescue during periods of drought supplementing the fodder supplies.

(b) Rehabilitation of waste and abandoned lands may be used for forage production through introduction of drought tolerant varieties of range grasses (anjan grass, yellow anjan and sevan grass) and legumes and top feed fodder trees. Similarly, perennial grasses like guinea grass, Hybrid napier, bhabhar grass, marvel grass, rhodes grass etc. should also be established on the bunds. These grasses will supplement the existing feed resources.

(c) Early grazing and overgrazing are common in drought affected areas and as a result the palatable species are not allowed to regenerate. Under these situations regulated grazing plays an important role in sustainable use of grazing lands. The controlled or regulated grazing depends on the support of community or panchayats. Participatory management of grazing and common forage resources in the village will help in great way the meeting scarcity of forages at local level. Seasonal and annual droughts, common in most arid and semi-arid zones, can have a serious impact on rangelands under these circumstances. Herbage quality and quantity, affected by the

Table 1 Suitable forage species for drought affected areas

Components	Suitable species
	Arid region
Grasses	*Lasiurus sindicus, Cenchrus ciliaris, C. setigerus*
Legumes	*Atylosia scrabaeoides, Lablab purpureous, Cholria ternatea*
Trees	*Acacia tortilis, A. nilotica, Prosopis cineraria, P. juliflora, Eucalyptus terminalis, E. camaldulenis, Azadirachta indica, Dichrostachys nutans*
	Semi-arid region
Grasses	*Cenchrus ciliaris, C. setigerus, Pennisetum pedicelatum, Chrysopogon fulvus, Sehima nervosum*
Legumes	*Stylosanthes humilis, S. gracilis, Stizlobium deeringeanum, Macroptilium atropurpureum, Atylosia scarabaeoides, Alysicarpus monilifer*
Trees	*Acacia tortillis, A. nilotica, A. catechu, Albizia lebbeck, A. amara, Dalbergia sissoo, Dendrocalamus strictus, Zizyphus spp., Emblica officinalis, Eugenia jambolana*

fluctuations of rainfall, vary considerably over time, and the regeneration of grazed vegetation is strongly hampered after extended periods of drought. Appropriate use of rangelands to minimize the risks associated with drought must focus on three components: protection, rehabilitation, and proper management.

(d) Alternative sources of feed must be found to supplement feed resources from natural pastures and rangelands, particularly in periods of drought, when major efforts should be devoted to the development of the livestock feed base.

(e) Local forage and pasture species and ecotypes, which have developed characters to cope with erratic rainfall, must be given priority.

(f) In addition, the possibility of using non-conventional water resources to irrigate forage has proved to be practical.

2.4.2 *Utilization of Non-Competitive Lands*

Field boundaries bunds etc may be used for raising drought tolerant perennial grasses, bushes and multipurpose fodder trees. The priority should be for plantations of genus *Prosopis* and *Opuntia*. These plantations serve as bio-fence, alternate source of fodder, fuel and food during severe drought situations, alteration of microclimate ultimately moderate the drought effect and acts as shelterbelts and wind breaks.

2.4.3 *Promoting Fodder Production in the Areas Having Irrigation Facilities*

Efforts have to be focused on the promotion of storable-forage production on arable land, particularly by including high quality fodder crops into rotations. Promoting fodder production activities in the adjoining districts where irrigation facilities are well developed will help in augmenting the fodder supply during periods of drought. An organized trade in fodder production, marketing and transportation will evolve over the years to sustain the livestock industry in the drought-affected region simultaneously, increasing the livelihood opportunities in the peripheral districts of the drought prone region. Fodder production activities may be promoted through demonstration, supply of quality seeds and other inputs besides facilitating the marketing of fodder.

2.4.4 *Utilization of Under-Exploited Naturally Growing Plant Resources*

In these regions naturally growing plants have tremendous potential to be exploited as food, forage and fuel such as Henna (*Lawsonia inermis*), Nirgundi (*Vitex regundo*), Jaffre (*Bixa orellona*), Neem (*Azadirachta indica*), perennial peanut, lablab bean (*Lablab purpureus*), snapmelon, *kachri*, ashgourd, tropical Kudzu (*Pueraria phaseoloides*), perennial soybean (*Nenotonia wightii*), *Desmodium* spp., *Indigofera* spp., *Lasura, Salvadora, Tephrosia* can be propagated in the existing cropping systems. Lasoda (*Corda myxa*) fruits can be used to make pickle. Edible cactus (*Opuntia ficus indica*) also needs to be promoted in these regions, which can be extensively used as fruit, forage, vegetable and medicinal crop. Exploiting the use of under-utilized plant resources as drought mitigation approach and stressed are to be focused on: (a) to generate an inventory of under-utilized/un-exploited plants and maintain their germplasm for productivity and quality improvement, (b) identification of ecological limits and optimal management practices for promising species and their evaluation as monoculture/mixed communities or even as under-storey crop with trees, (c) initiation of network projects on such species particularly on edible cactus, and with already existing international networks on such plants and (e) standardization of post harvest processing techniques for value addition and marketing opportunities.

3.0 AGRONOMIC MANAGEMENT

3.1 Selection of Crops and Agronomic Practices During *Kharif*

(a) By cultivation of short-duration and drought tolerant varieties of cultivated forage crops after the drought situation is relieved by some rainfall. The suitable crops are Maize cv. J-1000, Cowpea cv. Bundel lobia -1 and Bundel lobia-2; Sorghum cv. PC-6, HC-136; UPChari-1,2 Sorghum multicut, lablab bean cv. JLP4: Bajra cv. Giant bajra, Raj bajra chari 2 and Guar. Grain crops like Maize cv. Ganga 11, Pigeonpea, Black gram, Lentil may also be taken which supplements fodder in the event of midterm drought.

(b) Reduced plant density in case of mid season drought. In case of sorghum, if drought occurs at very early stage, re-sowing is advisable with subsequent rain rather that allowing poor plant density.

(c) Application of nitrogen can be reduced by 40% under rainfed conditions and it should be applied at sowing or planting time.

3.2 Catch Crops after Rains

Rapeseed, mustard, chinese cabbage, gobhi sarson and maize may be sown in last week of September or first week of October for fodder purpose wherever feasible. The crops may be harvested by November to facilitate *Rabi* sowing. Senji and Lucerne may be preferred over Berseem cultivation as per agro climatic suitability. Oats may be grown in October as multi cut fodder to ensure availability of green fodder for longer period or dual purpose crops like barley (var BH 75) may be sown in October and one cutting may be taken for fodder, after 50–60 days of sowing and subsequent one left for grain production. The area which received some showers in August and September, catch crop of maize, bajra, cowpea, bajra + cowpea, maize +cowpea and toria may be taken.

3.3 Contingent Planning for Forage Production

The drought condition which may prevail in the country during *Kharif* requires a contingency plan for meeting the needs of vast populations of humans and the animals. While, the requirements of grains for the human consumption can be met during the period, animals are likely to suffer due to non-availability of fodder and crop residues. Due to the failure of crops and uncertainties of the monsoon it is of utmost importance to emphasize growing of fodder crop even if there is slight availability of moisture so that, the animals do not suffer and they survive. The technologies, varieties and feeding management for perennial grasses are proposed in Table 2 to meet the demand:

Table 2 Contingent planning for forage production in different phases of drought

Crop	Varieties	Seed rate (kg/ha)	Row spacing (cm)	Fertilizer (N:P:K kg/ha)	Phase I	Phase II	Phase III
Bajra (Fodder)	Rajkoo, Giant Bajra	10	25	80:40:00	√	√	√
Bajra (Grain)	L–74, K–677 & Anand selection	5	50	60:30:00	√	√	×
Sorghum	UP–chari, Pusa Chari	40	25	60:30:00	√	√	√
(fodder)		15–20	50	60:30:00	√	×	×
Sorghum (Grain)	HC–136 & HD–2						
Maize (Fodder)	African tall, Kisan composite, Moti,	60	25	60:30:00	√	√	√
Maize (Grain)	Manjari & BL–7	25	50	90:40:30	√	√	√
Teosinte (Fodder)	Improved Teosinte	40	25	80:40:00	√	√	√
Cowpea (Fodder)	IGFRI–S–457, C–26, C–30, Russian Giant,	40–50	25	20:60:20	√	√	√
Cowpea (Grain)	Kohinoor & S–540	25–30	50	20:60:20	√	√	√
Guar (Fodder)	Durgapur safed, SFG–119, IGFRI–214,	30	25	20:60:00	√	√	√
Guar (Grain)	Guar No. 2 and FS–277	20–25	50	20:60:00	√	×	×
Rice bean (Fodder)	KRB–1	30	25	20:40:00	√	√	×

Phase I – upto August 10, Phase II – up to August 20, Phase III – up to August 31.
√ Crop may be grown successfully, × Crop may not be successful

4.0 FLOOD AND FORAGE RESOURCES

Floods are recurrent phenomenon in eastern and north eastern part of India. During last five years floods are affecting millions of land and property in nontraditional flood hit areas like Rajasthan, Gujarat, Punjab etc. In India about 6.0 m ha of cultivated land suffer due to water logging of which 3.4 m ha are subjected to surface flooding and remaining 2.6 m ha have high water table. The agricultural lands are also converted in to submerged and partially submerged conditions due to developmental activities like roads, canals, railways, urbanization. Forage resources are badly hampered due to all these situations. In order to overcome this situation, certain strategies are to be framed for fulfilling the feed and fodder to the livestock of flood/submergence affected areas on sustainable basis. Planting of fodder species tolerant to partial or full submergence in the river banks, embankments, elevated sites, bunds and receding ends of floods may provide a substitute under this scarcity (Table 3).

Table 3 Suitability of crops and multipurpose tree species (MPTS) under submerged and semi-submerged situations

Depth of water table (cm)	Crops	MPTS
150	Peas, Barley	*Tamorix articulate*
125–150	Bajra, Cowpea	*Syzyium cumini*
125	Mustard, Wheat	*Dallbergia sissoo*
75	Sorghum, Maize	*Terminalia arjuna*
60	Groundnut	*Syzyium cumini*
25	Toria, Sugarcane	*Casurina* sp.
20	Rice, Sesbania	*Casurina* sp.

Under continuous water stagnation *Tamarix articulate, Eucalyptus rostrata, Salix balyonia, Morus alba, Delbergia sissoo, Casurina equesetifolia* can be planted under agroforestry system in sub tropical India. *Lagerstroemica fleriginae,* and *Syzium cumini* are suitable species for tropical climate. Besides this the *Bracharia* sp., *Vitivaria zizanodis, Medicago denticulate*, *Vicia* sp., *Lathyrus* sp. and *Pisum* sp. can be grown on raised sunken emerged bunds as well as skip and staggered planting of vegetative species. There are large river beds are affected areas in Uttar Pradesh and Bihar called Diara lands. These areas should be provided with permanent vegetation cover of economic fuel-cum-fodder trees along with suitable grasses (Table 4). The tolerance limit of stagnating water differs from species to species (Table 5).

Table 4 Suitable species for fodder in diara lands and river-bed affected areas

Components	Suitable species
Grasses	*Pennisetum pedicelatum, Chrysopogon fulvus, Dicanthium annulatum, Eulaliopsis binata*
Legumes	*Stylosanthes gracilis, Calapogonium mucunoides*
Trees	*Delbergia sisoo, Acacia catechu, A. nilotica, Salix* spp. *Zizyphus* spp., *Psidium gujava*

Table 5 Tolerance limit of crops for standing water

Crops	Period of disposal
Fish	> 10 days
Rice	>10 days
Sugarcane	7 days
Banana	7 days
Cotton	3 days
Sorghum	3 days
Maize	2 days
Vegetables	1 day

5.0 WASTELAND AND FORAGE RESOURCES

India has a geographical area of 328.73 million hectares (mha). The reported area of 304.88 mha includes 142.82 mha net sown area of which 87.68 mha is rainfed, and 68.75 mha forest land of which 31.00 mha is degraded forest. About 63.8 million ha is estimated to be wastelands (Table 6), which are unutilized or underutilized. Wasteland is degraded land, which do not fulfill their life sustaining potential. Wasteland can result from inherent/imposed disabilities owing to the location, environment, chemical and physical properties of the soil or financial or management constraints.

Table 6 Areas and categories of wasteland in India (area in sq. km.)

S. No.	Category	Total wastelands	Percent of total geographical area covered
1	Gullied and/or ravinous land	20553.35	0.65
2	Land with or without scrub	194014.29	6.13
3	Waterlogged and marshy land	16568.45	0.52
4	Land affected by salinity/alkalinity-coastal/inland	20477.38	0.65
5	Shifting cultivation area	35142.20	1.11
6	Under utilized/degraded notified forest land	140652.31	4.44
7	Degraded pastures/grazing land	25978.91	0.82
8	Degraded land under plantation crop	5828.09	0.18
9	Sands-inland/coastal	50021.65	1.58
10	Mining/Industrial wastelands	1252.13	0.04
11	Barren rocky/stony waste/sheet rock area	64584.77	2.04
12	Steep sloping area	7656.29	0.24
13	Snow covered and/or glacial area	55788.49	1.76
	Total wasteland area	**638518.31**	**20.17**

Note: 1,20,849.00 sq. km. in Jammu & Kashmir is not mapped and hence not considered for calculating the percentage.

Source: 1,50,000 scale wasteland maps prepared from Landsat Thematic Mapper/IRS LISS II/III Data

Regeneration of degraded forest is also important for restoration of ecological balance. In addition there are also socio-economic priorities, as livelihood of the rural poor depends to a great extent, on adjoining forest areas that derive fuel, fodder and income though minor forest products (Norman *et al.*, 1999; Hazell and Chaherli, 2001). Historically wastelands are not wastelands. Such lands might have been wastelands for revenue-based colonial administration, as these lands did not contribute revenue to the crown. In a welfare state, the so-called wastelands should become productive assets, not only for landless households and marginal farmers, but also for the entire rural community. Wastelands need special attention both from ecological and socio-economical considerations. Besides reclamation, the trend of degradation should be arrested and reversed. It should be emphasized that wastelands, especially the common property resources, provide biomass lifeline for the rural poor. Reclamation of such lands and equitable distribution of biomass produced on these should receive the highest priority.

The idea behind wasteland development is to develop vegetative cover on non-arable lands, helping conserve soil and water in the uplands thereby converting wastelands into productive lands. *In situ* conservation of soil and water, making optimum use of trees and crops is the basic principle behind this approach. Livestock have been an integral part of farming systems. The rain-fed agriculture-animal system is characterized by a few animals raised by each household, often used as draught animals in addition to meat or milk. About one third of the total feed intake by the ruminants is by way of grazing on permanent pastures, current and permanent fallows, waste lands, tank bunds, problem lands, on road sides and cultivated field margins where green herbage is available. Permanent pastures constitute 3.6% of geographical area in the country. Grazing facilities last only 3 to 4 months after the northeast monsoon. During off-seasons, the animals graze in harvested fields.

5.1 Action Plan for Facilitating Vegetative Cover in Degraded Lands

5.1.1 *Suitable Forage Species Suitable for Waste Lands*

Identification of suitable forage species and their varieties for different wasteland situations is very important component of rangeland improvement programme. During natural calamities, the wealth of important species of grasses, legumes and trees has been suggested by Singh (1990). Category wise species are given as under:

5.1.1.1 *Cultivable Wastelands*

There are about 17 million ha of cultivable wastelands in the country which could be utilized for growing fuel-cum-fodder trees along with palatable grasses. This will go a long way in solving the problems of forage and fuelwood shortage.

Grasses: *Cenchrus ciliaris, C. setigerus, Pennisetum polystachyon, Panicum antidotale, Sehima nervosum, Chrysopogon fulvus, Dicanthium annulatum*

Legumes: *Stylosanthes* spp., *Macroptilium atropurpureum, Glycine javanica, Centrosema* spp., *Clitoria ternatea*

Trees: *Albizia* spp., *Hardwickia binata, Leucaena leucocephala, Acacia* spp., *Sesbania* spp., *Dichrostachys nutans*

5.1.1.2 *Salt Affected Lands*

There are about 8 million ha of land affected by saline and alkali soils in the country. Utilization of these problematic lands can be considered by growing salt tolerant species.

Grasses: *Cynodon dactylon, C. plectostachyon, Paspalum notatum, P. dilatatum, Choloris gayana, Lasiurus sindicus, Brachiaria mutica, Sporobolus marginatus, Urochloa mossambicensis*

Legumes: *Glycine javanica, Macroptilium* spp., *Phaseolus junatea, Stylosanthes humilis,*

Trees: *Acacia tortilis, A. nilotica, P. juliflora, Dendrocalamus strictus, Salvdora* spp., *Zizyphus* spp., *Sesbania* spp. *Albizia amara*

5.1.1.3 *Acidic Soils*

The acid soil can also be suitable utilized by growing tolerant species.

Grasses: *Pennisetum polystachyon, P. pedicellatum, Paspalum notatum, Pennisetum clandestianum*

Legumes: *Centrosema pubescens, Desmodium intortum, Stylosanthes guianensis*

Trees: *Ficus numeralis, Albizia chiensis, Morus serrata, Alnus nepalensis, Bucklandia populnea*

5.1.1.4 *Ravine Lands*

In India, there are about 4.0 million ha of ravine lands, a major portion of which is confined to Uttar Pradesh, Madhya Pradesh, Rajasthan and Gujarat. These lands come under capability class VI and VII and are far below their economic utilization. It is suggested that small to medium gullies can be put to silvipasture or hortipasture.

Grasses: *Cenchrus ciliaris, C. setigerus, Pennisetum pedicellatum, Chrysopogon fulvus*

Legumes: *Stylosanthes humilis, S. gracilis, Stizolobium deeringeanum, Macroptilium atropurpureum, Atylosia scarabaeoides, Alysicarpus monilifer*

Trees: *Acacia tortilis, A. nilotica, A. catechu, Albizia lebbek, A. amara, Dalbergia sissoo, Dendrocalamus strictus, Zizyphus spp., Emblica officinalis, Eugenia jambolana*

Posopis juliflora and *Acacia nilotica* (particularly for the head region of the ravines), *Syzygium cumini, Cassia siamea* are fit for the lower part of the ravines. Steep slopes are to be stabilized with *Saccharum munja* and *Eulaliopsis binata* grasses. Another successful tree for developing ravines is *Cassia auriculata*.

5.1.1.5 *Lands Affected by Shifting Cultivation*

In the hilly terrain of Assam, Orissa, Madhya Pradesh and Rajasthan, shifting cultivation has been practiced for centuries. This results in mass soil erosion and loss of land productivity. This has been recognized as a complex problem and specialized research efforts are urgently needed to tackle this challenging problem on priority. Introduction of agro-silvipastoral management will not only reduce the soil erosion hazards but will also be able to improve the socio-economic status of the tribal people.

a. North East Region

Grasses: *Cenchrus ciliaris, C. setigerus, Pennisetum pedicellatum, Chrysopogon fulvus*

Legumes: *Clitoria ternatea, Lablab purpureus*

Trees: *Leucaena leucocephala, Hardwickia binata, Acacia nilotica, A. tortilis, Dalbergia sisso, Albizia* spp., *Anogeissus pendula*

b. Assam and Orissa

Grasses: *Pennisetum pedicellatum, Dicanthium annulatum, Cynodon plectostachyon*

Legumes: *Stylosanthes hamata, Macroptilium atropurpureum*

Trees: *Albizia lebbek, A. procera, Leucaena leucocephala, Ficus spp.*

5.1.2 *Grazing in Forest Land*

In India, the forest cover is to the tune of 22.6% of which over 85% are under protection and conservation. These lands used to be a major source of feeds and fodder for the livestock rearing communities dwelling within and nearby the forests. Additionally grazers collect an estimated 175 to 200 million tons of green fodder annually. This results in overgrazing and over-extraction of green fodder, leading to forest degradation through damages to regeneration and compaction of soil. A sample survey of Forest Survey of India estimates that impact of grazing affects 78% of the country's forests, of which 18% suffers high incidence and 31% medium incidence. Grazing occurs even in protected areas. In another survey, 67% of the national parks and 83% of the wildlife sanctuaries surveyed reported grazing incidence. The ecological costs of this nearly ubiquitous form of land use can be dramatic. Examples of such costs include loss of biodiversity; lowering of population densities for a wide variety of plants; disruption of ecosystem functions, including nutrient cycling and succession; change in community organization; and change in the physical characteristics of both terrestrial and aquatic habitats. Because livestock congregate in forest ecosystems, which are among the biologically richest habitats in arid and semiarid regions, the ecological costs of grazing are magnified in these sites (Fleischner, 2002).

Controlled grazing generally refers to the degree of control or level of management applied to grazing animals through the use of such grazing systems as rotational stocking, continuous stocking, and strip grazing. The method or combination of methods used to achieve the appropriate level of control will vary with each livestock farm (Mosley, 1993). Regardless of the grazing method(s) utilized, the goal is to provide the amount and quality of the forage required by the particular class of grazing animal, while maintaining or improving the vigour of the plants being grazed. This system is highly sustainable for efficient utilization and harvest of forage by grazing animals. The number of animals in the grazing system (stocking rate) can usually be increased by 30 to 50 percent. Gain per acre can also be increased by ensuring that high-quality, fresh, and unsoiled vegetative growth is available throughout the grazing system (White and Wolf, 1996).

5.1.3 *Development and Management of Common Property Land Resources (CPLRs)*

The common land broadly covers three categories, namely, forest land, community lands and other government lands. CPLRs exist in the form of grazing grounds, village woodlands, catchments areas, dumping and threshing grounds, village ponds, rivers streams, banks and beds and other common wastelands. These resources constitute an important element of the production economy of rural areas. At present most of the CPLRs are in a state of degradation. Their size has shrunk over the past decades. The state-wise areas of common lands, based on the traditional land use classification statistics, are given in Table 7 below. From the table it appears that while nearly one-fifth of the land in the country is under forest cover, about one-tenth constitutes non-forest common land. In five states of Andhra Pradesh, Gujarat, Madhya Pradesh, Maharashtra, and Rajasthan, non-forest common lands exceed thirty lakh hectares. In U.P. this area is of the order of about 14.19 lakh hectares. Most of the common lands belong to the state government. Even where these are vested in panchayats, generally, the power of resumption is with the governments. Community management, barring exceptions, has remained, as yet, an unfulfilled dream (Bagchi and Philip, 1986).

It is estimated that since 1960 reductions in area of CPLRs have varied from 30 to 50 percent in the different states of India. Many factors have contributed to these shrinkages and to the incessant degradation of such lands. Among these are the high rate of population growth and consequent anthropogenic pressures, interventions of the state, operation of market forces, pressures generated by the development processes themselves, allotment of the lands to the landless and for other purpose, attenuation of village institutions, encroachments, land grabbing etc. If such lands are to be converted into common property resources then the most vital need is to establish and promote appropriate institutional management mechanisms at the local levels representative of user communities. These type of land resources become common property when the group of people who posses rights to their collective use is well defined and there is clarity in laws, rules and procedures for the use of such resources which are accepted and observed by the collectivity of users.

5.1.4 *Aerial Seeding of Range Grasses*

Aerial seeding is often the largest cost associated in worst drought affected areas but holds promise for rehabilitation of wastelands and grazing lands. Aerial seeding, usually grasses but occasionally also legumes, is one of the quicker methods to increase vegetative cover on an overgrazed site during the first few years. It is typically done where population pressure for exiting forage resources is high and native plant seed bank is believed to have been destroyed or severely reduced by the grazing. Seed is applied by fixed-wing aircraft or helicopter. Effectiveness depends on timeliness of seed application, choice of seed, pilot skill, protection from grazing, and luck in having gentle rains to stimulate seed germination before wind or heavy rains blow or wash soil and seed away. Proper timing of seed application depends on location. In some areas it is best to drop seed directly into dry ash, before any rain falls, to take advantage of the fluffy seedbed condition, while in others seed is best applied after the first snow so that it will germinate in the spring. Both conditions also reduce loss to rodents. Choice of seed determines how easily it can be applied – some grass species with long awns tend to clog in seeder buckets, and light seeds drift more than heavy ones – and how well it will grow, how long it will persist, and what impact it will have on natural regeneration. In general, legumes have not been found to be particularly effective at producing cover (Dewald and Wiedemann, 1983). A skillful pilot will apply the seed evenly, rather than in strips with unseeded areas in between them, providing better ground cover once the seed germinates. The government and other agencies should include aerial seeding as a drought proofing strategy in the selected niches of the country.

Table 7 Extent of common lands in India (000 hectares)

State	Reporting area*	Forest	Common lands		
			Barren and uncultivable land	Permanent pastures and other grazing lands	Total of columns 4 & 5
1	2	3	4	5	6
Andhra Pradesh	27440	6172	2296	899	3195
Assam	7852	1985	1541	184	1725
Bihar	17330	2736	964	136	1100
Gujarat	18826	1965	2502	845	3347
Haryana	4394	136	88	27	115
Himachal Pradesh	3113	809	144	1089	1233
Jammu & Kashmir	4506	2759	256	123	379
Karnataka	19050	1030	829	1260	2089
Kerala	3885	1082	86	5	91
Madhya Pradesh	44211	14015	2369	2820	5189
Maharashtra	30758	5312	1722	1584	3306
Manipur	2211	602	1419	–	1419
Meghalaya	2249	812	231	17	248
Nagaland	4099	286	–	–	–
Orissa	15540	6640	265	560	825
Punjab	5033	216	90	4	94
Rajasthan	34235	2151	2892	1836	4728
Sikkim	729	295	209	97	306
Tamil Nadu	13001	2030	500	155	745
Tripura	1048	578	120	–	120
Uttar Pradesh	29748	5120	1120	299	1419
West Bengal	8846	1184	121	4	125
Union Territories	9006	7276	28	12	40
All India	**304110**	**67191**	**19882**	**11956**	**31838**

* Excludes area under the illegal occupation of China and Pakistan
Source: National Wasteland Development Board, 1991

5.1.5 *Community Participation in Forage Resources Management*

Mobilization of people for the restoration of wastelands and for arresting the further progress towards land despoliation is now recognized as important management strategies of forage resources in periods of natural disaster (Dhar, 1997). It involves a number of factors that go in to organizing communities for planned action. The main responsibilities are concentrated in land use contexts, contacts of village leaders, community opinion makers, representatives of women, and youth groups as, also NGOs working in the field. A two pronged strategy seems to be called for. First, the reclamation of already deforested or overgrazed lands and second, the prevention of damage by deforestation and overgrazing itself (Dyson, 1991). Community-based forage resource management is the process of involving local communities to identify problems as well as solutions towards formulating and implementing a course of action involving the vegetation, livestock, water and land resources of their villages or panchayats, with emphasis on linkages between the upstream and downstream parts and the respective human and bio-physical endowments.

A Case Study of Rajasthan: Village community pastures (*Charagah or Charnot*), spread over 1.194 million ha are the major source of fodder in Rajasthan. However, over 40% areas are almost barren and unproductive due to excessive grazing, heavy soil erosion caused by cutting of shrubs and trees and complete neglect by the community. These pastures, managed by the caretakers of the Princely States, were transferred to Village panchayat. The villagers who are dependent on this resource to maintain their livestock were unwilling to share the responsibility. These community pastures pose a threat to the livelihood of the local poor. The situation worsens in drought periods. RRIDMA (Rajasthan Rural Institute of Development Management) an associate organization of BAIF had established a cattle development centre in Asind in 1979 with the support of the District Rural Development Authority, Bhilwara (BAIF, 2006). The villagers agreed to bring 40 ha under silvipasture development and the Swiss Development Cooperation provided financial support. A Committee was formed for community pasture management by identifying 11 members representing 10 major castes. Management of stray animals was controlled by cordoning off the area through the establishment of trench-cum-mounds followed by live hedge, using a popular fence plant locally known as *Thor (Euphorbia)*. It was also felt that the pasture should have woody perennials at 150–200 trees of Gum Acacia, *Azadirachta indica* (Neem) and *Prosopis cineraria* (Khejri) per ha, to improve the micro-climate and increase the soil fertility. Seeds of *Dhaman* grass (*Cenchrus setigerus*) and Stylo (*Stylosanthes hamata*) were sown after tilling the land by a tractor to ensure ground cover. With an annual rainfall of 680 mm, about 65% of the saplings survived and there was plenty of grass, ready for cutting. The committee arranged seed collection and decided that all the families could cut the grass and deposit 50% with the committee which was later sold to the needy members. Apart from the sale of the grass and forage seeds, the committee also permitted the members to cut thin branches of the trees for forage and fuel wood in autumn by paying Rs 10 per tree. This system provided additional biomass while inducing better growth of tree and grass and for building *Gramkosh* (Village development corpus). During subsequent years, the average income generated from the community pastures was between Rs. 1500–2000 per ha, of which 80–85% was contributed grass and the rest by sale of seeds and fine on stray cattle , etc. Recognizing the success of this programme, the Government of India provided financial support under Swarnjayanti Gram Swarojgar Yojana (SGSY) special projects to replicate this programme in 76 villages spread over 2520 ha. Having seen the benefits of community pasture development, the villagers from these villages, took greater interest and completed the development in the first year itself. Based on this initiative major keys to success are suggested as:

- Do not bring entire pasture land under protection in the first year.
- Sustainable unit of silvipasture is minimum 40 ha owned by an individual village.
- Link the community developed pasture with family based development.
- Watershed development and livestock improvement have supplementary benefits.
- People's participation representing different socio-economic groups is essential.

From such case study it may be concluded that lessons learnt from success stories for mitigating the natural disaster are to be replicated in different parts of the country. These approaches would utilize the local resources with close harmony in socio- economic terms having the role of public and private along with participation of local people. Forage resources in the drought and flood prone areas would be possible to manage and restore by bottom up approach.

APPENDIX

Appendix 1 - List of forages with their botanical names (in the text)

Common/English name	Botanical name
A. Cultivated forages – I Annual	
Maize	*Zea mays* (L.)
Sorghum	*Sorghum bicolor* (L.) Moench.
Pearl millet	*Pennisetum americanum* (L.) Leeke
Teosinte	*Zea mexicana* Schrad.
Foxtail millet	*Setaria italica* (L.) Beauv.
Coix	*Coix umbellate* Jobi.
Finger millet	*Eleusine corocana*
Oat	*Avena sativa* (L)
Barley	*Hordeum vulgare* (L) Emend. Bowden
Cowpea	*Vigna unguiculata* (L) Walp
Ricebean	*Vigna umbellate* (Thumb.) Ohwi & Ohashi
Lablab bean	*Lablab purpureus*
Horse gram	*Dolichos biflorus*_Lam.
Clusterbean	*Cyamposis tetragonoloba* (L) Taub.
Velvet bean	*Mucuna deeringiana*
Pigeonpea	*Cajanus cajan* L.

(Contd...)

(Contd...)

Common/English name	Botanical name
Broadbean	*Vicia faba*
Mung	*Vigna radiata*
Moth	*Phaseolus aconitifolius*
Berseem	*Trifolium alexandrinum* L.
Lucerne	*Medicago sativa* L.
Chinese cabbage	*Brassica pekinensis* (Lour.) Rupr.
Japan rape	*Brassica compestris*
Pea	*Pisum sativum*
Turnip	*Brassica rapa*
Deenanath grass	*Pennisetum pedicellatum* Trin.
A. Cultivated fodder – II – Perennial	
Napier bajra hybrid	*Pennisetum purpureum* Schum. X *Pennisetum americanum* (L) Leeke
Guinea grass	*Panicum maximum* Jacq.
Thin napier	*Pennisetum polystachyon, Pennisetum clandestianum*
B. Range grasses/ legumes – I Perennial grasses & legumes	
Buffel grass	*Cenchrus ciliaris* L.
Yellow buffel grass	*Cenchrus setigerus* Vahi.
Setaria grass	*Setaria anceps* Stapfex Massey
Doob grass	*Cynodon dactylon, C. plectostachyon*
Bahia grass	*Paspalum notatum, P. dilatatum*
Para grass	*Bracharia mutica* (Forsk.) Stapf.
Signal grass	*Bracharia brizantha* (Hochst ex. A. Rich) Stapf.
Congosignal grass	*Bracharia ruziziensis* Germain & Evard
Dharaf grass	*Chrysopogon fulvus* (Spreng) Chiov
Sewan grass	*Lasiurus hirsutus* (Forsk.) Boiss
Rhodes grass	*Chloris gayana* Kunth
Marvel grass	*Dicanthium annulatum* (Forsk.) Stapf.
Little para grass	*Urochloa mosambicensis* (Hack) Dandy
Sehima grass	*Sehima nervosum*
Blue panic grass	*Panicum antidotale*
Cocksfoot grass	*Dactylis glomerata*
Italian Rye grass	*Lolium multiflorum*
Tall Fescue grass	*Festuca arundinacea*
Perennial rye grass	*Lolium perene*
Timothy grass	*Phleum pretense*
Pangola grass	*Digitaria decumbans*
Guatemala grass	*Tripsacum laxum*
Bhabhar grass	*Eulaliopsis binata*
Rhodes grass	*Chloris guana*
Stylo	*Stylosanthes hamata* (L) Taub, *S. guianensis* (Aubl.) SW.*S. humilis* H.B.K., *S. gracilis*

(Contd...)

(*Contd...*)

Common/English name	Botanical name
Stizlobum	*Stizlobum deeringeanum*
Siratro	*Macroptilium atropurpureum* (DC) Urb.
Chatta ki ghas	*Alysicarpus monilifer*
Butterfly pea	*Clitoria ternatea* L.
Poa	*Poa annua* L.
Centro	*Centrosema pubuscens*
Wild groundnut	*Calapogonium mucunoides*
Hedge lucerne	*Desmanthus virgatus* (L) Wild.
White clover	*Trifolium repens*
Red clover	*Trifolium pretense*
Glycine	*Glycine wightii* (R. Grah.) Verde, *Glycine javanica*
Desmodium (Green leaf)	*Desmodium intortum* (Mill) Urb.
Desmodium (Silver leaf)	*D. uncinatum* (Jacq.) DC
C. Fodder shrubs/trees	
Subabool	*Leucaena leucocephala*
Australian keekar	*Casurina equisetifolia*
Atylosia	*Atylosia scrabaeoides*
Shevari/Agathi	*Sesbania egyptica, Sesbania sesban*
Babul	*Acacia nilotica* (black), *A. arabica* (white)
Videshi babul	*Acacia tortillis*
Kattha/ Kair	*Acacia catechu*
Khejri	*Prosopis cineraria* (native), *P. juliflora* (exotic)
Eucalyptus	*Eucalyptus terminalis, Eucalyptus camaldulensis*
Nutans	*Dichrostachys nutans*
Tamorix	*Tamorix articulate*
Jamun	*Syzyium cumini*
Pendula	*Anogeissus pendule*
Neem	*Azadirachta indica*
Madhuca	*Madhuca latifolia*
Gliricidea	*Gliricidia sepium*
Tamarind	*Tamarindus indica*
Arjun	*Terminlaia arjuna*
Flemingia	*Flemingia congesta*
Bhimal	*Grewia optiva*
Nagphani	*Opuntia* spp.
Bamboo	*Dendrocalamus sikkimensis, Dendrocalamus strictus, Bambusa balcoa*
Salvadora	*Salvadora* spp.
Anjan	*Hardwickia binata*
Siris	*Albizia amara, A. lebbek*
Shahtoot	*Morus serrata*
Nepal alder	*Alnus nepalensis*
Malayan aspen	*Bucklandia populnea*

(*Contd...*)

(*Contd...*)

Common/English name	Botanical name
Pink bauhinia	*Bauhinia porpurea*
Shisham	*Dalbergia sissoo*
Chinese ber	*Zizyphus mauritiana*
Wild ber	*Zizyphus nummularia*
Aonla	*Emblica officinalis*
Jamun	*Eugenia jambolana*
Mulberry	*Morus alba*
Himalayan elm	*Ulmus wallichiana*
Australian teak	*Acacia mangium* L.
Ficus	*Ficus religiosa, F. bengalensis, Ficus numeralis*
Calliandra	*Calliandra calothyrsus*
Guava	*Psidium gujava*

6.0 REFERENCES

Bagchi, K.S. and Philip, M. 1996. Wastelands in India: Untapped Potential, Uplabidhi, Trust for Developmental Initiatives, New Delhi, pp. 11–15.

BAIF, 2006. Community Pasture development in Rajasthan, BAIF/Rajasthan Rural Institute of Development Management (RRIDMA) Institute BAIF Communication, 4, BAIF Development Research Foundation, Warje, Pune (Maharashtra).

Begg, J.E. and Turner, N.C. 1976. Effect of drought on physiological aspects of cereals. *Advances in Agronomy*, **28**: 161–217.

Crafts, A.S. 1968. Cui, Y., Pandey, D.M., Hahn, E.J. and Paek, K.Y. 2003. Effect of drought on physiological aspects of Crassulacean acid metabolism in Doritaenopsis. *Plant Science*, **167(6)**: 1219–1226.

Crafts, A.S. 1968. Water deficits and physiological, processes. *In*: Water Deficits and Plant Growth (*Eds.*) Kozolowski, T.T., Academic Press, London.

Dewald, C. L. and Wiedemann, H.T. 1983. Improving rangelands seeding success. *Texas Agril. Exp. Sta. Consol. Prog. Rep*. **3665**: 98–99.

Dhar, T.N. 1997. Wastelands, common property resources and environment – Challenges and opportunities. *In*: Advances in Wastelands Development, (*Eds.*) Sarkar, T.K., Varish, R.C., Verma, V. and Gawande, S.P., Soil Conservation Society of India, New Delhi. pp. 12–27.

Dyson–Hudson, N. 1991. Pastoral production systems and livestock development projects: An eastern African perspective. *In*: Cernea, M.M. (*Ed.*), Putting People First: Sociological Variables in Rural Development, 2nd edn., pp. 219–56. A World Bank publication, Oxford University Press, Oxford, UK

Fisher, R.A. and Maures, R. 1978. Drought resistance in spring wheat cultivars. *Australian Journal of Agricultural Research*, **29**: 897–912.

Fleischner, T.L. 2002. Ecological costs of livestock grazing in Western North America. *Conservation Biology*, **8(3)**: 629–644.

Foyer, C.H., Valadier, M.H., Migge, A. and Beeker, T.W. 1998. Drought induced effects on nitrate reductase activity and MRNA and on the coordination of nitrogen and carbon metabolism in maize leaves, *Plant Physiology*, **117**: 283–292.

Hazell, P., Oram, P. and Chaherli, N. 2001. EPTD Discussion Paper No. 8 Environment and Production Technology Division, International Food Policy Research Institute 2033 K Street, N.W. Washington, D.C. 20006 U.S.A FAO, 2002. Production Year Book. Rome.

Henckel, P.A. 1964. Physiology of plants under drought. *Annual Review of Plant Physiology*, **15**: 363–86.

Hurd, E.A. 1971. Water deficits and physiological, processes. *In*: Drought Injury and Resistance in Crops (*Eds.*) Larson, K.L. and Eastin, J.D. CSSA, Madison. pp. 77–78.

Johnson, D.A., Rumbaugh, M.D. and Asay, K.H. 1981. Plant improvement for semi arid rangelands: Possibilities for drought resistance and nitrogen fixation, *Plant and Soil*, **58(1–3)**: 279–303.

Levitt, 1972. Responses of plants to environmental stress, Academic Press, New York

Mejia, E.G., Resendz, V.M., Tostado, E.C. and Loarcapia, 1998. Effect of drought on polyamine metabolism yield, protein content and *in vitro* protein digestibility in tepary (*Phaseolus aconitifolius*) and common been (*Phaseolus vulgaris*) seed, *Journal for Science of Foods & Agriculture*, **83(10)**: 1022–30.

Mosley, J. 1993. Grazing management during and other extended drought. Pub. No. 15, Department of Animal and Range species, Montana State University, Bozeman.

Norman, A., Eilers, J.J., Stuth, W. and Doug, T. 1999. Enhancing Conservation Management via Nutritional Profiling of Livestock on Grazing Lands: The NRCS National Evaluation of Forage Quality and Animal Well Being Program. Proc. VI International Rangeland Congress, Townsville, Queensland, Australia, pp. 378–9. People and Rangelands Building the Future

Pathak, P.S. and Kumar Sunil 2004. Forage and grazing resources in different agro-climatic regions in India. *In*: Kundu, S.S., Mishra, A.K. and Pathak, P.S. (*Eds.*). Buffalo production under different climatic regions. International Book Distribution Co. Lucknow: pp. 1–41.

Quizenberry, J.E. 1982. Drought resistance in plants. *In*: Breeding Plants for Less Favourable Environments (*Eds.*) M.H. Christianson and C.F. Lewis, John Willey & Sons, Inc., pp. 193–212.

Ramos, M.L.G., Gordon, A.J., Minchin, F.R., Sprent, J.I. and Parson, R. 1999. Effect of water stress on nodule physiology and biochemistry of drought tolerant cultivars of common bean (*Phaseolus vulgaris* L.), *Annals of Botany*, **91**: 143–153.

Rathore, M.S. 2005. State Level Analysis of Drought Policies and Impacts in Rajasthan, India, pp. 40. Colombo, Sri Lanka: Working paper 93: Drought Series Paper No. 6. IWMI.

Sharma, H.C. and Mohan, K. 1985. Effect of water stress on plant growth, *In*: Efficient Management of Dryland Crops. (*Ed.*) Venkateswarlu, J. CRIDA, Hyderabad. pp. 19–33.

Singh, P. 1990. Forage production for sustainable mixed farming. *In*: Agronomic Research Towards Sustainable Agriculture (*Eds.*) Singh, K.N. and Singh, R.P.,

Indian Society of Agronomy, Division of Agronomy, IARI, New Delhi pp. 96–118.

Singh, S.S. 1988. Dry farming. *In*: Crop management under irrigated and rainfed conditions, Kalyani Publishers, New Delhi- Ludhiana pp. 354–408.

Statyer, R.O. 1969. Comparative photosynthesis, growth and transpiration of two species of Atriplex. *Plants, (Berlin)* **93**: 175–89.

Turner, N.C. 1979. Drought resistance and adaptation to water deficits in crop plants. *In*: Stress Physiology in Crop Plants (*Eds.*) Mussel H. and Staples, R.C., Wiley & sons, New York.

White, H.E. and Wolb, D.D. 1996. Controlled grazing of virginea pastures, Virgina Cooperative Extension, Pub. No. 418–012, Virginia State University.

7

ENSURING FORAGE AVAILABILITY FOR LIVESTOCK DURING NATURAL DISASTERS: OPTIONS AND STRATEGIES

M.M. Roy

ABSTRACT

Of several reasons for devastating losses to animal herds during disasters, especially drought, lack of fodder is an important one. During disasters, many a times it is not feasible to actually produce fodder from the lands which other wise is meant for this purpose. The available options to produce quality fodder from arable lands and grasslands, management of grasslands, browse and top feed resources and to suggest long as well as short term strategies to maintain forage supplies for livestock in the affected areas are discussed. At present Indian grazing lands are considered as fragile ecosystems and ranked as class IV and V in land capability classification. The stocking rates are 1-51 adult cattle unit (acu) per hectare against the carrying capacity of 0.20-1.47 acu / ha in semiarid regions. Similarly in arid regions stocking rates are 1-4 acu / ha against the carrying capacity of 0.2-0.5 acu / ha. The degradation of Indian grazing lands is ascribed to many factors like large bovine population, free grazing practices, lack of management, and extremes of temperature, steep slopes, variable precipitation and scarcity of moisture in arid and semiarid regions. The range restoration techniques are based on ecological approaches e.g. protection for vegetation recovery, soil and water conservation, reseeding, improved range management techniques and grazing management. Region specific mechanisms are required to ensure adequate fodder supplies during the disasters with the aim of reducing losses to a great extent and the poor families are able to retain the herd. The strategies are required on region as well as local basis after taking into consideration available fodder resources and prospects of its production, conservation and utilization. The strategies should be long as well as short term. Implementation of these strategies through

Indian Grassland and Fodder Research Institute, Jhansi (UP), India

government and non government organizations and also community will lead towards effective combating of the disasters.

1.0 INTRODUCTION

Some or the other kind of natural calamities affect nations all over the world. In India on account of the large geographical area various natural calamities like floods, cyclones and drought occur fairly frequently in various pockets. At times, the same area is subjected to floods and drought situation in successive seasons or years. Calamities like earthquakes, hailstorms, avalanches, landslides, etc. occur quite suddenly but they are restricted in their impact in terms of time and space. Similarly, though floods and cyclones occur with some element of warning yet their occurrence is confined in duration. Drought, on the other hand, spans over a much longer time-frame and its adverse impact on the economic activities and life of an area is of a more lasting nature. The measures required to meet the threats posed by different calamities, therefore, differ considerably in terms of disaster preparedness and amelioration of the economic and social life of the affected people.

In drought affected areas, the most damage is in the food and agriculture sector. The livestock that forms an integral part of livelihood of small and marginal farmers faces the immediate impact of drought – be it health, lack of feed and fodder or drinking water. According to an estimate in drought areas in seventeen states, almost one third of the cattle were killed (Bhasin, 2002). Large scale migration of livestock – mainly cattle and sheep take place from these areas, resulting in various hardships to farmers. The farmers in areas not affected by drought or any other disaster do not like to receive these livestock as they damage their crops. Often there are clashes between the local and migrating farmers (Negi, 2002).

Feed and fodder scarcity becomes very acute during the disasters. The livestock owners need fodder supplies round the year for maintaining their animals. The common sources include crop residues, grazing lands, fodder crops and purchased fodder. During droughts and other disasters, the farmers tend to accept inferior options. However, roughages are usually not available during these periods in sufficient amounts. During droughts, the grass growth and water resource of livestock declines drastically. Purchase of fodder also becomes an extreme costly affair in these areas. All this leads to livestock mortality, distress selling of livestock and migration of livestock from the areas.

In order to improve this scenario for the benefit of livestock owners, it is very important to ensure forage availability in such affected areas. This paper is an attempt to highlight the available options to produce quality fodder from arable lands and grasslands, management of grasslands, browse and top feed resources and to suggest long as well as short term strategies to maintain forage supplies for livestock in the affected areas.

2.0 FODDER RESOURCES

2.1 Grazing and Grass/Legume Resources

Grazing based animal husbandry is still a major source of livelihood for landless and small and marginal farmers in various regions of the country,

Table 1 List of important range grasses for forage in different regions of tropical India

Botanical name	Common name	Climatic preference	Special situation /tolerance
	Perennial Grasses		
Andropogon gayanus	Gamba	Semiarid–humid	–
Brachiaria brizantha	Palisade	Subhumid–humid	Acidic
Brachiaria decumbense	Signal	Semiarid–humid	Waterlogged (light)
Bracharia mutica	Para	Semiarid–humid	Waterlogged
Brachiaria ruziziensis	Congo signal	Semiarid–humid	Waterlogged (light)
Bothriochloa intermedia	Fulkara	Semiarid–subhumid	–
Bothriochloa pertusa	Phulwa	Semiarid–subhumid	–
Cenchrus ciliaris	Anjan	Arid–semiarid	–
Cenchrus glaucus	Blue anjan	Arid–semiarid	–
Cenchrus setigerus	Dhaman	Arid–semiarid	–
Chloris gayana	Rhodes	Semiarid–subhumid	Saline/alkaline
Chrysopogon fulvus	Dhavalu	Semiarid	–
Dichanthium annulatum	Kail	Semiarid–subhumid	Saline
Dichanthium caricosum	Kartah	Semiarid	–
Digitaria decumbense	Pangola	Semiarid–humid	–
Diplanche fusca	Karnal	Semiarid–humid	Saline/alkaline
Heteropogon contortus	Lampa	Semiarid–subhumid	–
Iseilema laxum	Machhari	Semiarid–humid	–
Lasiurus sindicus	Sewan	Arid	–
Melinis minutiflora	Molasses	Semiarid–subhumid	–
Panicum antidotale	Blue panic	Arid–semiarid	–
Paspalum dilatatum	Golden crown	Semiarid–subhumid	–
Paspalum notatum	Bahia	Semiarid–subhumid	–
Pennisetum clandenstinum	Kikuyu	Semiarid–humid	–
Panicum maximum	Guinea	Semiarid–subhumid	Shade conditions
Pennisetum polystachyon	Thin napier	Semiarid–subhumid	–
Phragmites karka	Reed	Subhumid–humid	Waterlogged (running)
Sehima nervosum	Sen	Semiarid	–
Setaria sphacelata	Nandi	Semiarid–subhumid	–
Sporobolus marginatus	Usraili	Arid–semiarid	Saline/alkaline
Thysanolaena maxima	Broom	Semiarid–humid	Waterlogged
Tripsacym dactyloides	Guatemala	Semiarid–humid	–
	Annual Grasses		
Brachiaria miliformis	Cori	Subhumid–semiarid	Sandy soils
Themeda quadrivalvis	Gunkar	Semiarid	–
Pennisetum pedicellatum	Dinanath	Arid–semiarid	–

Table 2 List of promising tropical pasture legumes for wastelands in India

Species	**Preference**	
	Rainfall (mm)	**Soil**
Arachis glabrata (Perennial forage peanut)	800–2000	Well drained soils (sandy loam to clays), tolerates acidic (pH 4.5) and slight alkali soils
Arachis pintoi (Pinto peanut)	800–2000	Red sandy loam and river bottom soils, tolerates water logging
Atylosia scaraboides (Ban kulthi)	250–1000	Wide range, tolerates salinity
Calopogonium mucunoides (Calopo)	> 1500	Wide range of soils in the pH range of 4.5–6.5
Centrosema pubescens(Centro)	> 1500	Alluvial to medium fertile soils
Clitoria ternatea (Butterfly pea)	400–1500	Wide range, tolerates salinity
Desmanthus virgatus (Hedge lucern)	250–2000	Wide range including calcareous and saline soils
Desmodium intortum (Desmodium)	900–1300	Light to clay loams, adapted to poorly drained/water logged conditions
Glycine wightii (Glycine)	750–1500	Heavy black soils with good drainage
Indigofera astragalina (Hairy indigo)	900–1700	Moderately acidic and low fertile sandy loam soils
Lablab purpureus (Sem)	500–1500	Wide range, preferring deep sand to heavy clay (pH range 5.0–7.5)
Lotonois bainesii (Lotononis)	> 900	Well drained acid sandy soils, tolerates temporary water logging/flooding
Macroptelium atropurpureum (Siratro)	600–1800	Wide range, light textured to heavy clay (pH 4.5–8.0)
Macroptelium lathyroides (Wild pea bean)	750–2000	Tolerates soils that are acidic. Saline or alkaline and even with poor drainage
Macrotyloma axillaris (Archer dolichos)	> 900	Variety of soils, sand to clay but with good drainage
Neonotonioa wightii (Glycine)	750–1500	Heavy black soils with good drainage
Pueraria phaseoloides (Kudzu)	> 1500	Wide range, sand to clay (pH 4.0–5.0)
Stizolobium deeringianum (Velvet bean)	650–2500	Wide range, sandy to sandy loam and even on acidic soils
Stylosanthes guianensis (Tropical lucerne)	> 1200	Wide range from coarse sand to sandy loam but not heavy clays and saline soils
Stylosanthes hamata (Carribean stylo)	500–1300	Wide range of soils
Stylosanthes humilis (Townsville stylo)	500–1300	Prefers sand and sandy loams and may grow on acid and also saline soils
Stylosanthes scabra (Shrubby stylo)	300–1200	Wide range of soils having good drainage, tolerates salinity
Stylosanthes seabrana (Caatinga)	400–1200	Wide range of soils, preferably clay loams, loams and sandy earths
Vigna luteola (Vigna)	> 1500	Wide range, tolerates salinity and poor drainage conditions

including the disaster prone areas. Livestock rearing is also an important subsidiary activity among other farming communities. Here also grazing supplements animal's forage requirements. Importance of grazing depends on the other sources of available fodder and extent of area available for grazing and therefore, it varies in different regions of the country. In several states like Rajasthan, Gujarat, Madhya Pradesh, Andhra Pradesh and Maharashtra large areas are available for grazing. Most states in Indo-Gangetic Plains like West Bengal, Bihar, Uttar Pradesh, Haryana, Punjab and some other states like Assam, Tamil Nadu, Karnataka and Kerala have few grazing resources but are potentially rich in fodder production from arable lands. The hilly states in Western Himalayan region like Himachal Pradesh and Jammu & Kashmir and northeast region have large grazing areas (Roy and Singh, 2008).

Grasses are a cheap source of quality forage on account of high productivity on marginal lands with very low input, palatable herbage and good persistence (Table 1). Besides grasses a number of herbaceous leguminous species are available for wastelands (Table 2). These species also help in build up of soil fertility. Besides using them as green, they may be conserved in form of hay or silage as reserve for use in scarcity periods (Roy and Singh, 2009).

2.2 Crop Resources

In a diverse climate of India a variety of forage crops are used but area under cultivated fodder is stagnating around 8.5 million ha (4.6 % of the

Table 3 Area under fodder crops in major states of India

State	Area (million ha)	Share in sown cropped area (%)
Andhra Pradesh	0.152	1.20
Assam	0.009	0.20
Bihar	0.015	0.20
Gujarat	1.205	10.70
Haryana	0.545	9.30
Himachal Pradesh	0.004	0.40
Jammu & Kashmir	0.042	3.90
Karnataka	0.077	0.60
Kerala	0.002	0.10
Madhya Pradesh	0.805	3.30
Maharashtra	0.806	4.00
Punjab	0.770	10.10
Rajasthan	2.850	14.30
Tamil Nadu	0.188	2.70
Uttar Pradesh	1.003	3.90
West Bengal	0.003	0.05
All India	**8.469**	**4.60**

Note: Compilation based on the information available with State Departments of Agriculture gross cropped area) only on account of pressure of human population for food and other cash crops (Kelley and Parthasarathy, 1994). Table 3 presents an account of area under fodder crops in selected states of India. This includes area under guar that is grown for fodder as well as grain production in Rajasthan, Gujarat and Haryana. The area under major fodder crops is depicted in Table 4. Considerable regional variation exists in the area under fodder crops. For instance such crops are more important in Rajasthan, Punjab, Haryana and Gujarat where they account about 10 percent of the gross cropped area.

Table 4 Area under major fodder crops in India

Crop	Area (million ha)
Berseem	2.00
Oat	0.20
Lucern	1.00
Sorghum	2.60
Maize	0.90
Bajra	0.90
Others (Miscellaneous)	0.87
Total	**8.47**

Note: Compilation based on study reports and information from State Departments of Agriculture

There are many other crop species like cassava, fodder beet, turnip, mustard etc. that provide quality fodder in some or the other form in there respective areas. In the above crop resources, besides using them as green forage; there are opportunities to make hay and silage for lean periods.

2.3 Tree/Shrub

The trees and shrubs are an important fodder resource for livestock in various parts of the country (Table 5). Fodder is available from leaves and pods. It is usually rich in proteins, vitamins and minerals like calcium. Their importance increases with the severity of drought and other disasters (Roy, 2003). The advantage with these resources is that besides fodder they are very important in conservation of land, serve as windbreaks, useful as shade for livestock and a number of other products and complementary benefits (Sincere and Ranjhan, 1989). Of various uses to which fodder trees are put, perhaps the most significant is that dealing with soil fertility maintenance as it raises the chances of cultivation of fodder trees in production systems. Leguminous fodder trees such as *Gliricidia sepium*, *Leucaena leucocephala* and *Acacia albida* are the examples (Singh and Roy, 1993).

Table 5 List of important fodder trees/shrubs and their multiple uses

Species	Uses
Acacia albida Del.	1, 2, 10, 7
Acacia farnesiana (L.) Willd.	1, 10, 11, 7
Acacia ferruginea DC	2, 11
Acacia nilotica (L.) Willd ex. Del.	1 ,2, 10, 11
Acacia planifrons Wight & Arn	1, 2, 11
Acacia senegal (L.) Willd.	1, 2, 10, 11, 7, 6
Acacia tortilis Hayne	1, 2, 5, 7, 8
Acacia cordifolia Hook. F	9, 2
Ailanthus excelsa Roxb.	3, 2,1, 6
Albizia amara (Roxb.) Boiv.	1, 2
Albizia lebbeck (L.) Benth.	1, 2, 3, 6
Albizia odotissima Benth.	2, 11, 4
Albizia procera (Roxb.) Benth.	1, 2,
Albizia stipulate Boiv.	2, 11, 4
Anogeissus pendula Edgew.	2, 11, 4
Artocarpus heterphyllus Lamk.	9, 1, 2, 6
Artocarpus lakoocha Roxb.	1, 2, 4, 9
Azadirachta indica Adr. Juss.	1, 2, 6, 3, 4
Bauhinia purpurea Linn.	1, 11
Bauhinia variegata Linn.	9, 2, 10
Celtis australis Linn.	2, 1, 9, 6
Dalbergia sissoo Roxb. ex. DC.	1, 2, 7
Dendrocalamus strictus Nees	2, 7
Dichrostachys cinerea Wight & Arn	1, 5, 7
Ficus bengalensis Linn.	4, 6, 14
Ficus glomerata Roxb.	9, 6, 4
Ficus hookeri Miq.	1, 15, 6
Ficus religiosa Linn.	4, 6, 15
Gliricidia sepium (Jacq.) Walp.	1, 3, 2, 9
Grewia optiva Drummond ex. Burret	6, 2, 12
Hardwickia binata Roxb.	2, 12, 1
Holoptelia integrifolia Planch.	1, 2
Leucaena leucocephala (Lam) De Wit.	1, 3, 2, 8, 4, 13, 9
Mangifera indica Linn.	9, 2, 1
Moringa oleifera Lam.	1, 13, 9
Pithecellobium dulce Benth.	9, 1, 7, 10, 11
Pongamia pinnata (L.) Plerre	1, 2, 6
Prosopis cinerarea Druce	2, 1, 9, 6, 7
Prosopis juliflora DC.	1, 2, 7, 5
Pterocarpus marsupium Roxb.	1, 2, 10
Quercus semicarpifolia Sm.	1, 2
Quercus leucotrichophora Camas ex. Bahadur	1, 2, 7, 10
Sesbania grandiflora Pers.	9, 7, 1
Sesbania sesban Merrill	1, 2, 7, 6
Zizyphus jujuba Lamk.	9, 1, 2, 6, 5
Zizyphus nummularia (Brum. F.) W & A.	9, 1, 2, 6, 5

Note: 1 = fire wood/charcoal; 2 = timber/pole; 3 = mulch; 4 = shade; 5 = fence; 6 = medicine; 7 = soil conservation; 8 = windbreak; 9 = food/fruit; 10 = tannin/resin; 11 = gum; 12 = fibre; 13 = pulp; 14 = leaf plates; 15 = latex

2.4 Crop Residues

Large quantities of various crop residues are fed to livestock and in many parts it constitutes a major part of livestock diet. They include coarse straws, fine straws, leguminous straws, pulse straws and sugar cane tops. The major crops include wheat, rice, sorghum, maize etc. Total crop residue production in the country has been estimated to the tune of 393.39 million tonnes (GOI, 2006). The nutritional attributes of the main residues of wheat and rice are in fact not even sufficient to support maintenance requirements of growing and or lactating animals. However, there are opportunities to enrich such residues with various treatments to increase their protein value.

2.5 Concentrates and Other Miscellaneous Sources

Among the concentrate feed ingredients, coarse grains occupy the primary position. The production of coarse grains such as maize, sorghum, bajra and other millets amount to 37.76 million tonnes of which around 10 percent is used as livestock feed. Cereal by products such as brans and polish and various oil meals including ground nut cake, mustard cake, coconut cake, soybean meal, cotton seed meal and sesame cake amount to about 15.71 million tonnes. Various agro-by-products and other unconventional products like molasses, silk worm pupae, distillery waste, meat scrap, fish meal etc. are also put into use for livestock feeding. The others include vegetable and food processing residues and bagasse (GOI, 2006).

3.0 AVAILABLE OPTIONS

3.1 Improve Grazing Resources

At present Indian grazing lands are considered as fragile ecosystems and ranked as class IV and V in land capability classification. The stocking rates are 1–51 acu/ha against the carrying capacity of 0.20–1.47 acu/ha in semiarid regions. Similarly in arid regions stocking rates are 1–4 acu/ha against the carrying capacity of 0.2–0.5 acu/ha. The degradation of Indian grazing lands is ascribed to many factors like large bovine population, free grazing practices, lack of management, extremes of temperature, steep slopes, variable precipitation and scarcity of moisture in arid and semiarid regions. The condition in Himalayan pastures is even more alarming on account of severe pressures of sedentary, semi-migratory and migratory grazers (Pandeya, 1988).

Improvement of grazing resources and grassland restoration is the key factor in the improvement in livestock productivity that is sustainable and also the overall land productivity. Basically, the range restoration techniques are based on ecological approaches *e.g.* protection for vegetation recovery, soil and water conservation, reseeding, improved range management techniques and grazing management.

3.1.1 *Vegetation Recovery*

The rangeland degradation due to overgrazing leads to a situation where nutritionally low grade and unpalatable species dominate the scene. The vegetation becomes sparse and soil working in such situation is likely to further accentuate the erosion process. It has been found that just by mere protection through some sort of fencing a lot of benefit in terms of vegetation recovery can be obtained. A study revealed increase in herbage yield from 0.1 t/ha–3.5 t/ha within three years. The plant population of desired perennial grasses increased from 11 to 397 plants /m^2. However, time required for natural recovery of vegetation varies from one habitat to another and from one climatic zone to another. Once the desired level of vegetation recovery has been achieved, the rangelands should be released for grazing in the deferred rotational grazing management system by compartmentalizing with the help of inner fencing into paddocks (Shankar, 1986).

3.1.2 *Soil and Water Conservation*

Measures related to soil and water conservation in range and pasture management in dry areas are of crucial importance. The measures are especially important on degraded sloppy lands. Most useful such measures are described below:

3.1.2.1 *Bunding*

Along the contour lines and periphery bunds are made, either manually or using tractor mounted implements. The details *viz.*, cross section and interval of location of bunds are decided according to topography. In dry degraded areas, bunds are required to be stabilized by sodding with suitable grass species *viz.*, *Dichanthium annulatum, Cenchrus ciliaris* and *Panicum antidotale*. The excess runoff may be safely dispersed off through grassed waterways or pipe outlets (Srinivasan *et al.*, 1962). In areas having better soil and rainfall, technique of graded bunds along with grassed waterway and drop structure for safe disposal of water has been found to be better than contour bunds.

3.1.2.2. *Water Storage/Discharge Structures*

The construction of water storage/discharge structures *viz.*, checkdams, anne-cuts and embankments, of desired engineering specifications, in areas like ravines and watersheds etc. are required to be constructed for efficient conservation and utilization of already scarce water resources in dry areas. In some ravinous hilly areas gully plugging may also require erecting structure of various engineering specifications.

3.1.2.3 *Biological Measures*

Various grass and legume species are known to play an important role in soil and moisture conservation. The runoff and soil loss under such systems is reduced considerably in every agro-climatic region. In excessive windy areas a combination of trees may be planted as wind/shelter belt, in specific designs, to check the wind impact. The spreading of dry grasses or shrubby growth on bare ground along the contours or terraces has the potential to check soil and water loss. Various chemicals may also be used for mulching effect and soil stabilization (Singh, 1990a). Some examples are as under:

- Contour furrows (60 cm wide and 25 cm deep) at a distance of 10–15 m across the slope increased forage production of *Cenchrus ciliaris* considerably (130 percent over non-furrow plots).
- In inter row water harvesting system 30 cm wide ditch alternated with 70 cm wide raised beds, the yield of pasture grasses and legumes raised on the edge of ditch was higher by 124 and 66 percent, respectively when compared to the conventional system of planting.
- Use of Staggered Pitter Discer, a tractor attached device significantly increased forage yield, 80–115 percent in natural pasture and 50–70 percent in seeded pasture.
- In natural grasslands, contour furrows made at an interval of 6 m was most effective in soil moisture conservation when compared to 3 and 12 m spaced furrows.

3.1.2.4 *Reseeding*

The grasslands/grazing lands are usually of poor quality, containing less than seven percent CP for most of the time. Also, because of excessive grazing pressure on most of these grasslands, the productivity of such lands has deteriorated to considerable extent. In order to increase the pasture productivity, it is essential to replace low yielding annual grasses with high yielding perennial grasses that are adaptable to the prevailing condition of that region. As the use of fertilizers in these lands is of little practical value on account of high costs, introduction of suitable pasture legumes may be one of the best ways to improve the quality.

The important species identified for reseeding is provided in Tables 6 and 7. Selection of such species depends on the site conditions. *Dichanthium annulatum* and *Cenchrus ciliaris* have wide range of climate and soil in areas having rainfall up to 750 mm. In such situation suitable legumes include *Alysicarpus rugosus*, *Stylosanthes hamata*, *Styzolobium deeringianum*, *Indigofera endecaphylla* and *Lablab purpureus*.

Pre-monsoon broadcasting of *Stylosanthes* seeds in *Sehima-Heteropogon* grasslands was found best in terms of legume establishment. In case of *S. hamata* broadcasting of seeds after harrowing gave maximum establishment as compared to broadcasting alone or line sowing at 50 cm apart. Also, use of pelleted seeds gave better establishment. However, in arid situations pelleting of seed has not proved to be of any specific advantage except in case of *Laisurus sindicus*. Significant increase in nitrogen fixation is possible by appropriate selection of *Rhizobium* strain and by appropriate seed pelleting practice. The guide to arrive at required quantity of *Rhizobia* culture for use in inoculation of pasture legumes is given in Table 8. By using a seed rate of 4–6 kg/ha after required scarification pre-treatment and pre-monsoon sowing/broadcasting have provided the nitrogen equivalence to the tune of 40–60 kg/ha in natural grassland (Rai *et al*., 1980).

Table 6 Important grass species for reseeding in arid and semiarid regions

Condition	Species	Strains/varieties	Seed rate (kg/ha)
Extreme arid situations	*Lasiurus sindicus*	CAZRI–317, 319, 351; M 30–7, 20–5 and 295	4–5
Sand dunes	*Panicum antodotale*	CAZRI–331, 333, 379, 347	2–3
Arid and semi–arid –sandy loam soils	*Cenchrus ciliaris*	CAZRI–358; Marwar Anjan (75); IGFRI–3108	4–5
Arid and semi–arid –sandy loam soils	*Cenchrus setigerus*	Marwar Dhaman (CAZRI–76, 296, 175, 415); Pusa Yellow Anjan	4–5
Arid and semi–arid red gravelly soil/sloppy lands	*Chrysopogon fulvus*	IGFRI–1	3–4
Semi–arid clay soil with higher moisture status	*Iseilema laxum*	–	3–4
	Panicum maximum	Hamil, PGG–1, PGG–9	5–6
Arid and semi–arid– loam soils with higher moisture status	*Dichanthium annulatum*	CAZRI–490, 485 ; IGFRI–495–1; Marvel–8	2–3

Source: adapted from Roy, 2003

Table 7 Recommended seed rates for promising tropical pasture legumes

Species	Seed rate (kg/ha)	Species	Seed rate (kg/ha)
Atylosia scaraboides	6–10	*Macrotyloma axillaris*	2–2.5
Calopogonium mucunoides	6–8	*Neonotonioa wightii*	4–6
Centrosema pubescens	4–5	*Pueraria phaseoloides*	3–6
Clitoria ternatea	15–20	*Stizolobium deeringianum*	25–30
Desmanthus virgatus	2–3	*Stylosanthes guianensis*	4–5
Desmodium intortum	2–2.5	*Stylosanthes hamata*	5–6
Indigofera astragalina	3–5	*Stylosanthes humilis*	5–6
Lablab purpureus	20–25	*Stylosanthes scabra*	4–5
Macroptelium atropurpureum	6–12	*Stylosanthes seabrana*	3–4
Macroptelium lathyroides	2–3.5	*Vigna luteola*	8–12

Source: Skermann *et al*., 1998; Trivedi, 2004

Table 8 A guide to quantity of Rhizobia culture for use in inoculation of pasture legumes

Seed size group (no. seeds/kg)	Culture treatment (g/kg)	Selected examples
> 2.2×10^6	155	*Lotonois bainesii*
> 6.6×10^5-2.2×10^6	30	*Indigofera spicata, Lotus carniculatus*
> 1.3×10^5-6.6×10^5	10	*Desmodium intortum, Macrotyloma axillaries, Stylosanthes guianensis*
> 2.2×10^4-1.3×10^5	5	*Calopogonium mucunoides, Centrosema pubescens, Macroptelium atropurpureum*
2.2×10^4	2.6	*Lablab purpureus, Stizolobium deeringianum*

Source: Norris, 1970; Skerman *et al*., 1988

3.1.3 *Management Options*

3.1.3.1 *Bush Cleaning*

The undesirable bushes are required to be removed, either manually or by mechanical felling and removal of stumps or application of selective weedicides on the cut stamps to kill them to stop coppicing. The leaf fodder yielding shrubs may, however, be retained in appropriate proportion. It has been found that in rangelands of Indian deserts medium stock density level of *Zizyphus nummularia* (14 percent of land area covered by shrub canopy) was optimum for high forage as well as leaf fodder production (Kanodia and Patil, 1982).

3.1.3.2 *Application of Fertilizers*

Application of fertilizers on productivity of grasslands has been extensively studied in arid and semi-arid regions. The response is maximum in case of nitrogen doses followed by phosphorus, almost no response with potassium. The timing of fertilizer application is also important, maximum response has been obtained when applied about 30 days after the start of monsoon. Similarly, application of 60 kg N/ha in three split doses at 25 days interval was found to be superior (Rai and Kanodia, 1981). The forage yield increases were in the range of 40–70 percent over control.

3.1.3.3 *Controlled Burning*

Controlled burning, if used judiciously, is a powerful range/pasture management tool. It is preferable to use this tool in combination with moderate grazing management. Burning is recommended during wetter periods with longer intervals, about three years (Shankar and Trivedi, 1988).

3.1.3.4 *Grazing Rotations*

In pasture grazing systems, a period of rest is essential for the grasses to recoup and rejuvenate. In *Sehima* dominated grasslands in Jhansi deferred rotational grazing system has been found to be superior in providing more number of animal days (2925) as compared to continuous grazing system

(2097 days) with a body weight gain of 191 g/heifer/day, without being detrimental to the range health. Similarly, in *Lasiurus* dominated arid grassland the former system was superior in terms of average body weight gain and calving of "Tharparker" breed of heifers over continuous grazing system (IGFRI, 1990).

A mixed pasture of *Cenchrus ciliaris* + *Stylosanthes hama* has been adjudged best for body weight gain (342.5 g/animal/day) in case of calf, while in case of heifers *Cenchrus ciliaris* + *Macroptelium purpureum* pasture was the best. Studies on management practice for "Mandya" sheep indicated that it is be well maintained throughout the year on natural grasslands, if these animals are allowed to graze on stylo - siratro mixed pasture for about two hours in addition to six hours grazing in natural grasslands (IGFRI, 1990).

3.1.3.5 *Grazing Economics*

Grassland and rangeland degradation is the greatest challenge to the rangeland users in India. In a long-term study biomass and returns obtained under different grassland management systems is depicted in Table 9. The introduction of stylo added substantially (40%) to the biomass in the second year but reduced to almost negligible after second year of grazing. It showed the necessity of introducing stylo every third year. Natural grasslands, however, proved more economical in terms of output per rupee invested. Body weight gain in goats was similar in sown and improved pastures but it decreased in natural grasslands. Sheep showed best performance in sown pastures. Cattle gave better performance in improved pasture (Singh, 1995).

Table 9 Biomass yield and economics under three different grassland management systems in semiarid region

System	Yield (DM t/ha)						Return
	Yr 1	Yr 2	Yr 3	Yr 4	Yr 5	Average	(B/C)
Sown pasture	2.7	10.9	8.3	4.3	3.6	6.8	1.27
Improved pasture	2.3	9.7	8.0	3.7	6.1	6.2	1.17
Natural pasture	1.3	6.1	6.8	6.0	3.7	5.1	2.65

3.2. Agroforestry Practices

Woody perennial based forage production systems are capable of yielding leaf fodder, wood and food at the same time besides conserving and rehabilitating the ecosystem. Such systems have the ability to cope with the disasters in a more efficient way (Singh *et al.*, 1994).

3.2.1 *Agro-Silvipastoral Systems*

Such systems integrate crop, pasture and/or animals with trees. Here again woody perennial, preferably of fodder value, are introduced deliberately and systematically and managed scientifically (Rai, 1999). Such systems can be used for food production and soil conservation besides providing fodder and firewood. These systems hold promise especially in highland humid tropics (Pillai *et al.*, 1987). It may be tree-livestock-crop mix around homestead, woody hedge rows for browse, green manure, soil conservation or an integrated production of pasture, crops, animals and wood.

3.2.2 *Agri-Silviculture*

Such systems integrate crops and trees (Ramaprasad and Prasad, 1991). For fodder production, forage crops may be grown in association with woody perennial, preferably of fodder value. These systems are very promising in arid and semi-arid regions under irrigated situation. The positive associative effect of *Leucaena leucocephala* and *Sesbania aegyptica* has been found in forage crops like hybrid napier (IGFRI-3), lucerne, oat and several other cereal forages (Singh, 1990b). Agri-silvicultural technology for fodder production in *terai* belt of Uttar Pradesh in irrigated conditions involving poplars and berseem has given a fodder production of 50 t per ha in five to six months.

3.2.3 *Alley Cropping*

Alley cropping, also called hedge row intercropping, integrates the benefits of fallow period directly into the cropping period. Crops are sown in alleys between rows of trees, usually leguminous. The trees are pruned at regular intervals, at about 0.6 m above ground level, and the pruning are used for mulch or animal feed. It is a quite flexible technology that benefits crop and livestock activities and can, through a modification of tree management techniques, provide firewood for the household. It is a system that can be adapted to meet particular priorities of individual farmers. *Leucaena leucocephala* and *Gliricidia sepium* have been most widely used for alley farming. Now, use of two or more species is recommended to reduce the possibility of pest or disease completely destroying the production as psyllid pest did to *Leucaena* in much of the Asia and the Pacific.

In semi-arid tropics water competition between trees and adjacent food crops may be the dominant factor. But fodder shortage during late part of dry season increases the attractiveness of legume forage to livestock owners. However, *Leucaena* and *Gliricidia* may not be ideally suited to semi-arid tropics everywhere. There is a need of investigation on suitability of other tree species, their spacing and other management different environments.

3.2.4 *Hortipastoral Systems*

Such systems integrate fruit trees with pastures. The main objective is to obtain fruits. Trees become focus of the system and activities to establish pastures or produce animals become secondary. In such systems it is important to check that introduction of forage system is beneficial or at worst neutral to tree crop. There may be competition between trees and pasture but the extent and effect will depend upon spacing (Dasthagir and Suresh, 1990).

3.2.5 *Boundary Plantings*

Farm boundaries, banks, river banks etc. may also be suitably utilized for lean period fodder. At IGFRI Jhansi, *Leucaena leucocephala* at 2 m spacing along farm roads harvested at 3–4 years gave 11 and 1 t per row km of firewood and fodder respectively. The same plantation after 4 to 5 years in the second rotation produced 27.4 and 2.6 t per row km firewood and forage.

3.2.6 *Live Hedges*

These are lines of trees or shrubs planted on farm boundaries or on borders of home compounds, pastures, fields or animal enclosures. The purpose is different from other agro-forestry technologies based on trees planted in lines such as boundary plantings, contour strips or hedge row intercropping. Besides their main function to control human and animal movement - they provide fodder, fuel-wood and food; act as wind breaks; and enrich the soil depending on the species used.

The fodder species selected for this purpose should be adaptable to local agro-climatic conditions, of fast growth rate, of high coppicing ability and should provide higher economic returns. Depending on the site conditions species like *Gliricidia sepium*, *Carissa* sp., *Zizyphus mauritiana*, *Prosopis juliflora*, *Dichrostachys cinerea*, *Acacia* sp. may be used.

3.3 Use of Improved Varieties

The fodder cultivation is practiced in areas where rearing of improved cattle and buffaloes is common. It is desirable that available improved varieties are used for enhancing fodder production per unit area (Table 10). There is plenty of scope to introduce suitable varieties for a particular agro-climatic zone. Crops like sorghum, maize, pearl millet, cowpea and guar may be grown in *kharif* and oats, berseem, lucerne, barley in *rabi* season. The species like guinea grass, napier bajra (NB) hybrid, *Setaria* etc. may be grown as a perennial fodder resource.

Crops like NB hybrid, berseem, oats, maize are suitable for intensive production areas under irrigation while sorghum, pearl millet, lucerne, guinea grass etc. are able to tolerate moisture stress conditions. Although a number of varieties have been released in many fodder crops, the selection of a variety should be based on its merit and suitability for a particular situation.

Table 10 List of promising varieties in selected fodder crops for cultivation on arable lands

Crop	Varieties
Sorghum	Pusa Chari-6, Pusa Chari-9, SL-44, Hc-136, MP Chari, Meethi Sudan, IS-4776, Jawahar Chari-6, Jawahar Chari-69, JS-20, JS-263, GFS-4, SSG-988, 855 F, PCH-106, pantchari-3, AS-16, MFSH-3, LS-250
Pearl millet	Giant Bajra, K-677, Raj Bajra Chari-2, L-72, Rajko, TNSC-1
Maize	African Tall, Vijay composite, Jawahar, Moti composite, J-1006, Manjari composite
Cowpea	Russian Giant, NP-3, CO-1, UPC-287, UPC-5286, Gujarat Lobia-3, Lobia-88, Bundel Lobia-1, Bundel Lobia-2, Haryana Lobia-88, UPC-8075, UPC-52877, C-30, Kohinoor, Shweta,
Berseem	Mescavi, Wardan, BL-1, BL-2, BL-10, BL-22, Bundel Berseem-2, Bundel Berseem-3, JB-2, UPB-103, UPB-110
Lucerne	Type-9, Anand-2, LL composite-3, LL composite-5, Anand-3, RL-88, CO-1, SS-627
Guar	HFG-156, Fs-277, Bundel Guar-1, Bundel Guar-2, Bundel Guar-3, HFG-119, Guara-80
Oats	Kent, OS-6, UPO-212, OL-125, UPO-94, OL-9, JHO-810, Bundel Jai-822, JHO-851
Hybrid Napier	Pusa Giant, NB-21, Co-1, IGFRI-3, IGFRI-6, PBN-83, CO-2, CO-3, IGFRI-7, IGFRI-10, Yeshwant, CO-4
Setaria	PSS-1, Tall Fescue Grass, Him-1
Guinea grass	Macuenii, Hamil, PGG-1, PGG-19, PGG-101, PGG-3, PGG-9
Dinanath grass	PS-2, Bundel-1, Bundel-2, IGFRI-3808, TNDN-1

3.4 Design Crop Sequences

In view of limited scope of extending area under fodder crops it is important to (i) integrate forage crops in traditional cropping systems; (ii) identify new niches for fodder production and (iii) design productive and remunerative crop sequences (Patil and Alagundagi, 2006). The dry lands often fail to support two successive grain crops in rotation because of their long rotation. However, short duration forage crops may be included to precede or succeed the food crops to increase the cropping intensity on black soils. The forage sorghum – lentil sequence is more productive and remunerative. The fallows during *kharif* season may be made use of by growing cover-cum-forage crops like cowpea for enhancing land productivity as well as soil protection.

The red soils are traditionally mono-cropped in many parts as the residual moisture is low to support winter crops in sequential system. Intercropping of short duration forages in long duration rainy season food crops such as sorghum and pigeon pea, which are widely spaced, is a good option. The legume forage component in such sequences apart from providing nutritious fodder; controls the weeds and enriches the soil. Potential intercropping schemes are depicted in Table 11. In these schemes the forages are harvested at 50 days and thereby allowing main crop to grow free of competition and produce grain on residual soil moisture.

Table 11 Potential intercropping scheme for red soils

Main crop	Fodder crops
Grain Sorghum	Cowpea, sunhemp, moth bean, *Sesbania sesban*
Grain Pigeon pea	Forage sorghum, teosinte, maize, bajra, sudan grass, cowpea, cluster bean, sunhemp

Ratooning is another means of increasing productivity of mono-cropped dry land areas for providing fodder as well as food. The select cultivars of pearl millet possess high regeneration capacity and so could be rationed after 30–40 days of growth and the top dressed with 20 kg N/ha. The yields are – 19.2 t green forage/ha followed by 7.4 q grain/ha and 3.05 t stover/ ha). Over seeding of leguminous forage *Stylosanthes* in maize, sorghum and pearl millet increases grain yield (6–26%) and yield nutritious fodder as well. Lay farming sequences (fodder-food) are known to have remarkable improvement in soil physical and chemical properties. Continuous cropping of either pure *Stylosanthes hamata* or a combination of *Cenchrus ciliaris* + *Stylosanthes hamata* for three years; before taking sorghum as a grain crop increases the yields remarkably.

There are several crop sequences for irrigated areas like (i) sorghum-berseem-pear millet, (ii) maize + cowpea-berseem-pearl millet and (iii) NB hybrid intercropping with berseem. Staggered planting of cropping sequences in rotational fashion by sub dividing forage fields in 2–3 plots and conserving excess fodder as hay or silage is expected to solve many fodder related problems in a drought affected region.

3.5 Assured Seed Availability

Shortage in availability of quality forage seeds is one of the major bottlenecks in (i) realizing higher forage yields per unit area and (ii) rehabilitation of grazing lands by reseeding. There is a huge gap (80–90%) in demand and availability of seeds of range grasses and legumes both in tropical and temperate regions (Mishra and Bhatt, 1996). This is attributed to several reasons *viz.*, non-synchrony in maturity of grass seed, seed shedding, poor seed productivity of high forage yielding varieties and difficulties in harvesting.

There are several options available to tackle this issue through technological interventions *viz.*, use of growth retardants/hormones, foliar spray of mineral nutrients, agronomical and management practices. There is a need of developing a strong national seed network for ensuring the issue of forage seed availability.

4.0 STRATEGIES

In the event of any disaster, especially drought that prolongs for much a larger duration, there is a great concern for fodder. Our system should ensure adequate fodder supplies during the disasters in order to reduce losses to a great extent and the poor families are able to retain the herd. In India, fodder for livestock is met through various sources and there is a great deal of variability across different regions. During disasters, many a times it is not feasible to actually produce fodder from the lands which other wise is meant for this purpose. Similarly, productivity of the natural areas either declines or simply not accessible.

In view of many available options, including the technological ones; it is suggested to have strategies based on short (immediate) as well as long term (drought proofing) contingency plans (KNNB, 2006).

4.1 Immediate Measures

4.1.1 *Feeding Strategies*

- The grain crops like maize, sorghum, bajra and minor millets are harvested at flowering stage for fodder.
- *Kharif* sorghum is harvested for fodder and ratoon allowed for grain production.
- Minimize wastage (30–40%) of fodder by chaffing.
- Tree/shrub leaves (lower branches) are pruned and used to feed the livestock.
- Left over dry roughages after feeding may be stored and re-fed after enrichment.

4.1.2 *Use of Non-conventional Fodder*

- Chopped banana leaf and stem is fed to livestock to the extent of one-third of its daily requirement.
- Grasses from forest and other areas are harvested and fed to livestock.
- Feeding of sugar cane tops and bagasse.

4.1.3 *Other Measures*

- The dry fodder like *jowar kadabi*, wheat *bhusa* etc. available in other areas is procured and distributed in the affected areas by some agency/ community.
- Establish fodder banks near forest areas to procure and store available forage and then supply to the affected areas by some agency/ community.
- Grow range grasses and legumes in wasteland and forest areas by some agency so that additional supplies are ensured.
- Encourage planting of multipurpose fodder trees/shrubs in social and farm forestry programmes for developing a fodder resource for disaster periods.

- An agency is made responsible for keeping adequate stocks of fodder seeds for contingency planning. Sowing of dual/grain types of even hybrids (F_2s) should be done in event of delay in procuring specific varieties.
- Assess fodder requirement in drought affected districts and locate areas where shortages are likely to occur and arrange for supplies from near by districts.
- Monitor the prices of fodder in selected places/markets and arrange to procure fodder from surplus States.
- State forest departments arrange for the cutting and bailing of grasses in the forests, wherever possible to meet the demand from fodder deficit districts.
- Encourage fodder cultivation, wherever feasible.
- Obtain necessary pre-mired feed and urea-molasses-bricks from NDDB and other sources.

4.2 Medium to Long Term Measures

As all natural calamities cannot be predicted and prevented, a state of preparedness and ability to respond quickly to a natural calamity can considerably mitigate loss of livestock resource and restore normalcy at the earliest. It is, therefore, of paramount importance to take long term measures based on a plan of action for dealing with contingencies that arise in the wake of natural calamities. It is required to be periodically updated as well.

The basic principle of long term measures for tackling the issue of fodder should aim at (i) investing in a good year for a bad, and (ii) initiatives that emphasizes on self reliance of communities and they are less dependent on government. The emphasis should be towards:

- Developing drought resistant varieties of crops and propagating their spread.
- Completing quick maturing irrigation schemes and ensuring water flow in the medium term, say, within 3 years.
- Taking up watershed development activities and developing a cropping system based on such watershed approach.
- Afforestation of all exposed hill-slopes as per 'Ridge to Valley' watershed development approach and increasing forest canopy cover in order to ensure conservation of rain water.
- Incorporation of fodder trees in various afforestation schemes to build a strong top feed reserve resource.
- Encouraging programmes on improvement of grazing lands and adoption of silvipastures.

- The State Governments should interact with local/state level NGOs and communities in disaster management activities.
- The State Governments should try to construct check dams in the drought affected areas with actual participation of beneficiaries/NGOs so that the stored water could be used for various purposes during drought.
- Training and education of officers responsible for fodder development and farmersin the area of grassland/fodder production technologies.

5.0 CONCLUSIONS

Since the livestock are a core asset for poor farm families or landless; the loss in this resource during disasters is critical. Of several reasons for devastating losses to animal herds during disasters, especially drought, lack of fodder is an important one. During disasters, many a times it is not feasible to actually produce fodder from the lands which other wise is meant for this purpose. Similarly, productivity of the natural areas either declines or simply not accessible.

In India, fodder for livestock is met through various sources and there is a great deal of variability across different regions. So region specific mechanisms are required to ensure adequate fodder supplies during the disasters with the aim of reducing losses to a great extent and the poor families are able to retain the herd. The strategies are required on region as well as local basis after taking into consideration available fodder resources and prospects of its production, conservation and utilization. The strategies should be long as well as short term. Implementation of these strategies through government and non government organizations and also community will lead towards effective combating of the disasters.

6.0 REFERENCES

Bhasin, N.R. 2002. Drought and livestock. *Indian Dairyman*, **54(8)**: 33–35.

Dasthagir, M.G. and Suresh, K.K. 1990. Fodder production under trees in hortipastoral systems. *My Forest* **26**: 8–12.

KNNB, 2005. Community fodder banks – a guideline. Drought Proofing Cell, Kutuch Nav Nirman Abhiyan (KNNB), Bhuj. pp. 23.

GOI, 2006. Report on Working Group on Animal Husbandry and Dairying for the Eleventh Five Year Plan (2007–2012). Planning Commission, Government of India, New Delhi. pp. 29–33.

IGFRI, 1990. *Forage Production Technology*. Indian Grassland & Fodder Research Institute, Jhansi. pp. 43.

Kanodia, K.C. and Patil, B.D. 1982. Pasture development. *In*: *Desert Resources & Technology* (*Ed*. Alam Singh). Scientific Publishers, Jodhpur. pp. 263–305.

Kelley, T.G. and Parthasarathy Rao, P. 1994. Trends in cultivated green fodder crops in India. *In: Proc. Workshop Quanity and Quality of Fibrous Crop Residues* (*Eds.* Joshi, A.L., Doyle, P.T. and Oosting, S.J.) BAIF, Pune. pp. 21–29.

Mishra, L.P. and Bhatt, R.K. 1996. Forage seed: research and technology. *In*: *Transfer of Forage Production Technology* (*Eds.* Bhag Mal and Hagger, R.J.). Indian Grassland and Fodder Research Institute, Jhansi. pp. 91–99.

Negi, S.S. 2002. Drought and livestock farming. *Indian Dairyman*, **54(8)**: 29–32.

Norris, D.O. 1970. The intelligent use of inoculants and lime pelleting for tropical legumes. *Tropical Grasslands*, **1**: 107–121.

Pandeya, S.C. 1988. *Status of the Indian Rangelands*. RMSI, Jhansi. pp. 213.

Patil, S.A. and Alagundagi, S.C. 2006. Forage production and feeding during scarcity. *In*: *Livestock Feeding Strategies for Dry Regions* (*Eds.* Pathak, P.S. and Kundu, S.S.). International Book Distributing Company, Lucknow. pp. 1–14.

Pillai, G.R., Bai, M.M. and Lakshmi, S. 1987. Coconut based fodder production systems in Kerala. *In*: *Forege Production in India* (*Ed.* Singh, P.). IGFRI, Jhansi. pp. 97–102.

Rai, P. 1999. Silvipastoral systems: growing greener and increasing production from wastelands. *Wasteland News* **14(3)**: 48–50.

Rai, P. and Kanodia, K.C. 1981. Effect of split application of nitrogen on the quality and quantity of *Sehima-heteropogon* grassland. *Forage Res.*, **7**: 83–85.

Rai, P., Kanodia, K.C., Patil, B.D., Velaydhan, K.C. and Agrawal, R. 1980. Nitrogen equivalence of range legumes introduced in natural grasslands. *Indian J. Range Mgmt.*, **1**: 97–101.

Ramaprasad and Prasad, R. 1991. Under planting of stylo legume crop in teak seed orchards. *Vaniki Sandesh* **15**: 1–5.

Roy, M.M. 2003. Grassland and pasture management systems in arid and semiarid regions. *In*: *Sustainable Animal Production* (*Eds.* Jakhmola, R.C. and Jain, R.K.). Pointer Publications, Jaipur. pp. 112–134.

Roy, M.M. and Singh, K.A. 2008. The fodder situation in rural India: future outlook. *International Forestry Review*, **10(2)**: 217–234.

Roy, M.M. and Singh, K.A. 2009. Tropical pasture legumes in sustainable livestock production. *In*: Legumes for Ecological Sustainability (*Eds.* Ali, M., Gupta, S., Basu, P.S. and Naimuddin). ISPRD, Kanpur. pp. 480–490.

Shankar, V. 1986. Vegetation recovery under protection. *In*: *Desert Environment: Conservation and Management* (*Eds.* Shankarnarayan, K.A. and Shankar, V.), CAZRI, Jodhpur. pp. 25–28.

Shankar, V. and Trivedi, B.K. 1988. Grazing burning rotation for the management of *Sehima-Heteropogon* grasslands in Central India. *In*: *Proc. 3rd International Rangeland Congress* (*Eds.* Singh, P., Shankar, V. and Srivastava, A.K.), RMSI, Jhansi. pp. 382–385.

Sincere, M.N. and Ranjhan, S.K. 1989. The place of fodder trees and shrubs in grassland systems. *In*: *Proc. XVI International Grassland Congress. Association.* Francaise pour la Production Fourragere, Nice (France). pp. 1761–1768.

Singh, K. 1990a. Grassland and pasture development in the arid region of Rajasthan. *In*: *Proc. Rehabilitation of Degraded Land Ecosystem (UNESCO-ICAR)*, CAZRI, Jodhpur. pp. 1–8.

Singh, P. 1990b. Agrisilviculture systems in India. *In*: *Shrubs and Tree Fodder for Farm Animals* (*Ed*. C. Devendra), IDRC, Ottawa (Canada). pp. 186–195.

Singh, P. 1995. Improved methods in rangeland production and utilization systems for sustainable development in India. *In*: *Proc. V. International Rangeland Congress*. Society for Range Management, Denver. pp. 516–517.

Singh, P. and Roy, M.M. 1993. Silvipastoral systems for ameliorating productivity of degraded lands in India. *Ann. For.*, **1**: 61–73.

Singh, P., Pathak, P.S. and Roy, M.M. (*Eds.*). 1994. *Agroforestry Systems for Sustainable Land Use*. Oxford & IBH, New Delhi: pp. 28.

Skerman, P.J., Cameron, D.J. and Riveros, F. 1988. *Tropical Forage Legumes*. FAO, Rome. pp. 692.

Srinivasan, V., Bonde, W.C. and Tejwani, K.G. 1962. Studies on grass and their suitability to stabilize and maintain bunds in the ravine lands of Gujarat. *J. Soil Wat. Conserv. (India)*, **10**: 72–78.

Trivedi, B.K. 2004. *Grasses and Legumes for Tropical Pastures*. IGFRI, Jhansi. pp. 47.

8

FEEDING STRATEGIES: MEETING THE CHALLENGES DURING NATURAL CALAMITIES

M.M. Das, S.K. Mahanta and A.K. Misra

ABSTRACT

Long spell of drought leads to an adverse effect on feed resources as dried roughages, (which are in short supply) alone account for over 50 to 80 percent of dry matter and TDN required for ruminants (viz., cattle, buffalo, sheep, goats and camel). Further, during drought the nutritional quality of available forages (including green) particularly crude protein (CP) and TDN deteriorated due to the stunted growth of plant. Information on nutrient requirement of different classes of livestock is given. The deficiency of nutrients that arises during natural calamities and the requirement of protein, energy, mineral and vitamins etc. has been provided. The feeding strategies for sustaining life, formulation of diets from locally available feed stuffs, the nutritive value of non conventional feed stuffs, their incorporation in the diet, the role of top feeds during scarcity are discussed. Urea treatment of straws is so far the only chemical treatment with practical potential under field conditions. Urea-treated straw can save concentrate in the ration, increases daily milk yield by 1-2 litres/animal, offers better economic returns to the farmers. The chapter also provides guidelines on the preparedness strategies to counter act the feed scarcity that arising during calamities by adopting suitable technology viz. complete feed block, feed and fodder bank etc. Fodder banks are the buffer stocks of fodder grasses, straws of paddy, wheat etc which are to be used for feeding livestock during scarcity and drought. The minimum size of fodder bank should be between 10% to 20% of the required feed and fodder required annually in a particular zone, state or district. The making of block reduces the transportation cost of roughages drastically.

Indian Grassland and Fodder Research Institute, Jhansi (UP), India

1.0 INTRODUCTION

Feeding of livestock is important for the survival of existing stock and for different production purposes. The feed of livestock can be classified in to three major categories *viz.* a) dry roughage (straws like paddy straw, wheat straw, stover, leguminous straws etc.), b) green fodders (berseem, cowpea, maize, sorghum etc.) and c) concentrate (mustard cake, groundnut cake, cotton seed cake, maize grain, barley grain etc.). While feeding to livestock, the prime consideration is to ascertain and to meet up the total nutritional requirements in terms of i) dry matter, ii) digestible crude protein (DCP), and iii) total digestible nutrients (TDN). The requirement of the quantity of dry matter depends on the body weight of the animal and also on the nature of its production (milk/meat/egg/wool). Cattle generally consume 2.0 – 2.5 kg dry matter for every 100 kg of body weight daily while buffaloes and crossbred animals consume 2.5 – 3.0 kg per 100 kg live weight daily. As DM intake is the limiting factor, all the requirements like carbohydrate, protein, fat, minerals or vitamins of the animal should come from the total dry matter consumed. Further the bulk in the diet of ruminant is essential for animals satiation, the dry matter allowance is generally divided into different proportions *viz.*, dry roughage (approximately 66%) and concentrate 34%. The amount of dry roughage can be changed as per availability of the green roughages. The requirement of DCP and TDN depends on the physiological needs or the purpose for which animals are kept *i.e.*, whether the animal is just to maintain itself or in addition to maintenance, it has to meet out the demand of pregnancy or milk production. The requirement of DCP and TDN for all such purposes for different categories of livestock is given in (Tables 17 to 26).

However, such ideal nutritional requirement cannot be met out in areas, which is severely affected by natural calamities like drought and flood every year. The scarcity of animal feed caused by floods is of a temporary' nature, however, severe shortage of animal feed especially roughages is encountered during drought. Long spell of drought leads to an adverse effect on feed resources as dried roughages, (which are in short supply) alone account for over 50 to 80 percent of dry matter and TDN required for ruminants (*viz.,* cattle, buffalo, sheep, goats and camel).

Further, during drought the nutritional quality of available forages (including green) particularly crude protein (CP) and TDN detoriated due to their stunted growth of plant. Animal fed on such forages will not be able to fulfill its normal requirement of protein, energy and other nutrients like minerals and vitamins. Special measures can be taken to make up these nutrient deficiencies.

2.0 SUPPLEMENTATION DURING SCARCITY

2.1 Urea Molasses Supplement

Urea molasses preparation (containing 5 kg urea and 25 kg molasses dissolved in 100 to 200 litres of water), or the use of a ready-mixed commercial product, sprayed at a rate of 1 litre of mix per animal over a thick stand of grass. Treating poor-quality roughage with the same mixture is highly effective in promoting intake, as well as improving its nutritional value.

2.2 Urea Molasses Mineral Block (UMMB) - As Licking Material

Sometime during scarcity period only straw is available which can not be enriched *in situ* due to non availability of molasses. Under such circumstances, reduction in milk production and weight losses of animals are common features, which culminate in substantial economic losses to the farmers. The use of urea molasses mineral blocks (UMMB) for supplementing crop residues based diet of livestock has been well documented in ruminants and has the potential to increase the viability of livestock production, to increase dry season milk supplies and increase household income. Use of UMMB is more convenient than the use of urea molasses liquid diet (UMLD), especially for small farmers during scarcity condition during drought and flood. Urea molasses mineral blocks could be fed to the affected ruminants as supplement for protein, energy and minerals. Hungry animals must not be allowed sudden, free access to licks containing high levels of urea. In addition, animals not fed urea for a period exceeding 48 h, must first be adapted by gradually increasing the levels of urea in the supplement.

As a feed supplement, it provides essential nutrients such as protein, energy, and minerals usually deficient in most forages and crop residues. Urea-molasses-mineral block (UMMB) is prepared in a solid, hard, compact form, but it is water-soluble. It is given to ruminants as a lick.

The block can be prepared using several formulations depending on the supply and price of the needed ingredients. The main ingredients and their proportion in the formulation are indicated below:

The composition of UMMB is given below

Ingredients	Parts/100 kg of mixture
Molasses	38%
Urea	10%
Wheat/rice bran	40%
Portland cement	10%
Mineral mixture	1%
Salt	1%
Total	**100%**

The amount of UMMB to be given depends on the type of animal and its weight. A block of 5 kg may last for 7~10 days when given to a cow or water buffalo weighing 350–400 kg. However, such blocks cannot be used as a supplement for animals younger than 6 months, or by animals which have not eaten anything for the whole day. Consumption of too much UMMB must be prevented. Animals must always be provided with clean drinking water. The block must be protected from rainwater so that it does not soften.

2.3 Protein

Protein requirement of a cow will be 300 to 500 g per day for maintenance. Protein requirement therefore be met through supplementing oil cakes, tree leaves, pods, urea and poultry litter. Poultry litter, can be fed to mature animals @ 1.5–2.0 kg/d in addition to low quality roughage. Dry cows should be provided approximately 250–400 g of supplemental crude protein and lactating cows with 500–600 g of supplemental crude protein per day.

2.4 Energy

Several options are available for supplying energy to animals on drought-stressed pasture. Wheat bran, grain, and crop processing byproducts can be used to supply energy to grazing animals. Grain should be supplemented @ 0.2% of body weight. Corn and oats can be fed whole but may be better utilized if coarsely rolled before feeding. It is desirable to crack the hull of barley so that rumen microorganisms and digestive enzymes can access starch in the kernel. Fine grinding and rolling which results in excess fineness and dust should be avoided. In addition, extremely dusty supplements are unpalatable. Straws can be ammoniated to improve their digestibility and available energy content.

2.5 Minerals

During drought phosphorus supplementation is crucial. A mixture of 50% trace mineralized salt and 50% dicalcium phosphate should be supplied free choice to the cowherd. The salt mixture should be placed close to stock watering locations.

2.6 Vitamins

Vitamin A content of forages during drought and hay produced from drought-affected forages are low. Therefore Vitamin A should be supplemented in the diet and vitamin A and D booster shots should be given approximately 30 days prior to calving and their calves should receive vitamin A and D at birth.

During calamities the strategies for emergency feeding depends on whether animals have to be saved from starvation only or whether their production of milk, meat etc. are also to be maintained. It has been estimated that 43% of the total livestock in India are perpetually affected by drought. Drought of 2002–03 affected 62% bovine population (156 million) in 18 states. Due to inadequate nutritional support, milk production declined in many states, average daily milk procurement was reduced by 7% in Tamil Nadu, 8% in Maharashtra and 22% in Rajasthan.

Besides the availability, transport of feed is also a major problem during natural calamities. The bulkiness of the feeds is a serious constraint during transportation and increases the cost of transport. The use of high quality feeds (concentrate), or the densification of straws is options to ease transportation problem. If transportation of feeds is not feasible, the available nutrients should be diverted from the feeding of smaller number of producing animals, to the feeding of a larger number of animals at (sub) maintenance levels of production. In such situation, the survival of animals becomes the most essential form of animal production. During drought, two situations may arise, one relates with lack of fodder availability and the other with lack of feed quality. In both the situations, three programmes need to be followed depending on the severity of the drought or flood situation. 1) Feeding strategy for maintenance of animals to ensure its survival; 2) Feeding strategy to maintain minimum level of production and growth; 3) Feeding strategy to maintain desired production by purchasing supplements from outside.

3.0 RATIONS FOR SUSTAINING LIFE OF LIVESTOCK DURING CALAMITIES

During drought situations when the availability of conventional roughages is scarce, the following livestock rations can be used for meeting the immediate nutritional requirements.

3.1 Urea Molasses Liquid Diet (UMLD)

Urea molasses liquid supplements can be fed during natural calamities along with small amount of roughage to maintain the livestock health. Urea molasses liquid feed can be prepared by mixing 93 kg molasses with 2.5 kg urea, 2.5 kg water, 1.5 kg salt, 0.5 kg mineral and vitamin supplements. The mixture of urea molasses should be condensed by heating till its DM content reached around 60–70%.

Lactating cows producing 8 kg of milk can be sustained on liquid urea-molasses diets supplemented with 1 kg of DM (5–6 kg green) through green fodder like oats, maize etc or from tree leaves (*Leucaena leucocephala,*

Sesbania sesban etc.) according to availability per 100 kg of body weight, 0.8 kg of cereal grain like maize, barley etc and 1 kg of groundnut cake or soybean cake, cottonseed cake per head/day. The animals can consume about 6 kg of liquid diets (Ranjhan and Khera, 1976). However, calves of 3–4 months of age should not be fed such liquid diet.

3.2 Procedure of Feeding to Adult Animal

Animal should be fed 5–6 kg dry roughage on first day. On third day, 4–5 kg dry roughage with 1 kg urea molasses liquid feed should be offered to the animal. Subsequently in the next seven days, the quantity of urea molasses liquid feed should be increased to 3–4 kg along with 3–4 kg dry roughage in the diet. From 11th day onward, liquid diet should be offered after consumption of 3 kg dry roughage. It is important to provide drinking water at all times. In places where there is acute shortage of forages, the following composition of UMLD may be used (Table 1).

Table 1 Ingredient composition of UMLD (without roughage)

Ingredient	(g)
Molasses	840
Concentrate	100
Urea	30
Mineral mixture	20
Phosphoric acid	10

During scarcity, sugarcane bagasse which is produced in the sugar factory can be used as an alternate roughage resource along with urea and molasses to form a life saving diet for livestock. Based on the research conducted at Indian Veterinary Research Institute (IVRI), Izatnagar (U.P.), a technology was developed for use in sugarcane growing areas (Table 2). It was successfully tested in 1973 in Maharashtra when drought prevailed in that state. More than 34000 animals including 5000 crossbred animals were fed urea-baga-molasses formulation under the guidance of Bhartiya Agro-Industry Foundation (BAIF).

Table 2 Composition of the ration of different categories of livestock

Ingredient	Milk producing adult cattle (150–300 kg B.W)	Adult cattle (150–300 kg B.W)	Growing cattle (50–150 kg B.W)
Bagasse (kg)	2.0	4.0	2.0
Molasses (kg)	0.4	0.8	0.8
Sugar cane tops (Chaffed) (kg)	8.0	3.0	3.0
Urea (g)	22.0	20.0	40.0
Common salt (g)	30.0	20.0	20.0
Mineral mixture (g)	50.0	50.0	30.0

Preparation: The bundles of bagasse are spread out in layers of about 15–20 cm high. The mixture of urea and molasses is sprayed over the materials can now be offered to animals. Three days ration can be prepared at one time.

4.0 FEEDING STRATEGIES DURING CALAMITIES

4.1 Complete Feed Mix (CFM)

During natural calamities, optimum utilization of the roughage and concentrate should be ensured. Thus the available feed resources in a particular area may be fed in the form of complete feed. Ruminants are generally given wheat bhoosa, paddy straw and stovers during scarcity, but such materials are poorly digested. Nutritive value of these roughages can be improved by mixing with urea and other ingredients like molasses, mineral mixture etc. Complete feed system is advantageous against conventional system of feeding as it reduces labour cost, maintains uniform roughage concentrate ratio, uniform feed intake favoring uniform supply of nutrients for maintenance and conducive for microbial growth. This system of feeding is well suited to our country as it helps in utilizing locally available crop residues, agro-industrial by products and non-conventional feeds during scarcity.

4.1.1 Straw Based Complete Feed Mix

Straw based complete feed can be prepared for drought or flood situation. To prepare one ton of feed (10 q), the ingredients can be mixed in the following proportion.

a) Straw of wheat, paddy, bajra, jowar etc. 9 q
b) Urea 10 kg
c) Molasses 85 kg
d) Mineral mixture with salt 5 kg
e) Vitamin supplement 50 g

Procedure: Dissolve urea thoroughly in 10–15 litres of water and then mix it thoroughly in above-mentioned quantity of molasses. The mineral and vitamin supplement should also be mixed with this solution at the same time. The straw should be chaffed and spread on pucca floor/polythene sheet. The solution should be sprayed with sprayer or hose used in gardens. Constant stirring of solution and the straw is essential for through mixing. This mixture can be fed as such to all categories of livestock except young calves in the following quantity.

Adult (non lactating)	8–10 kg
Lactating (2–3 kg milk)	*Ad lib*
Heifers	6–8 kg
Bullocks and bulls	*Ad lib*

Table 3 Composition of complete feed mix for growing animals (% DM basis)

Ingredients	Ration I	Ration II	Ration III
Wheat straw	30.0	30.0	40.0
Prosopis juliflora pods	20.0	–	10.0
Rice polish	6.5	20.5	–
Mango seed kernel	–	–	10.0
Wheat bran	5.0	18.0	8.0
Groundnut cake (deoiled)	25.0	18.0	18.5
Urea	0.5	0.5	0.5
Molasses	10.0	10.0	10.0
Salt	1.0	1.0	1.0
Mineral mixture	2.0	2.0	2.0

http://www.icar.org.in/drought/drought2.htm

Table 4 Composition of complete feed mix for milch animals (% DM basis)

Ingredients	Ration I	Ration II
Wheat straw	45.0	45.0
Groundnut cake (deoiled)	25.0	–
Prosopis juliflora pods	13.0	13.0
Babul seed chuni	3.5	3.5
Mustard cake (deoiled)	–	25
Urea	0.5	0.5
Molasses	10.0	10.0
Salt	1.0	1.0
Mineral mixture	2.0	2.0

4.2 Utilization of Straws/Stovers Submerged in Flood

1. Recover the straw from the flood water
2. Wash it with clean water (if available)
3. Spread in the open air for drying after spraying 1% common salt solution. This will prevent the growth of moulds and fungi.
4. Use as early as possible preferably within a month
5. Straw can be ensiled with chaff green fodder (tree leaves, grasses, aquatic plants etc.) and molasses with urea.

4.3 Use of Unconventional Feed Resources

Some feed and fodder, which are not generally used for animal feeding (mainly due to un-palatability), can also be used during scarcity (Tables 5 & 6).

4.3.1 *Flood Prone and Coastal Areas*

4.3.1.1 *Cher Leaves and Fruits (Avicennia officinalis)*

Cher grows near the seashore in stagnated seawater. During scarcity, cher leaves are available which can be used for feeding animals. Ripen cher fruits floating on water can also be used for animal feeding. It is the experience of livestock owners that cher leaves and fruits increase the milk fat content.

4.3.1.2 *Aquatic Plants*

Aquatic plants available in river, pond and other water logged areas can be used for feeding farm animals. Water hyacinth, aquatic spinach, stalks and leaves of lotus plants, pistia, aquatic weeds and jangli paddy etc. are common aquatic plants tested for animal feeding.

i) *Water hyacinth*: The chemical composition of water hyacinth varies from species to species and climatic conditions. It contains 10–15% crude protein, 1–2% ether extract, 15–25% crude fiber, 40–50% nitrogen free extract and 13–15% ash.

ii) *Water spinach* (*Ipomoea reptans*): It is fast growing aquatic plant containing 90–92% water, 2–2.5% crude protein and 3–4% carbohydrates. It can be fed as a part of ration to supply about 20–25% of dry matter.

4.3.2 *Drought Prone Areas*

4.3.2.1 *Cactus*

Cactus is primarily found in deserts hence it is easily available during scarcity also. As such it is not used for feeding animals but during scarcity it can be used. It contains 3.3% crude protein, 3.8% crude fat and 56% carbohydrates. It also provides water to animals. The thorns should be removed/burnt before feeding cactus. 8 to 10 kg cactus can be fed daily to an adult animal along with other roughages.

4.3.2.2 *Agaves (Ketki)*

Agaves also grows in desert and semi desert areas and hence during scarcity it is available. It is not fed to animals routinely but can be used during scarcity. It contains 4.5% crude protein, 3.3% crude fat and 48% carbohydrates. Animals can be maintained on it along with other concentrates during scarcity.

4.3.2.3 *Neem Seed Cake*

Neem seed kernel cake is a protein rich (34–40% CP) agro-industrial by-product and available to the tune of 0.7 MT annually. It's ensiling with 2.5% fertilizer grade urea for 5–6 days followed by sun drying and grinding resulted in improvement of cake. Such processed cake can be used in the feeding of buffalo calves, fattening lambs and kids, broiler poultry and rabbits without adversely affecting the growth, nutrient utilization, blood biochemical profile. Thus, urea ammoniated neem seed kernel cake can be used as a wholesome protein substitute in the feeding of various farm livestock to spare costly oil cakes.

4.3.2.4 *Sugar Cane Bagasse*

Sugarcane bagasse is produced in the sugar factory can be used as an alternate roughage resource in the diet of livestock during scarcity. Country is producing nearly 383 MT of sugarcane bagasse. By treating bagasse with urea the nutritional quality can be enhanced in terms of protein content, palatability and digestibility. Sugarcane bagasse is not palatable as such but mixed ration containing sugarcane bagasse, molasses, 1% urea, 1% salt and 2% mineral mixture is quite palatable to animals after adaptation and animals can be maintained on it.

4.3.2.5 *Sugarcane tops*

Sugarcane tops can also be used as basal roughage during scarcity period. Cane tops are low in N content and unable to support optimum fermentation in the rumen. Therefore it should be fed with some nitrogen rich supplements like urea and tree leaves. Cane tops contain 5.3% CP, 2.3% DCP and 49% TDN with low phosphorus content (0.1–0.2%) and high in silica. It should therefore be supplemented with calcium, phosphorus and molasses to improve its palatability.

4.3.3 *Both Drought and Flood Prone Areas*

4.3.3.1 *Paper Waste*

Paper waste is rich in carbohydrate (cellulose) and it helps in satisfying the hunger of animals. Ground paper waste (6 kg) supplemented with molasses (4 kg), salt (50 g), and mineral mixture (50 g) per day is sufficient to maintain an adult animal. Complete feed with 30% ground paper waste and other concentrates can be effectively used to maintain the animals during scarcity.

4.3.3.2 *Sawdust*

During scarcity when nothing is available, saw dust can also be fed to animals. Complete feed containing 30% saw dust, 32% maize bran, 31% molasses, 4% urea, 2% salt and 1% mineral mixture can be used as maintenance ration.

4.3.3.3 *Fruit Factory Waste*

The waste material like pine apple waste, orange peel, tomato pomace can be used as apart of livestock diet after processing through ensiling.

4.4 Use of Top Feeds During Scarcity

During drought, the green fodder is almost unavailable. During natural calamities severe shortage of dry roughages is also encountered. Under this situation, the use of tree leaves in the diet is the best option to sustain

Table 5 Some unconventional feeds and fodders and their nutritive value (% DM basis)

Feeds/Fodder	DCP	TDN
Tree leaves		
Neem	8	53
Saura	11	60
Tara	5	55
Aquatic plants		
Water hyacinth	7	50
Lotus plant	6	50
Deoiled salseed cake	0	45
Neem seed cake	8	59
Sugar cane bagasse	0.5	40
Sugar cane tops	1.5	45
Tea waste	8	45
Tapioca leaf meal	15	45
Mango seed kernel	6	70
Animal organic waste		
Rumen fill	6–12	40–45
Bovine excreta	8–10	35–45
Digested gobar slurry	6–8	50–55

Table 6 Optimum level of important unconventional feeds in the diet of animals

Name of feed	Species/stage of animal	OptimumLevel of inclusion (%)
Ambadi cake	Calves	20
Babul seed/pods	Calves	30
Castor bean meal	Sheep	10
Dhupa cake	Calves	20–30
Guar meal	Calves	15–20
Karanje cake (deoiled)	Calves	24
Kosum cake	Calves	35
Mahua seed cake (water steeped)	Calves	20
Mango seed kernel	Bullocks	40
Neemseed cake	Calves	12.5
Rubber seed meal	Milch cattle	30
Salseed meal	Sheep	20
Salseed meal (alkali treated)	Cows	30
Water hyacinth meal	Calves	10–20
Warai bran	Calves	30

Gupta, 1991

animal production. Tree leaves serve as a concentrate to goat and sheep reared by poor farmers at a very low cost (only the labour cost of harvesting fodder). Trees and shrubs are an effective insurance against seasonal feed shortage or against the risk of drought and floods. Tree leaves have protein content (18–26%), supplement deficient nutrient in the diet, enhances microbial growth and digestion of cellulosic biomass in the rumen, provide undegradable protein, vitamins and minerals.

Leaves of *Gliricidia, Erythrina, Prosopis cineraria, Zizyphus, Flacurtia indica*, neem, mango, banyan, pipal, babul, subabul, mahuva, Sesbania, etc. can be used as green fodder. *Flacourtia indica* are used traditionally as fodder in semiarid region of central India and is quite palatable and may be fed without any harmful effect. *Zizyphus* leaves of different species are also widely used as animal feed particularly in Rajasthan and Central India. Khejri (*Prosopis cineraria*) leaves are an important animal feed resource in arid and semi arid regions of India. (FAO, 1992). Ardu (*Ailanthus excelsa* Roxb.) is a lofty deciduous tree, though it is widely distributed in the country.

Complete feed prepared from 50 kg tree leaves, 5 kg groundnut cake, 25 kg vilayati babul pods (*Prosopis julifora* pods), 15 kg molasses, 1 kg urea and 2 kg mineral mixture is palatable to animals and it can form a good maintenance ration. The complete feed prepared with fallen teak leaves (*Techtona grandis*) 40%, maize 15%, groundnut cake 20%, deoiled rice polish 23%, mineral mixture 1% and common salt 1% can be used as a growth ration. Leaves of *M. alba* are highly palatable to sheep and can meet the maintenance requirement of adult sheep. The vegetable leaves and creepers like cabbage, cauliflower and potato can also be used as animal feed during scarcity.

Table 7 Proximate principles (%DM) of tree leaves of some tropical species

Name of species	Local name	CP	CF	NFE	EE	Total ash
Albizia lebbek	Sirish	16.8–26	26–37	36–42	2.8–4.7	7–11.5
Azadirachta indica	Nim	12.4–18	11.4–23	43–66	2.3–6	7.7–19
Bauhinia variegata	Kanchan	10.7–16	25–33	41–52	1.3–3.9	6.3–12
Ehretia laevis	–	13.5	18	51.2	6	11.3
Ficus bengalensis	Bot	7.7–11.2	13.9–30	37.6–51	2.5–5	9.3–17
Ficus glomerata	Gular	11–15	12–16.5	53–59	2.2–2.9	11.9–18.4
Ficus religiosa	Aswatha	13.8–14	19–21.4	46–50	2.2–3.8	14
Grewia optiva	Bihul	19	22	43.9	2.2	12.8
L. leucocephala	subabul	18.9–27	10–17	46–52.6	2.6–5.9	7.5–10.9
Morus alba	Mulberry	15–27.6	9–15	48–49.7	2.3–8	14–23
Prosopis cineraria	Shami	14.2	22.1	53.2	4.3	6.1
Shorea robusta	Sal	10	27.4	55.4	3.2	3.9
Tamarindus indica	Tintri	13.8	15.9	60.5	1.9	7.94
Z. mauratiana	Berboroj	16.9	13.5	56.1	2.2	11.5

Mondal, 1997

Table 8 Nutritional quality of some of the important fodder tree in tropical India

Tree species	Period of availability		Palatability		Crude Protein	
	Leaf	Pod	Leaf	Pod	Leaf	Pod
Acacia nilotica (Babul)	May–Feb	May–June	Good	Good	13.9	13.6
Ailanthus excelsa (Arya, Maha nim, Maharukh)	May–March	–	Good	–	19.8	–

(Contd...)

(*Contd...*)

Tree species	Period of availability		Palatability		Crude Protein	
	Leaf	Pod	Leaf	Pod	Leaf	Pod
Albizia amara (Tugli, Chikreni)	June–March	–				
Albizia lebbek (Sirish)	April–Oct	–	Moderate	–	29.2	–
Albizia procera (Safed Sirish)	July–March	–				
Bauhinia variegata (Rakta Kanchan)	April–Nov	–	Fair	–	13.3	–
Dalbergia sisso (Sishu, Sisham)	Feb.–Dec.	–	Fair	–	16.6	–
Dichrostachys cinerea (Kunlai, Sigamkati)	July–April	Aug–Dec.	Good	Good	15.2	11.4
Ficus glomerata (Gular, Umar)	Dec.–Aug	March–July	–	–	–	–
Hardwickia binanta (Anjan)	April–Jan	–	–	–	–	–
Leucaena leucocephala (Subabul)	July–Dec	–	Good	–	21.4	–
Parkinsonia aculeate (Vilayati Kikar)	Throughout year	Dec.–March	Moderate	–	13.1	–
Pitheceolobium dulce	Throughout year	April–June	–	–	–	–
Prosopis cinerea (Shami)	Throughout year	June–Aug	Good	–	13.9	–
Prosopis juliflora	–	March–May	Fair	Good	21.4	30.8
Sesbania grandiflora (Basna)	July–Nov	–	Good	–	26.2	–
Ziziphus nummularia (Kokan ber, Bhor, Paragi)	Throughout year	Dec–Jan	Good	–	11.7	–

Table 9 Digestibility coefficient of nutrients of some important top feeds

Tree	Animal	DM	OM	CP	CF	EE	NFE
Ailanthus exelsa (Arya, Mahanim, Maharukh)	Sheep	72.1	75.1	81.3	55.1	31.7	81.7
Zizyphus nummularia (Kokan ber, Bhor, Paragi)	Goats	52.4	–	41.9	43.5	37.5	66.4
Moringa oleifera (Sainjna, Sujina)	Sheep	64.3	65.3	77.3	34.9	65.8	65.1
Moringa oleifera	Goats	56.6	61.4	65.8	29.6	61.5	65.0
Symingtomia populnea	Goats	64.7	66.5	27.9	44.3	61.3	76.5

(*Contd...*)

(Contd...)

Tree	Animal	DM	OM	CP	CF	EE	NFE
Gliricidia ssp.	CB calves	56.7	–	62.2	50.1	52.8	65.3
Pithecellobium dulce	Sheep	–	62.2	76.9	57.7	57.2	57.6
Vicus cunia	Goats	40.2	51.1	52.0	24.4	36.8	61.1
Morus alba (Kala Shehtoot)	Goats	75.5	76.8	74.1	72.3	55.5	82.9
Morus alba (Mulberry)	Sheep	72.9	77.3	83.2	82.3	52.3	79.0
Thespesia populnea (Parash, papal)	Sheep	71.8	76.5	73.4	48.3	43.8	85.5
Thespesia populnea	Goats	73.0	76.7	73.3	51.5	40.5	84.7
Acacia leucopholoea (Rinj, Arinj)	Calves	46.7	49.6	49.7	43.9	54.0	51.1
Boswellia serrata (Tad, Salai, Saler)	Sheep	70.2	75.9	58.4	63.9	62.0	73.1
Boswellia serrata	Goats	71.1	75.7	77.5	63.1	63.9	77.0
Leucaena leucocephala (Subabul)	Sheep	57.7	74.5	56.3	58.9	64.7	59.4

Table 10 Chemical composition of forbs (% DM basis)

Name	DM	CP	NDF	ADF	Lignin	Ash	Hemi cellulose	Cellulose	EE
Forbs									
Digera muricata	16.62	15.67	52.15	26.16	7.65	12.11	25.99	17.98	2.56
Ipomea pestigridis	16.55	13.78	53.58	37.79	9.35	10.72	15.79	27.70	1.98
Celosia argentea	11.38	12.55	48.78	28.43	7.30	11.97	20.35	20.35	1.78
Phyllanthus urinaria	45.05	7.46	52.15	36.52	7.91	6.55	15.63	27.79	2.32
Commelina benghalensis	20.39	19.04	52.81	26.05	5.74	10.50	26.70	18.49	5.23
Leucas aspera	16.78	14.09	47.00	31.76	6.04	10.28	15.24	21.46	2.18
Sida acuta	32.25	20.30	44.21	20.05	3.66	8.79	24.16	15.76	2.44

Table 11: Important top feeds for different agro–climatic zones of India

Tree species	Region	Fodder availability period		Preference to livestock
		Leaf fodder	Pod fodder	
Acacia albida	SAR	May–Feb	Mar–June	Goat, Sheep, Camel
Acacia nilotica (Babul, Kikar)	AR, SAR, IGP, CZ, COZ	May–Feb	Apr–June	Cattle, Goat, Sheep, Camel
Acacia senegal (Khor, Kumta)	AR, SAR	Apr–Dec	May–June	Goat, Sheep, Camel

(Contd...)

(*Contd...*)

Tree species	**Region**	**Fodder availability period**		**Preference to livestock**
		Leaf fodder	**Pod fodder**	
Acacia tortilis	AR, SAR	May–Feb	May–June	Goat, Sheep, Camel
Ailanthus excelsa (Arua, Maharukh)	AR, SAR	May–Mar	–	Goat, Sheep
Albizia lebbek (Indian walnut, Siris)	AR, SAR, IGP, CZ, NER	Apr–Nov	Aug–Nov	Goat, Sheep, Cattle
Albizia procera (Safed siris)	SAR, CZ, NER	Jul–Mar	Jan–Mar	Cattle, Goat, Sheep
Anogeissus latifolia (Dhaura)	SAR, AR	Apr–Oct	Jan–Mar	Cattle, Goat, Sheep and Camel
Artocarpus lakoocha (Dhew, Lakuch)	IGP, NER, STZ	Apr–Nov	–	Goat, Sheep, Cattle
Azadirachta indica (Neem)	AR, SAR, IGP, CZ, NER	Throughout the year	July–Aug	Goat, Sheep, Cattle
Bauhinia purpurea (Krairwal, Gairal)	STZ, CZ, NER, IGP	Apr–Nov	–	Goat, Sheep, Cattle

During flood there is scarcity of basal roughages for feeding livestock. In some areas like northeastern region, tree leaves are available in plenty. So these tree leaves can constitute the roughage portion of the diet.

Table 12 Ration for non–producing, growing and lactating cattle supplemented with tree leaves

Category	**Body weight (kg)**	**Tree leaves (kg)**	**Paddy/wheat straw (kg)**
Non producing cattle	100–200	5–7	1.0–2.5
	250–300	8–10	2.0–3.0
Growing cattle	50–100	2–3	0.5–1.5
	150	3.5	2.0
Lactating cows Producing 2–3 kg milk/day	250–300	4–6 plus 1.5 kg concentrate	2–3

Besides roughages, 30 g common salt, 20 g bone meal and 0.5 g mineral mixture should provided to each animal.

5.0 PREPAREDNESS TO COUNTERACT THE FEED SCARCITY OF NATURAL CALAMITIES

Acute shortage of feedstuffs is observed just after the recede of the flood and during drought. The preparedness is required to be under taken by the farmer, village panchyat, local administration and NGOs to meet out the challenges of feed scarcity. The following technologies can be adopted before calamities happen so that sufficient feedstock is available to counter act the feed scarcity during calamities.

5.1 Conservation of Urea Ammoniated Straws

Urea treatment of straws is so far the only chemical treatment with practical potential under field conditions. Urea-treated straw can save concentrate in the ration, increases daily milk yield by 1–2 litres/animal, offers better economic returns to the farmers. The process of urea ammoniation of straw is as follows:

Straw, kg	1,000
Urea, kg	40
Water (l)	400
Space requirement	180 × 150 × 150 cm

First urea is dissolved in water and then the straw in stacks of 20–30 kg is sprayed mechanically with a solution of fertilizer-grade urea (40 kg urea/tonne of straw, as DM basis). Polyethylene sheet are placed at the bottom of the stack to prevent the leakage of ammonia produced after treatment due to reaction of urea with plant urease. A 2–3 m strip of free plastic sheet is left to facilitate sealing of the stack. Special care is taken in folding the corners. The two plastic sheets (bottom and top) are then covered with soil. The stacks can be opened after 1–4 weeks depending on environmental temperature for feeding of livestock.

5.2 Silage

The basic principle of silage making is to convert the sugars in the ensiled fodder into lactic acid, this reduces the pH of the silage to about 4 or lower depending on the type of process. Crops and plant material rich in soluble sugars such as maize, sorghum, oats, sugarcane tops, hybrid Napier grass and other grasses are highly suitable for ensiling. The dry matter concentration of the forage at the time of ensiling should be around 15–30%, but higher is possible. Chaffing of the material for ensiling increases its compactness, thus eliminating the air space to the maximum extent. Green to semi-green forage, which may use the oxygen present for respiration, results in high quality silage. The silo should be air tight for filling. Fermentation starts within hours after closing the silo, and accelerates over the next 2 to 3 days. It terminates in about three weeks. Organic acids, primarily lactic and acetic acid, ethanol and gases such as CO_2, CH_4, NO_2 and NH_3 are produced during the fermentation process. Due to the production of acid, the pH of the biomass is reduced to a level below 4, resulting in the termination of all biological activities, after which the material remains conserved under anaerobic conditions.

Ensiling paddy straw with fruit factory waste and poultry droppings

- Fruit by products are rich source of soluble carbohydrates. So paddy or wheat straw can be mixed with fruit waste and poultry droppings in the ratio of 50:25:25 for a period of 4 weeks. After ensiling period, the material will be ready for feeding of animals.
- Paddy straw, poultry droppings, green grass and molasses in the ratio of 40:40:10:10 on dry matter basis form very good silage. This silage may be fed to non producing adult animal or growing animal for maintenance.

5.2.1 *Crops Suitable for Silage Making*

5.2.1.1 *Fodder Crops*

Maize, oats and sorghum are important fodder crops that are rich in carbohydrates. During periods of abundant green fodder availability, they can be chopped and ensiled to produce silage for feeding during scarcity periods. Sorghum and oats should be harvested at flowering stage when 50% ears have emerged, while maize should be harvested at its milk stage. The main stovers used for silage are from sorghum, though also crops like pearl millet and green maize can provide stover for silage. The prussic acid which is present especially in younger sorghum plants or quick regrowth, and which is dangerous if fed to ruminants in larger quantities, is destroyed completely by ensiling.

5.2.1.2 *Grasses*

During the monsoon season, cultivated as well as forest grass grows luxuriantly, and there is abundant availability of green fodder. These grasses can be harvested at their pre-flowering or flowering stage when their feeding value is still high. If the weather is too humid for hay making, these grasses may be conserved in the form of silage for feeding during scarcity periods.

5.2.1.3 *Aquatic Plants*

Water hyacinth is an aquatic weed, abundantly growing in Eastern India. In its fresh form ruminants do not like it. Its leaves are rich in protein but the plant contains 90–94% water. In order to prepare silage from water hyacinth, it is necessary to reduce the water concentration to about 70%. As the leaves dry faster than the petiole and the stems, they drop down during the drying process, and the nutritive value is reduced.

5.3 Feed and Fodder Bank

A fodder bank is an accumulation of feed to be used as emergency feed in times of scarcity. The principle is to preserve surplus feed produced in times of plenty; and to purchase emergency feed in times of plenty, when the unit price is lower than in times of need. The size of the fodder bank will increase with the aridity of the land. Acute fodder shortage is a common phenomenon during natural calamities every year in different parts of the country. The prices of fodder rise steeply during scarcity and small and marginal farmers are compelled to get rid of their livestock due to inability to feed them. In such areas, establishment of community fodder bank is a boon to sustain livelihood even under distress condition.

Fodder banks are the buffer stocks of fodder grasses, straws of paddy, wheat etc which are to be used for feeding livestock during scarcity and drought. Establishment of location specific buffer stock (fodder bank) of compact fodder and bulk fodder purchase from surplus regions (*e.g.* Punjab where 5 million tons of paddy straw and some quantity of wheat straw are destroyed by burning in the field) during peak production seasons can mitigate the fodder shortages during drought. In fodder banks, grasses which are easily available in monsoon season, can be harvested, sun dried and stored in the form of hay in specially constructed structure in which it can be stored for almost two years without much deterioration in the nutritive values. After testing for aflatoxin contents, pesticides and drug residues, the damaged feeds that are not fit for human consumption but suitable for use in ruminants or non-ruminants, may be spared for livestock use and stored in feed banks. Such feeds will be useful for all class of livestock. Grasses from forest areas or crop residues from regions with surpluses are commonly transported to scarcity regions during droughts. The possibility of transport depends on the nature of the calamity. Transportation of feed is only an option if the drought is regional. It has little or no value if infrastructure is highly damaged (floods) or if the calamity is of a much larger scale (severe droughts that occur in most parts of India).

5.3.1 *Reduction of Transportation Cost*

The viability of fodder bank is primarily depends on

- Cost of transportation of fodder from surplus to deficit region
- Storage cost of fodder

Table 13 Effect of densification on reduction of transportation cost

Feed materials	Bulk density (kg/m^3)	Capacity (quintal/ Truck)	Transportation cost (rupees/ quintal)			Reduction in transportation cost (%)
			100 km	200 km	400 km	
Grass loose	25	7.25	620	896	1379	94
Grass block	453	131.37	34	49	76	
Wheat straw loose	94	27.26	165	238	366	77
Wheat straw block	421	122.09	36	53	81	
Paddy straw loose	89	25.81	174	251	387	77

Samanta *et al.*, 2004

Some times the transportation cost is higher than the cost of feed. The only way to reduce transportation cost is to compress or densify the bulky feeds. Therefore, the making of block not only reduces the transportation cost of roughages drastically but also disposes the roughages from the production field judiciously.

5.3.2 *Assumption*

Size of truck: $8 \times 8 \times 16$ feet (29 m^3)

Transportation cost Rs. 4500/100 km, Rs. 6500/200 km and Rs. 10000/400 km

5.3.3 *Reduction of Storage Space*

Conventionally roughages are stored in heap (field) or in go down, which occupies huge space due to its low bulk density. Therefore, storing of roughages particularly for in fodder bank is very costly affair as it will occupy huge space for storage. The space requirement can be reduced by two ways

- Baling of fodder
- Block making of fodder

The block making tremendously reduces the storage space requirement because of increase in bulk density (Table 14). As a result it becomes cost effective.

Table 14 Effect of densification on reduction of storage space

Feed resources	Space required per quintal in loose form (cubic meter)	Space required per quintal in block form (cubic meter)	Reduction in space requirement (cubic meter /quintal)
Grass	3.89	0.22	3.67
Wheat straw	1.06	0.24	0.82
Paddy straw	4.04	0.25	3.79
Black gram straw	1.69	0.18	1.51
Groundnut haulm	1.78	0.17	1.61
Stylo straw	3.54	0.21	3.33
Sorghum stover	2.36	0.18	2.18
Leucaena leaves	1.20	0.13	1.07

Surplus production can also be stored in the block form economically (Fig. 1) which may be utilized in shortfall year during which prices of roughages may reach at any high. Due to increase in the bulk density (four times) in compressed complete feed block, the handling, storage and transportation are easy and economical (Verma *et al.*, 1996). Therefore, the complete feed block can constitute the part of 'Famine Feed Bank' in drought prone regions or during emergence of any natural calamities.

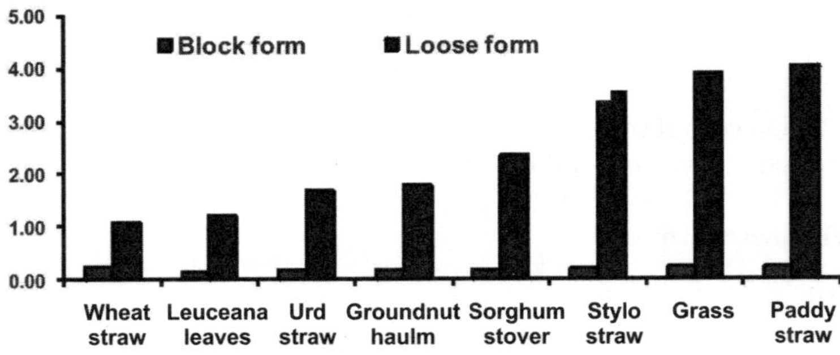

Fig. 1 Effect of densification on storage space (m^3 per 100 kg)

Chopped crop residue is mixed thoroughly with other concentrate ingredients. One of the ingredients should be molasses as it helps in better block making and also acts as a binder. Molasses could be avoided along with leguminous fodders, as these fodders have inherent qualities of blocking under pressure. Pressure application appeared to be most critical factor in process of block making. The block texture and compactness improved progressively on the application of increased pressure. A force of 15 to 20 tonnes (15 mpa) was considered satisfactory. The moisture content higher than 23% resulted into fungal growth and loose blocks.

During calamities fodder banks play a crucial role in supplying emergency feeds. But the question is what should be the size of fodder bank at district or state level? The size of the fodder bank should be such that it can hold the minimum amount of feed/forages required for maintenance of livestock of the affected zone or state. Assuming the degree of severity and length of flood and drought period of 2 and 4 months, respectively, the estimated sizes of fodder bank are given in Table 15.

Table 15 District level size of fodder bank as percent of dry matter required annually

Type of calamity	Size of fodder bank as % of feed DM required annually	Size of fodder bank as capacity to hold the material (million tons)*
Flood	10	0.15
Drought	20	0.33

*Assuming drought and flood period of 4 and of 2 months, respectively

The minimum size of fodder bank should be between 10% to 20% of the required feed and fodder dry matter annually in a particular zone, state or district. For example, if the estimated requirement of feed dry matter is 207.38 and 338.82 million tons for drought and flood prone zones of the country, respectively, then each district should have a minimum size of fodder bank of about 0.33 and 0.15 million tons capacity. The approximate cost of fodder bank is given in Table 16.

Table 16 Investment required for a typical fodder bank

S. No.	Particulars	Investment (Rs) in lakh
1.	Land and building	20.00
2.	Fodder procurement and distribution	20.00
3.	Training	1.00
4.	Plant and machinery	10.00
5.	Treatment	3.00
6.	Miscellaneous	11.50
7.	Total	65.00

Source: Rajasthan Infrastructure Agenda, 2025

5.3.4 *Advantages of Fodder Bank*

- Availability of fodder in dry span of year
- Availability of quality fodder throughout the year
- To stop the migration of animals during summer months
- Fodder banks would not only keep the livestock alive but also productive
- Availability of fodder at reasonable rate
- Additional income for rural families as fodder bank would be established through self-help group.

For regular availability of fodder, there is a need of commercial fodder production. In areas having fodder banks, government should encourage self-help group to cultivate the fodder in the nearby non-arable lands. Since grasses can grow with little water, it can be grown even during water scarcity. The government should also give incentive for the cultivation of fodder crops, as this would ensure continuous supply of fodder. The government should also encourage cultivation of fodder trees on the bank of canals, water reservoir, panchayat land, wasteland and other community land.

It is necessary to ensure full participation of stake holders, particularly farmers. It is also necessary to involve NGOs and CBOs to educate the farmer and to form Self Help Groups (SHG). The SHGs should be responsible for the operation of fodder bank. The role of various agencies will be required for the establishment of fodder bank

- Central Government should provide subsidy for the construction of fodder bank
- In absence of buyers, the fodder bank should dispose the degenerated fodder stored for long period. Therefore, it is necessary to provide financial assistance from time to time to compensate the loss.
- State Government should responsible for making the project and will submit for funding to the central government.

- Animal husbandry department should act as nodal agency at state level and will assess the number and location of such fodder bank to be set up across the state.
- Non Government Organizations and government agencies will motivate stake holders at village level about need for fodder bank and facilitate the formation of fodder bank.

Table 17 Nutrient requirement for maintenance of cattle and buffaloes

Live weight (kg)	DM (kg)	DCP (g)	TDN (kg)
250	4-5	140	2.20
300	5-6	168	2.65
350	6-7	195	3.10
400	7-8	223	3.55
450	8-9	250	4.00
500	9-10	278	4.45
550	10-11	310	4.90
600	11-12	336	5.35

ICAR, 1985

Table 18 Nutrient requirements for maintenance of sheep

Live weight (kg)	DM (kg)	DCP (g)	TDN (kg)
20	500	25	240
25	625	30	300
30	750	32	360
35	875	35	420
40	1000	40	480
45	1125	42	500
50	1250	45	540
55	1375	50	600
60	1450	54	640

Table 19 Nutrient requirements for maintenance of goats

Live weight (kg)	DM (kg)	DCP (g)	TDN (kg)
30	900	35	400
40	1200	40	500
50	1400	45	560
60	1600	55	670
70	1800	60	760

Table 20 Nutrient requirements for maintenance of camels

Live weight (kg)	DM (kg)	DCP (g)	TDN (kg)
500	12.0	500	5.5
550	13.0	525	6.0
600	14.0	550	6.5
650	15.0	600	7.0
700	16.5	650	7.5

Table 21 Nutrient requirements of growing cattle (Growth rate 550 g/d)

Live weight (kg)	DM (kg)	DCP (g)	TDN (kg)
70	2.10	259	1.39
80	2.33	282	1.53
100	2.78	328	1.80
120	3.23	373	2.07
140	3.67	419	2.34
150	3.90	442	2.47
160	4.12	465	2.61
180	4.57	510	2.88
200	5.02	556	3.14
220	5.47	601	3.41
240	5.97	647	3.68

Table 22 Nutrient requirements of growing buffaloes (Growth rate 450 g/d)

Live weight (kg)	DM (kg)	DCP (g)	TDN (kg)
70	1.97	156	1.24
80	2.20	177	1.38
100	2.65	198	1.64
120	3.10	279	1.90
140	3.56	300	2.18
160	4.01	264	2.45
180	4.46	288	2.72
200	5.10	312	2.90
220	5.60	345	3.10

Table 23 Nutrient requirements of growing lambs

Live weight (kg)	Daily gain (g/d)	DM (kg)	DCP (g)	TDN (kg)
10	50	400	35	220
	100	450	45	280
	150	500	55	340
15	50	500	45	300
	100	600	55	360
	150	700	65	450
20	50	700	50	400
	100	800	70	520
	150	1000	80	640
25	50	800	65	500
	100	1100	85	700
	150	1200	100	800

Table 24 Nutrient requirements of growing kids

Live weight (kg)	DM (kg)	DCP (g)	TDN (kg)
10	425	25	275
15	600	30	350
20	700	35	400
25	800	40	450
30	950	45	500

Table 25 Nutrient requirements of growing camels (Growth rate 100 g/d)

Live weight (kg)	DM (kg)	DCP (g)	TDN (kg)
200	5.0	250	2.0
250	6.0	325	2.5
300	7.5	350	3.0
350	9.0	470	3.8
400	10.0	600	4.7
450	10.5	650	5.0

Table 26 Nutrient requirements per kg of milk production

Fat (%)	DCP (g)	TDN (g)
3.0	40	270
4.0	45	315
5.0	51	370
6.0	57	410
7.0	63	460
8.0	69	510
9.0	75	560
10.0	81	600
11.0	85	700

6.0 REFERENCES

FAO, 1992. Legume trees and other fodder trees as protein source for livestock. Animal Production and Health Paper No.102.FAO, Rome.

ICAR, 1985. Nutrient Requirements of Livestock and Poultry.

Gupta, B.S. 1991. Toxic and incriminating factors in feeds and fodders and their detoxification from animal health and production. *Proceedings of First International Animal Nutrition Workers Conference* for Asia and Pacific held in Bangalore from September 23–28, 1991.

http:/www.icar.org.in/drought/drought2.htm.

Mondal, L. 1997. Nutritive values of tree leaves of some tropical species for goats. *Small Ruminant Research* **24**: 95–105.

Rajasthan Infrastructure Agenda 2025 http://ppp.rajasthan.gov.in/PWCreport/KD4/Agriculture/agri-vol-III.pdf

Ranjhan, S.K. and Khera, R.C. 1976. Feeding Farm Animal During Scarcity, ICAR, New Delhi.

Samanta, A.K., Singh, K.K., Das, M.M. and Kundu, S.S. 2004. Complete feed block: A balanced ration for livestock.

Verma, A.K., Mehra, U.R., Dass, R.S. and Singh, A. 1996. Nutrient utilization by Murrah buffaloes (*Bubalus bubalis*) from compressed complete feed blocks. *Animal Feed Science and Technology* **59**: 255–263.

9

MEETING WATER REQUIREMENT OF LIVESTOCK DURING NATURAL DISASTER

S.K. Mahanta, Sultan Singh and M.M. Das

ABSTRACT

Water is an important but often overlooked nutrient for livestock. It is the main constituent of the animal's body contributing 50 to 80 percent of the live weight. An animal can lose almost all of its fat and about 50 percent of its body protein and survive. However, the loss of 10 percent of its body water can be fatal. It becomes more important under different calamities like drought, earthquake etc when there is acute shortage of drinking water. During flood, cyclone and tsunami water sources become contaminated and unsafe for drinking by livestock. Drought not only causes the quantitative deprivation of drinking water but it also causes deterioration of water quality like high concentration of minerals (excess salinity), high nitrogen content (nitrates and nitrites), heavy microbial / blue-green algae growth and contamination of chemicals (pesticides, fertilizers and accidental spills of petroleum). Water requirements of different categories of animals, impact of natural calamities on water access of livestock, occurrence of water intoxication, safe levels of different potentially toxic nutrients and contaminants in water, common water borne diseases like blackleg, anthrax, malignant edema, tetanus, foot rot, mastitis, botulism, listeriosis etc during flood and their treatment / preventive measures etc. are discussed in this chapter. Natural calamity can also have impact on the water supply system like breaks in pipelines and damage of water shortage structures. Restoration of drinking water service is essential in ensuring the health and well being of humans as well as livestock and as such fulfills an important role in the development / rehabilitation process.

1. 0 INTRODUCTION

Water is an important but often overlooked nutrient for livestock. It is the main constituent of the animal's body, contributing 50 to 80 percent of the

Indian Grassland and Fodder Research Institute, Jhansi (UP), India

live weight, depending on age and degree of fatness. An animal can loose almost all of its fat and about 50 percent of its body protein and survive. However, the loss of 10 percent of its body water can be fatal. A good water supply is defined both in terms of quantity and quality of the water. In fact, unless livestock have access to an adequate supply of safe water every day, they may reduce feed intake, thereby lowering production and causing an economic loss to the livestock owner. The situations aggravated under different calamities like drought, earthquake etc when there is acute shortage of drinking water. During flood, cyclone and tsunami water sources become contaminated and unsafe for drinking by livestock.

The **main functions of water** in the body are:

- To help in elimination of waste products of digestion and metabolism (feces of healthy cattle often contain 75 to 85 percent water)
- Transport of nutrients and other compounds to and from cells
- To help in digestion and metabolism of nutrients
- To regulate blood osmotic pressure
- A major component of secretions (milk and saliva) as well as in the products of conception and growth
- In the body's thermoregulation as affected by evaporation of water from the respiratory tract and from the skin's surface
- Provision of fluid environment for the developing foetus

Body water is again divided into intracellular and extracellular compartments. Intracellular is the largest compartment, accounting for about two-thirds of water in the body. The extracellular fluid comprises water around cells and connective tissue, water in plasma, and transcellular water or water in the gastro-intestinal tract. Intestinal water accounts for 15–35% of body weight. Cows in early lactation had about 15% of their body weight in gastro-intestinal water, while cows in late lactation and in gestation had 10 to 11%. Resident time of water in the rumen was estimated to be 61 minutes in sheep and 62 minutes in lactating dairy cattle.

Loss of water from the body occurs through milk production, urine excretion, fecal excretion, sweat and vapor loss from the lungs. Water losses through milk of a good lactating cow are about 26 to 34 percent of total water intake (feed plus free water consumed). Fecal water losses are similar to those of milk (30 to 35% of total water intake), and urine losses are about half of fecal losses in lactating cows. Factors that affect the fecal water loss include dry matter intake, dry matter content of the diet, and digestibility of the diet. Urinary water excretion in cattle is variable at 4.5 to 35.4 litres/ day. Urinary water excretion is related positively to water availability, amount of water absorbed from the intestinal tract (total intake minus fecal loss), urinary nitrogen and urinary potassium excretion, and negatively related to dietary dry matter content. Sweat, salivary and evaporative losses combined account for about 18% of water loss.

2.0 WATER REQUIREMENT UNDER NORMAL CONDITIONS

As in the case of other nutrients, the water requirement/ consumption depends on many factors such as:

- Kind and size of animal
- Rate and composition of body weight gain
- Pregnancy
- Lactation
- Quality of water
- Type of diet
- Level of dry matter intake
- Level of activity
- Temperature of the water offered
- Surrounding air temperature

Assuming that drinking water needs are species, farm and management-specific, Table 1 provides an estimate of water consumed daily by different categories of livestock. Milk is composed of nearly 87% water. An adequate supply of quality water for dairy animals is extremely important. The dairy animal's peak water intake generally occurs during the hours of greatest feed intake. A cow yielding 10 litres per day secretes about 9.0 litres of water in the milk, and for every litre of milk produced 2 to 3 litres of water is said to be required, apart from what the animal needs for its body function. Young calves receiving milk diets or growing calves consume greater amounts of water in relation to the dry matter of their diet, than older cattle fed on dry feeds such as straw.

Again goats and sheep are regarded as having more efficient renal system (kidney function) which reduces water loss compared with temperate and Zebu cattle. In the Rajasthan desert during summer, the adapted goats are considered to be more efficient at economizing on water use than sheep from the same region. Many workers who have studied goats, that have access to shade, concluded that goats are better adapted to more arid conditions than sheep.

In swine, the introduction of three-site production and all-in/all-out facilities has altered the water usage patterns. Pigs in the growing phase are often grouped in segregated rooms or entire barns by age. They are kept in this facility until the target weight is reached by the majority of animals, when the room or barn is emptied and then restocked. The use of wet/dry feeders and liquid feeding systems has reduced the volume of drinking water required because of the higher moisture content of the feed ration and the reduction in spillage of water from these systems. Swine system specialists have estimated that, earlier approximately 50% of the water supplied to a hog pen in a day was lost as spillage. Today, more efficient feeding systems have cut these losses significantly.

Table 1 Water requirement of different categories of livestock

Livestock category	Water requirement range (L/day)
Dairy cattle	
Dairy calves (1–11 months)	3–14
Dairy heifers (12–24 months)	15–30
Dry cows	20–35
Milking cows with 5 kg milk	30–45
– with 10 kg milk	40–55
– with 15 kg milk	50–65
– with 20 kg milk	60–75
Dairy buffalo	
Buffalo calves (1–11 months)	5–14
Buffalo heifers (12–24 months)	15–30
Dry buffaloes	30–50
Milking buffaloes with 4 kg milk	40–60
– with 8 kg milk	50–70
– with 12 kg milk	60–80
Beef animals	
Feedlot buffalo (180–360 kg b. wt.)	15–40
Feedlot buffalo (360–600 kg b. wt.)	27–55
Lactating cows with calves	43–67
Dry cows, bred heifers & bulls	22–54
Sheep and goats	
Feeder lamb/ kids (10–25 kg b. wt.)	1.0–3.5
Ewe/doe & ram/buck (40 kg)	2.0–4.5
Lactating meat ewe/doe plus unweaned offspring (40 kg +)	4.5–6.5
Gestating dairy ewe/doe (45 kg)	2.5–4.5
Lactating dairy ewe/doe (45 kg)	4.5–6.0
Swine	
Weaner (7–22 kg)	1.0–3.2
Feeder pig with 23–36 kg b. wt.	3.2–4.5
36–70 kg b. wt.	4.5–7.3
70–110 kg b. wt.	7.3–10
Gestating sow/boar	13.6–17.2
Lactating sow	18.1–22.7
Horse	
Small size (225 kg b. wt.)	13–20
Medium size (450 kg b. wt.)	26–39
Large size (675 kg b. wt.)	39–59

3.0 IMPACT OF NATURAL CALAMITIES ON WATER ACCESS AND LIVESTOCK

3.1 Drought

Droughts, unlike other natural calamities, do not occur suddenly but are slow onset calamities resulting from insufficient rain or snow (temperate areas) over a period of months, and sometimes years. Its effects are principally seen in the decrease or extinction of sources of drinking water both for human and livestock. Surface water such as rivers and ponds usually suffers the effects of drought before ground water. Depending upon the characteristics of surface water sources and the type of drought, impacts could include a decrease in the normal volume of drinking water, which depending on its severity results in moderate to severe rationing of water for livestock.

3.1.1 *Dehydration*

The deprivation of drinking water, when continued under hot climates leads to dehydration of animals. However, two factors are involved in the pathogenesis (Fig. 1) of dehydration: i) the depression of tissue fluid levels with resulting interference in tissue metabolism, and ii) anhydraemia with reduction in the fluid content of the blood. The initial response to negative water balance is the withdrawal of fluid from the tissues and the maintenance of normal blood volume. The fluid is drained primarily from the intravascular compartment and the interstitial fluid space. Essential organs including the central nervous system, heart and skeleton contribute little and the major loss occurs from connective tissue, muscle and skin. In the goat, total body water may be reduced as much as 44% before death occurs. The secondary response is a reduction in the fluid content of the blood causing a reduction in circulating blood volume (oligemia) and an increase in the concentration of the blood (hemoconcentration). Because of the hemoconcentration, there is an increase in the viscosity of the blood which impedes blood flow and further exacerbates the peripheral circulatory failure. In addition, water is preserved by reduced fecal output and increased absorption, which results in dehydration of the contents of the rumen and large intestine which in turn results in dry scant feces.

Dehydration also exerts some important effects on tissue metabolism. There is an increase in breakdown of fat, then carbohydrate and finally protein, to produce water of metabolism. The increased endogenous metabolism under relatively anaerobic conditions results in formation of acid metabolites and the development of acidosis. Urine formation decreases because of the restriction of blood flow and this, together with the increased endogenous metabolism, causes a moderate increase in blood levels of non protein nitrogen. The body temperature rises slightly (dehydration hyperthermia) because of insufficient fluid to maintain loss of heat by evaporation. The onset of sweating in steers after exposure to high environmental temperatures has been shown to be delayed by dehydration. Thus dehydration may even cause death.

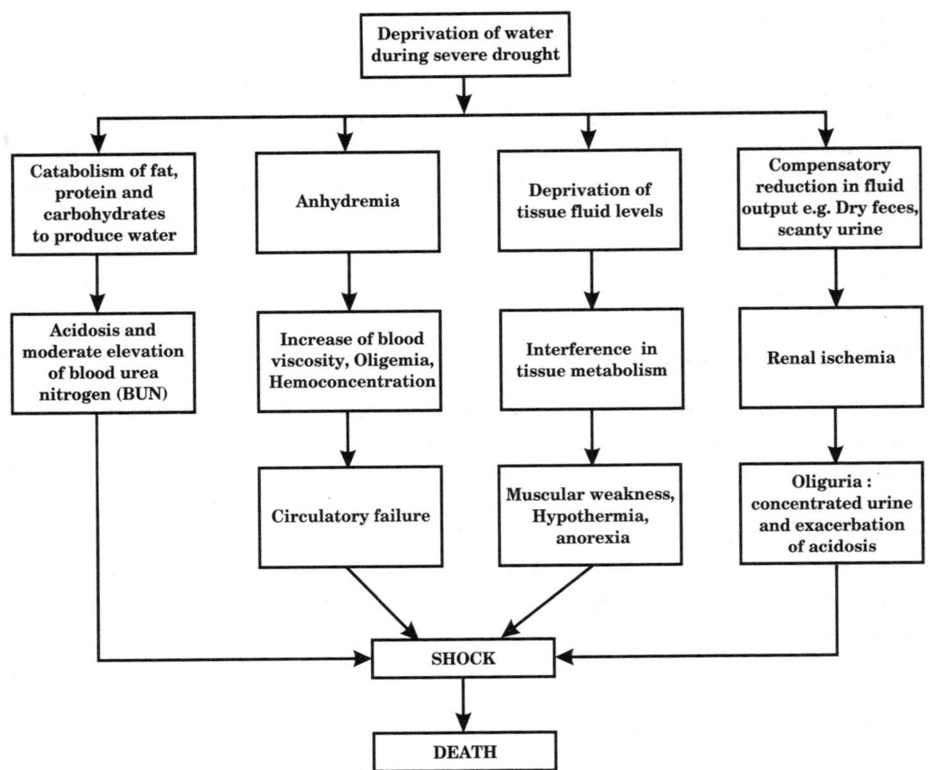

Fig. 1 Etio-pathology of dehydration in animals

3.1.2 *Signs of Dehydration*

The most important signs in dehydration are i) dryness and wrinkling of skin, giving the body and face a shrunken appearance. The eyeballs recede into the sockets and the skin subsides slowly after being picked up in to the fold. The skin of the upper eyelid and of the neck gives one of the best indications of the degree of dehydration, which on clinical basis is commonly assessed as a percentage of body weight. Some guidelines are as follows. The mildest form of dehydration at 4–6% of the body weight may be just barely detectable clinically but hemoconcentration may be present. At 6–8% dehydration, the eyes are sunken and the skin fold (tenting of the skin) will remain elevated for 2–4 seconds; at 8–10% dehydration the eyes are markedly sunken and the skin fold remains elevated for 6–10 seconds; at 10–12% dehydration the eyes are markedly sunken and the skin fold remains elevated for 20–45 seconds. ii) Loss of body weight also occurs rapidly. In horses deprived of water for 72 hours there is a mean body weight loss of about 15%. iii) There is muscular weakness and loss of appetite. iv) The degree of thirst present will depend on the presence or absence of toxemia. In water deprivation due to severe drought, dehydrated animals are very thirsty when offered water. v) Loss of milk production: Experimental restrictions of the water intake in lactating dairy cattle for up to 4 days may reduce milk yield by 75% and decrease body weight by 14%. A 10% reduction in water intake causes a drop in milk production which may be difficult to detect. vi) Behavioral changes are obvious; cows spend considerable time licking of water bowls. In cold climates, cattle are often forced to eat snow as a source of water. The snow must be soft enough so that it can be scooped up by the cattle and 3–5 days are necessary for the animals to adjust to the absence of water and become dependent on snow. During this time there is some loss of body weight. Lactating ewes relying on snow as a source of free water reduce their total water turnover by approximate 35%. vii) Urine excretion decreases and the urine become more concentrated and the renal insufficiency may accentuate pre-exciting acidosis and electrolyte imbalance hence the importance of restoring renal function. viii) Goats are more sensitive to water deprivation during pregnancy and lactation than during anestrus. Water deprivation for 30 hours causes a marked increase in the plasma osmolality and plasma sodium concentration in pregnant and lactating goats.

3.1.3 *Water Quality Problems*

Drought not only causes the quantitative deprivation of drinking water but it also causes deterioration of water quality. Following are the most common water quality problems affecting livestock production during drought-

- High concentration of minerals (excess salinity)
- High nitrogen content (nitrates and nitrites)
- Heavy microbial/ blue-green algae growth
- Contamination of chemicals (pesticides, fertilizers and accidental spills of petroleum)

3.1.2.1 *Salinity*

Salinity refers to salt dissolved in water and is expressed as parts per million (ppm) or as milligrams per litre (mg/L). The expression "total dissolved solids" (TDS) is often used to denote the level of water salinity. Salts commonly present include carbonate, bicarbonates, sulfates, nitrates, chlorides, phosphates, and fluorides. Highly mineralized waters (high solids) do not have much effect on health as long as there are no objectionable effects from specific ions, such as sulfate, and as long as normal amounts of water are consumed. One gram of sulfate per litre (1000 ppm) may result in scours (diarrhoea) and may also reduce copper availability in the diet. The limiting health concern is often sulfate, because the acceptable sulfate level generally exceeds before TDS levels are high enough to be a concern.

Salts, such as sodium chloride, change the electrolyte balance and intracellular pressure in the body, producing a form of dehydration. Salts also place a strain on the kidneys. Excess fluoride causes degeneration of the teeth. High salt concentrations that are less than toxic level may actually cause an increase in water consumption. Animals may refuse to drink high saline water for many days, followed by a period when they drink a large amount. They may then become sick or die. The tolerance of animals to salts in water depends on factors such as water requirements, species, age, physiological condition, season of the year, and salt content of the total diet, as well as the water. Animals have the ability to adapt to saline water. However, abrupt changes from water with low salt to water with high salt concentrations may cause harm while gradual changes do not. Table 2 gives the level of mineral content that either makes the water taste too bad to drink or causes a detrimental effect on health. It appears to make little difference whether the total quantity of dissolved salts or dissolved solids is made up of a single salt or a number of different salts. Table 3 shows recommendations for various animal species in relation to dissolved solids content.

Table 2 Guidelines for intake of problematic water (> 2500TDS) by livestock

TDS (ppm or mg/L)	Comments
2,500–3,000	Usually satisfactory for most livestock.
3,000–5,000	May not cause adverse effects to adult livestock. Growing/young livestock could be effected by looseness or poor feed conversion. At levels near 5,000 ppm the water is unacceptable for poultry.
5,000–7,000	Should not be used for pregnant or lactating females. Usually laxative and may result in reduced water intake.
7,000–10,000	Do not use for swine. Do not use for pregnant or lactating ruminants or horses.
10,000 or more	May cause brain damage or death.

Table 3 Recommended levels of dissolved solids for various animal species

	Dissolved solids (ppm)				
Species	**Excellent**	**Good**	**Fair**	**Poor**	**Limit**
Humans	0–800	800–1,600	1,600–2,500	2,500–4,000	5,000
Horses:					
– Working	0–1,000	1,000–2,000	2,000–3,000	3,000–5,000	6,000
– Others	0–1,000	1,000–2,000	2,000–4,000	4,000–6,000	10,000
Cattle	0–1,000	1,000–2,000	2,000–4,000	4,000–6,000	10,000
Sheep	0–1,000	1,000–3,000	3,000–6,000	6,000–10,000	15,000
Swine	(young pigs and market pigs appear to tolerate less than cattle)				

During drought and famine situations, desert sheep has the ability to conserve water by desiccation of faeces, concentration of urine, reduction of metabolic rate and evaporative water losses and reallocation of water in rumen. Animals in drought affected areas are compelled to consume unsuitable water which can aggravate other diseases, cause weight loss and even responsible for death of the animal due to toxicity. Most of the water sources in drought affected areas have increased salt and other toxic substances which make them unpalatable or dangerous to drink. Ground water in most of the desert areas is saline in nature which is also hazardous to animal health. Majority of sheep flocks in Thar desert depend on saline water for more than 8 months in a year. Total soluble salts (TSS) more than 3.5 g per litre causes diarrhoea, emaciation and depression of estrus cycle and increase incidence of abortion. Many districts of Rajasthan have higher concentration of nitrate and fluorine and this affect the reproductive performances.

3.1.2.2 *Nitrates and Nitrites*

The presence of nitrates or nitrites in water often indicates contamination of the water supply with fecal material or seepage from a septic field. Nitrates themselves are not very toxic, but when reduced to nitrites, problems can develop. Nitrites that get into the blood stream convert the red pigment, hemoglobin, to a dark brown pigment, methemoglobin. Hemoglobin is responsible for carrying oxygen from the lungs to other

Table 4 Guide to the use of waters containing nitrate for livestock

Nitrate content[1] (ppm nitrate nitrogen)	Comments
Less than 20	Nitrate levels above this level may be detrimental to poultry performance. Research evidence indicates that nitrate levels greater than 10 ppm may impact broiler performance as measured by poor weight gains, health problems or poor flock performance.
Less than 100	Experimental evidence to date indicates that this water should not harm cattle, sheep, swine and horses.
100 to 300	Water containing over 100 mg/L of nitrate is a potential health hazard with feed containing high nitrates. When feeds contain nitrates, this water can add greatly to the nitrate intake making it dangerous. Ruminants are the most susceptible because bacteria in the rumen convert nitrate to the much more dangerous nitrite. Pigs are less susceptible because this conversion doesn't occur to the same extent.
Over 300	This water, when fed with feed containing high nitrates, can cause typical nitrate poisoning in cattle and sheep. Its use for these animals is not recommended. This level of nitrate contributes significantly to salinity. The use of this water for swine, horses or poultry should also be avoided.

[1]Includes nitrite nitrogen

tissues of the body. Oxygen cannot be carried in the methemoglobin form. When about 50 percent of the hemoglobin is in the form of methemoglobin, the animal shows signs of distress suggesting a shortage of breath. Above this level, respiratory distress may result in death. At 80 percent or more, the animal usually dies from a type of suffocation. In ruminant animals, bacteria in the rumen convert nitrate in the feed or water to nitrite, which can diffuse into the blood stream. Nitrate toxicity from water is unusual, but the combination of nitrates in feed plus those in water should be watched to avoid problems. The conversion of nitrate to nitrite is not a major problem with monogastric animals. Table 4 categorizes the levels of nitrate in water.

3.1.2.3 *Microbial/Blue Green Algae Growth in Poor Quality Water*

There are many microorganisms in our normal water supply for livestock. Most are quite harmless but during drought or contamination from other sources, some of them do cause animal health problems. Coliform counts below 50 per millilitre of water are safe for all cattle. Other possible contaminants include coccidia, staphylococcus, streptococcus, virus, leptospira, etc. Green scum that builds up in livestock drinking troughs and tanks is algae. It cannot grow without sunlight. Some blue-green algae are toxic. No good method exists to predict whether toxins will be produced or not. Livestock behavior should be monitored closely during periods of heavy algal blooms. Copper sulfate ($CuSO_4$) or other commercial copper-containing products, often called bluestone, will kill the algae for a period of several months. In troughs or small tanks, a safe dosage is one level teaspoon of copper sulfate per 6500 litres of water. Thirty grams will treat 35000 litres of water. Generally, treatment is done only when algae growth is heavy or if a toxicity problem occurs. Hold livestock off the treated water source for at least 24 h. Signs of blue green algae poisoning are diarrhoea, lack of coordination, labored breathing, and death. During recovery, unpigmented skin may slough off. A suggested treatment for algae afflicted animals is large quantities of medical-grade charcoal and mineral oil, given orally. These are given to try to prevent any further absorption of the toxins by the affected animals. Contact your veterinarian for more information and assistance.

3.1.2.4 *Chemical (Pesticides/Herbicides) Contamination*

Many chemicals may be found in water, some of which could be detrimental to livestock. Safe levels of herbicides and pesticides in water for animals have not been determined. Table 5 gives guidelines for humans. These appear to be reasonable for livestock also.

Pesticides can enter into ground water or surface water supply from run-off, drift, and accidental spills. Provide adequate drainage around the water supply. Wells should be located on elevated ground to prevent surface run-off in to the well. Fish are much more sensitive to pesticides than other

Table 5 Maximum tolerable levels of various pesticides in water for humans

Pesticides	Maximum concentration (mg/L)
Chlordane	0.0020
Endrin	0.0020
Hephtachlor epoxide	0.0002
Hephtachlor	0.0004
Lindane	0.0002
Methoxychlor	0.0400
Toxaphene	0.0030
2,4-D	0.0700
2,4,5-T	0.0700

livestock. Table 6 lists the safe levels of potentially toxic nutrients and contaminants in water for cattle. These should be analyzed only when there is good reason to suspect their presence at excessive levels.

Table 6 Safe levels of potentially toxic nutrients and contaminants in water for livestock

Element	ppm or mg/L
Aluminum	5.00
Arsenic	0.20
Boron	5.00
Cadmium	0.05
Chromium	1.00
Cobalt	1.00
Copper	0.50
Fluorine	2.00
Lead	0.05
Mercury	0.01
Nickel	1.00
Nitrate-Nitrogen	100.00
Nitrite-Nitrogen	10.00
Selenium	0.05
Sulfate	1,000.00
Vanadium	0.10
Zinc	25.00

3.1.2.5 *Mitigation*

The water quality problems arises basically from the contamination of sources of drinking water, which occur due to-

- Decrease in the self-cleansing capacity of rivers or ponds because of reduced flow
- Increase in concentration of pesticides, insecticides or industrial wastes
- Decrease in free oxygen resulting in contamination from fish kill off
- Contamination cased by dead animals near intakes for drinking water

Hence, it may be necessary-

- To add chemical additives to lessen health risks or turbidity
- *Even alternative sources (emergency deep wells) may need to be constructed or put into operation*: Depending upon the duration of the drought and local hydrological characteristics, there can be new demands for on ground water for emergency drinking water supply for livestock as well as human beings. The resulting decrease in water table will reduce the productivity wells, and requires increased pumping to obtain required flow. This may entail an increase in operation costs for wells and decrease in productivity of pumps.
- To take over wells used for other purposes (industry, recreation or agriculture) to provide livestock with drinking water.
- *Development of springs and seeps*: Many areas still exist where springs and seeps can be developed. Any such water source, however small, is vital in a drought year—and it should prove helpful even in normal years.
- *Horizontal well*: A horizontal well consists of a pipe bored at a slight angle down into a hillside to tap small, seepy flows of water. The water is controlled with appropriate valve systems. As with springs, water usually is of high quality.
- Reliance on water from tank trucks/railway wagons, even with the loss of water quality and increase in cost

3.1.2.6 *Water Intoxication*

Again the ingestion of large quantities of water by a thirsty animal may result in water intoxication, particularly when there has been much loss of salt due to severe exercise or high environmental temperature. Under field conditions calves are most commonly affected. Cellular hydration occurs and in organized tissues the cells increase in turgor. This is especially noticeable in the brain where a condition analogous to cerebral edema occurs and causes nervous signs including muscle weakness and tremor, restlessness, ataxia, tonic and clonic convulsions and terminal coma. In unorganized tissues, particularly the erythrocytes, lysis occurs and the resulting hemolysis may cause severe hemolytic anemia and haemoglobinuria. Additional sign include hypothermia and salivation. However, water intoxication does not occur commonly and can be avoided by allowing thirsty animals to have only limited access to water. In cattle on pasture and deprived of water for up to 9 days and then give access to water there will be staggering, falling, convulsions and some death. This is similar to the syndrome of salt poisoning in pigs. Treatment of affected animals should include sedation, the administration of diuretics and in severe cases the intravenous injection of hypertonic solutions.

3.2 Flood

The most serious consequence of flood is large scale contamination of drinking water. In such situations water borne diseases (discussed below), usually associated with poor hygiene and sanitation, can affect livestock. The contamination of water can be caused by-

- Contamination of surface sources of drinking water due to animal cadavers near intakes, excessive increase in the turbidity of water or pollution from other types of contaminants.
- Flood levels that surpass the height of well head walls, or waters that flow directly over wells and other intakes.
- The rise of water levels in sewer outfalls can cause waste water to back up and flood the interiors of homes, lower levels of buildings, and public throughways. In homes this occur through toilet and washbasins; in streets it occurs through manholes and rainwater sinks.
- If fuels mix with flood waters, it will be more difficult for sterilization.
- Damage to pipelines and distribution channels of drinking water supply.

Because of the serious risk of occurrence of different diseases, methods of water treatment with chemical sterilization may also be considered for consumption by the livestock under emergency cases. However, cloudy water is harder to disinfect and we won't be able to disinfect it to the same degree as clear water. Generally, groundwater is the preferred source of drinking water. If it is not available and surface water must be used, avoid sources that contain floating material or have a dark color and/or odor. Filter mudy or colored water through clean cloths or allow particles to settle. Then draw off or decant the water, from the top that can be disinfected with the help of following chemicals.

- ***Chlorine Bleach*:** Common household bleach contains a chlorine compound that disinfects water. Mix it thoroughly and allow the treated water to stand for 30 minutes. The water should have a slight chlorine odor; if it does not, repeat the dosage and allow it to stand for another 15 minutes. If the treated water has too strong a chlorine taste, it can be made more palatable by allowing the water to stand exposed to the air for a few hours.
- ***Granular Calcium Hypochlorite*:** Add and dissolve one heaping teaspoon of high-test calcium hypochlorite (approximately 7.5 grams) for each 10 litres of water. This mixture produces a stock chlorine solution. To disinfect water, add the stock chlorine solution in a ratio of one part chlorine solution to 100 parts of water to be treated. To remove any objectionable chlorine odor, aerate the water.

- ***Chlorine Tablets*:** Chlorine tablets containing the necessary dosage for drinking water disinfection can be purchased in a commercially prepared form. These tablets are available from drug and sporting goods stores and should be used as stated in the instructions. When instructions are not available, use one tablet for each quart (around 1.0 litre) of water.
- ***Tincture of Iodine*:** Common household iodine available from the medicine shop or first aid kit may be used to disinfect water. Add five drops of 2 percent tincture of iodine to each litre of clear water. For cloudy water, add 10 drops and let the solution stand for 30 minutes.
- ***Iodine Tablets*:** Commercially prepared iodine tablets containing the necessary dosage for disinfecting drinking water can be purchased at drug and sporting goods stores. Use as stated in the instructions. When instructions are not available, use one tablet for each litre of water to be disinfected.
- ***Lime*:** Calcium oxide known as lime are also commonly used as disinfectant for drinking water of livestock.

4.0 COMMON WATER BORNE DISEASES PREVALENT DURING FLOOD

Animals surviving from natural calamities, especially animals that have been evacuated or displaced and relocated during flood, are under tremendous amount of stress. They also may be commingled with other livestock. This combination of stress and exposure to new diseases can lead outbreaks of different diseases including water borne diseases. Animal should be given extra care, particularly if they are stranded by flood water and have been off regular feeding schedules. We should also be cautious when wet feeds are given to livestock. First a few animals should be fed and watched them few days before giving wet feeds to all the animals. In addition, we should be watchful for signs of water borne and flood related diseases.

4.1 Blackleg

Blackleg, caused by micro-organisms (*Claustridium chauvoei*) spread over fields by standing water, is a potentially serious post-flood disease. It most commonly affects cattle of 6 to 24 months old, but it also affects sheep, goats and swine. Symptoms include acute lameness, depression, fever, and swelling in the hip, shoulder, chest, back, neck or throat muscles. Main symptom is black colouration of muscle. If untreated, blackleg is usually fatal within 24 h after onset. Treatment may be effective in the early disease stages. The best prevention against blackleg is inoculation of all unvaccinated young cattle before they are put out on pastures that have been flooded.

4.2 Anthrax

Anthrax is another disease which may breakout following flooding. All the animals which die suddenly following flooding should be necropsied by a competent veterinarian. It is caused by *Bacillus anthracis* micro-organism and leads to acute death in all species of livestock, but especially of cattle. Signs are hyperpyrexia, excessive bleeding from body orifices and sudden death. It is contagious to other animals and people. Proper carcass disposal is critical to prevent future cases. Anthrax spore vaccine (1 ml per head as sub-cut) is given as preventive measures.

4.3 Malignant Edema (Gas Gangrene)

Malignant edema is an acute wound infection caused by organisms of the genus *Clostridium*. There is an acute inflammation at the site of infection and profound systematic toxemia. Hot, painful swelling at point of infection, high fever, loss of appetite, decreased milk production, difficult breathing and convulsions followed by death are signs of malignant edema. This disease kills animals 1 or 2 days after symptoms appear. In its early stage malignant edema can be successfully treated by a veterinarian. Both specific and combined formalized vaccines are available for preventing the occurrence of the disease in enzootic areas.

4.4 Tetanus (Lockjaw)

Tetanus is a problem whenever animals have punctured wounds. This disease is highly fatal, infectious disease of all species of domestic animals caused by the toxin of *Clostridium tetani*. Symptoms include generalized stiffness caused by muscles contractions. Legs and tail are extended, the third eye-lid hovers over the eye when its head is raised. Animals can be vaccinated as preventive, and disease is treatable in its early stage. However, the response to treatment in horses and sheep is poor but cattle frequently recover. The main principles in the treatment of tetanus are to eliminate the causative organism, neutralize residual toxin, relax the muscle tetany to avoid asphyxia, and maintain the relaxation until the toxin is eliminated or destroyed.

4.5 Foot Rot

This is an infectious disease (being caused by *Fusobacterium necrophorum*) of cattle and sheep characterized by inflammation of the sensitive tissues of the feet and severe lameness. Constant exposure to mud and water softens tissues around the feet of cows and sheep, greatly increasing their susceptibility to foot rot. Lameness, a painful swelling of the hoof, and foul smelling dead tissues in the space between the claws are common symptoms of the disease. To prevent foot rot and other foot infections, walk cows through a solution of copper sulphate (2 kg ordinary commercial bluestone in 25 litres of water). As long as cow's feet are stained with the copper sulphate solution, they are reasonably protected against foot root. Antibiotics may be needed in severe cases.

4.6 Mastitis

Organisms like *Streptococcus agalactiae*, *Staphylococcus aureus* and *Escherichia coli* etc present in mud and muddy water can cause severe mastitis. Coliforma organisms are usually involved. They cause acute intoxication (septicemia) in the udder, and death of udder tissues (gangrene). To protect cows against mastitis, clean their teats thoroughly before milking and maintain hygienic conditions. In case of acute, severe and life-threatening mastitis of dairy cows generally requires veterinary assistance for treatment. Intramammary antibiotics can be administered, but only after sterilizing and thoroughly drying the teat ends. Injectable or oral antibiotics also may help.

4.7 Botulism

Botulism is a fatal, motor paralysis caused by the ingestion of the toxin of *Clostridium botulinum*, an organism proliferates in decomposing organic materials. Botulism toxin can proliferate in decaying animal carcasses or vegetables and can contaminate water supplies. Horses are particularly susceptible, but pigs appear to be resistant. This disease causes paralysis, which shows up first as weakness with difficulty in eating and swallowing. Treatment is supportive care, and prevention is avoiding exposure to stagnant water and decaying carcasses and vegetables. Vaccination with type specific or combined (bivalent C and D) toxoid is also practiced in enzootic areas. A single dose, precipitated toxoid is available which gives good immunity after 2 weeks and the immunity is solid for about 24 months.

4.8 Listeriosis (Circling Disease)

An infectious disease caused by *Listeria monocytogenis*. This disease affects the nervous system and causes circling, blindness and other abnormal behavior. It occurs in wet pasture when grass is grazed too close, in old and wet hay, and improperly ensiled maize, sorghum and other feedstuffs. Cattle and goats seem to be particularly susceptible. Antibiotics (strepto-penicillin and chlortetracycline in high doses) are necessary for treatment. Prevention includes removing animals from the source of *Listeria* bacteria.

5.0 CONCLUSIONS

Different regions of the country are exposed to a large variety of natural calamities. Droughts, floods, earthquakes, cyclones, landslides etc also affect the livestock and cause major disasters. There are number of deaths, injuries and animals seriously affected, damage to infrastructure, disruption of services including water supply and economic losses. Restoration of drinking water service is essential in ensuring the health and well being of humans as well as livestock and as such fulfills an important role in the development process. In fact, during emergency or disaster situations the basic services like drinking water supply are imperative for the rapid return to normalcy. The impact of a natural calamity can cause contamination of water, breaks in pipelines, damage to structures, water shortage, and collapse of the entire water supply system and needs proper preparedness for tackling of these problems with provision of minimal survival requirements (Table 7).

Table 7 Minimal survival requirements for livestock

Animals	Feed per day	Water per day (litres)	Space required per head* (sq ft.)
Dairy cow in production	6–9 kg hay	20–35	30–50
Dairy cow, dry	4–6 kg hay	15–25	20
Beef cow, dry	5 kg hay or 3 kg ground maize, oats or barley	20–30	30
Beef cow with calf	6 kg hay or 4 kg ground maize, oats or barley	30–40	150
Weaning calves	3–5 kg hay or 2 kg ground maize, oats or barley	10–20	30
Brood sow with litter	1–1.5 kg grain	10–15	40
Brood sow, dry	0.5–1.0 kg grain	4.5	20
Weaning pigs to market weight	0.5–1.5 kg grain	1–4	4–12
Ewe/Doe with lamb/kids	1.5 kg hay or 1.25 kg grain	3	32
Ewe/Doe, dry	1 kg hay or 0.6 kg grain	2	16
Weaning lamb/kid	0.25–0.5 kg grain	1	16

* Close quarters increase water and ventilation requirements

6.0 REFERENCES

Anonymous, 1998. Flood related diseases in poultry and livestock. The Disaster Handbook (National Edition). Institute of Food and Agricultural Sciences, University of Florida, USA.

Blood, D.C. and Radostitis, O.H. 1989. Veterinary Medicine: A text book of the diseases of the cattle, sheep, pig, goats and horses (7th edn.). ELBS, Bailliere Tindall Ltd., London, UK.

Dennis, B. 2008. Livestock water supply. Fact Sheet (N. 01/07), Australia. www.pir.sa.gov.au/factsheets.

Dunbar, J.R. and Miller, R.O. 2008. Assessing water quality for livestock under drought condition. Drought Tips (No. 92–30). www.drought tips.pdf.

McKague, K. 2008. Water requirement of livestock. www.omafra.gov.on.ca/english/engineer/facts/07–023.html

PAHO, 1998. Natural disaster mitigation in drinking water and sewerage systems: Guidelines for vulnerability analysis. Pan American Health Organization, Washington, DC.

10

ANIMAL CARE DURING DISASTER

N. Das[1] and K.S. Das[2]

ABSTRACT

The animal care varies from disaster to disaster; in cyclones, high winds and hurricanes physical injuries are predominate, whereas in drought and snow, starvation and dehydration may be problems. There are five primary emergency actions (evacuation, feeding and watering, identification of animals, treatment, prophylactic measures), which must be taken at disasters to prevent injuries and death. Evacuation of the trapped and injured animal from the damaged animal sheds should be taken up to prevent further injuries to the animals. Evacuated animals need to be housed in one place and sufficient feed, fodder and water should be offered after identification of each animal. The injured animals need to be treated appropriately. Vaccination and deworming in advance of any expected disaster (e.g. before rainy season in flood prone areas) can protect the animals post-disaster. Many small changes / precautions (like following construction norms, immediate completion of repair works, ensuring clear drainage channels, storing chemicals and toxic materials in secured containers etc.) can minimize the losses. Scattered animal carcasses in open are not uncommon during flood, tsunami, and earthquake and even in prolonged drought situation. Depending upon the epidemiological situations and with regards to infectious diseases, the safe disposal of carcasses can be categorized into two categories viz. high risk and low risk carcass. High risk carcasses should be disposed under the guidance of veterinarian. The standard guidelines followed for disposal of carcasses as well as disinfection of the surroundings are mentioned. The details of the disaster specific care and management are also described in this chapter.

[1]Indian Grassland and Fodder Research Institute, Jhansi (UP), India
[2]Central Institute of Research on Buffalo, Nava (Punjab), India

1.0 INTRODUCTION

The animal care varies from disaster to disaster; in cyclones, high winds and hurricanes physical injuries are predominate whereas in drought and snow, starvation and dehydration may be problems. Following fires, smoke inhalation and burn wounds will be issues that require attention of qualified veterinarians. *There are five primary emergency actions that must be taken at disasters to prevent injuries and death* (EMI, 1998).

1.1 Evacuation

Evacuation of the trapped and injured animal from the damaged animal sheds should be taken up to prevent further injuries to the animals. Evacuation of large number of animals particularly large ruminants (cattle, buffalo) needs necessary machineries and elaborates planning. Evacuations are coordinated with neighbors, friends, livestock associations, livestock development/veterinary office, animal welfare organization, local body (municipality, panchyats), local administration, fire brigades etc. Both the destination and the method of lifting and transportation need to be sorted out before the evacuation operation is started.

1.2 Feeding and Watering

Immediately after evacuating the livestock, they should be housed in one place and sufficient feed, fodder and water may be offered. Feeding diets that have moderate energy levels and meet the minimum nutritional requirement reduces the chance of illness.

1.3 Identification of Animals

In large-scale disasters when many animals are evacuated and kept in a common place, these animals should be uniquely identified immediately by marking (use oil paints or 1% silver nitrate solution). Ear tagging can be done later on for permanent identification.

1.4 Treatment

The injured animals need to be treated appropriately and a course of general antibiotics and supportive therapy is to be given to those animals which are susceptible to illness.

1.5 Prophylactic Measures

Vaccination and deworming in advance of any expected disaster (*e.g.* before rainy season in flood prone areas) can protect the animals post-disaster.

2.0 PLANNING AND ORGANIZING

a. Local veterinarians, community leaders, NGOs/local clubs and farmers should communicate each other once disaster has occurred and finalize disaster management action plan.
b. A local coordination committee should be formed and the committee immediately initiates the evacuation and other follow up action.
c. A safe shelter for rescued animal needs to be identified. Potential places are fairgrounds, playground, other farms, grazing land etc.
d. Animals should be evacuated and taken to the shelter.
e. Local administration and voluntary organizations should arrange appropriate vehicle for transport of injured and dead animals.
f. Stray animals moving in the roads should also be taken to the safer places and provided with feed, fodder and drinking water.
g. Ensure timely and adequate supply of food and water in the common shelter area so that feeding could be continued to the injured and evacuated animals.
h. Prepare I) one disaster kit which included an up-to-date list of all animals including their records (health, vaccination, feeding, production etc in summarized form), materials required for temporary identification (like ear tags, paints, marker pen etc.) and II) one basic kit that includes commonly used surgical instruments, antiseptics, antibiotic, bandage, analgesic and handling equipments like halters, cage, rope, bolt cutter instruments etc.
i. Facility for safe disposal of dead and decomposed carcasses and farm wastes need to be created.
j. Local veterinary hospital should stock adequate drugs (*e.g.* antibiotics, analgesics, antinflamatory, dewormers, ointments and emergency health care facilities for effective treatment of the injured animals.
k. In the flood prone areas, all the animals should be vaccinated against contagious diseases before the onset of monsoon.
l. Ensure the availability of vaccines that can be needed when disease outbreak is expected after the disaster.
m. The minimum requirement of feed and water for different categories of animals is given at Table 1.

2.1. Earthquake

i. Identify the safest place for shelters and be prepared to keep the animals survive for 2–3 days without any outside help.
ii. All the farm workers and animals should have vaccination against tetanus.

Table 1 Minimum requirement of water and feed for survival

Animal	Water per day (in litre)		Feed per day
	Winter	Summer	
Milk producing cows and buffaloes	30	40	5 kg legume hay + 5 kg straw
Dry cows	25	30	8 kg straw
Growing calves (above 6 months of age)	10	20	4–5 kg hay/straw
Pregnant cow and buffalo	25	30	5–7 kg legume hay
Sheep and goat (with unweaned lamb/kid)	5	5	2 kg hay
Growing lamb or kid (above 4 months of age)	1	2	1 kg hay
Adult male sheep/goat and dry female	2	3	1.5 kg hay
Sow with litters	15	20	4 kg grain
Pregnant sow	5	10	1 kg grain
Boar	5	5	1 kg grain
Horses	20	25	8–10 kg hay.
Layer birds (poultry)	20 (per 100 birds)	25 (per 100 birds)	8 kg/100 birds
Broilers	15 (per 100 birds)	20 (per 100 birds)	5 kg/100 birds
Turkey	40 (per 100 birds)	50 (per 100 birds)	20 kg/100 birds

iii. If the workers and animals are remain within the shed during earthquake, remain indoor and try to stand under the door and strong furnitures that will not break easily. The pet animals should be held on their legs.

iv. Keep away the animals from the objects (like tools, equipments etc) which can easily fall and cause serious injury to the animals.

v. Remember that animals can be frightened by an earthquake and become aggressive and may bite out of fear and stress.

2.2 Cyclone and Flood

i. The cyclones surveillance is done by satellite INSAT and powerful cyclone detection RADAR installed at Kolkatta, Paradeep, Vishkapatnam, Machhilipatnam, Chennai, Karaikal, Mumbai, Goa, Cochin and Bhuj. The vigil is so satisfactory that no cyclone can escape detection and it has been possible to issue timely warnings through cyclone warning centres located at Kolkatta, Bhubaneswar, Visakhapatnam, Chennai, Mumbai and Ahmedabad.

ii. Before the low-lying areas are flooded, move the animals to higher ground. But the animal evacuation routes must not interfere with human evacuation route. Places where animals are to be taken should be decided in advance and arrangements are made with the owners to accept their animals. Since all animals may not be evacuated at a time, owners should decide ahead of time which is the most important ones to save. Proper identification of every animal is must in resolving arguments of ownership in case animals get mixed up. Disasters attract individuals who gouge and prey on the misfortunes of victims.

iii. Keep enough stock of food and medicines. Items that can be obtained are hay stack, ropes, halters and other animal restraint equipments.

iv. In the coastal areas, prepare enough cyclone shelters for both human and animals. Although cyclones can be predicted with a great degree of accuracy, salvaging cattle has never been a priority. The animal shelter should have provision for stocking fodder, medicines and drinking water.

v. During cyclone, animal left outside may suffer less injury than those placed in shelter. This is because some of the animal sheds may not withstand the high winds. Animals are injured by collapsing structures and flying objects that may have been avoided on the outside. Flood waters that inundate a shed can trap animals inside causing them to drown.

vi. Local ponds and canals are regularly inspected and cleared off from the obstructions.

vii. Ensure that animals are vaccinated against tetanus and rabies. Endemic disease and chronic conditions like worm infestation or ticks require special attention.

viii. Ensure that outflow from manure pit does not contaminate the drinking water source.

2.3 Tsunami

i. Farmers in the coastal area, should be alert whenever any information of earthquake is reached from anywhere in the world.

ii. If tsunami warning is issued, animals should not be allowed to go near the beach and plan escape route and emergency shelter.

iii. Keep the animal loose.

iv. When waves are receded, evacuate the animals, check for injuries and seek veterinary care if necessary.

2.4 Drought and Extreme Heat

i. Undertake water conservation measures in the farm before the summer starts and during the time of water scarcity.

ii. Create artificial sheds in the pasture and humidifier in the shed to keep the animal cool.

iii. As and when an animal is found suffering from heat stroke, provide supportive measures like access to shade; supply drinking water, spraying water on the body of animals etc.

iv. Provide regular rest periods for working animals.

v. ***Water Resource Management***: Based on meteorological and hydrological conditional assessment, early warning may be issued for better preparation to manage the drought. Protect the existing water reservoir from misuse and prepare water budget, repair the tube wells, clean the water tank, large ponds, canals and local water catchments before the onset of monsoon.

vi. A food security system needs to be developed with active participation of panchayat, veterinary and agriculture department, local administration, NGOs and other voluntary organizations so that farmers get feed and fodder at subsidized/reasonable rate. Encourage fodder cultivation and tree plantation wherever possible and encourage supply of molasses and urea to the farmers for enrichment of low quality fodder.

vii. Overgrazing will increase the problems in drought affected area and should be avoided; this can be done through collective effort of all the villagers and livestock grazers.

viii. The quality and availability of feed in the pasture is often poor during hot dry season in arid and semi-arid region. Animals attempt to compensate by increasing the grazing hours. However high temperature and scarcity of water prevent the animals to graze longer hours. Provision of trees in field helps the animal to reduce heat stress. The provision of adequate shade for pregnant ewes and female goats (does) offers a practical and effective way to reduce the incidences of low birth weight and neonatal mortality in hot environments.

ix. There must be sufficient space provided under the shed to prevent overcrowding because it leads to humid unsanitary conditions that facilitate air borne infectious diseases.

viii. For cattle preservation in drought prone areas, it is essential that state authority should take stock of their cattle population and assess fodder requirement and locate areas where shortages are likely to occur. Whenever necessary, the local administration should arrange procuring fodder from surplus areas.

ix. Close monitoring of the fodder price is needed so that unscrupulous traders can not take the benefit of scarcity and increase the price of fodder. The forest department should undertake steps for the cutting and bailing of grasses in the forest whenever possible to meet the demand from fodder deficit districts.

x. Storage depots for fodder and concentrates along the lines of FCI as a means of buffer stock for livestock protection need to be considered as a suggestion for long term planning for drought mitigation.

xi. Among the domesticated animals, sheep is worst affected from the occurrence of drought. To reduce the losses, it is desirable that breeding/mating of ewes is avoided by keeping rams away from the female flock during the entire period of drought. Sheep in pregnancy and in early lactation can be managed on 3 watering per week and in rest of stages of production the animals can be reared on 2 watering per week. These animals can be managed on moderately underground saline water containing up to 4.0 g/l of total soluble salts without any harmful effects. The sheep can be raised on underground waters containing 0.7 g/l of nitrate and 4.5-9.0 part per million of fluoride with no apparent harmful effect on their productive processes. Further the underground water containing higher level of total soluble salt, nitrate and fluoride can be diluted with potable waters (Mittal, 2002).

xii. Leaves from trees can serve as life saving forage both during flood and drought. A massive tree plantation programme with the involvement of community will help avoid considerable loss to livestock.

3.0 PREVENTIVE MEASURES AT FARM LEVEL

Many small changes/precaution can be undertaken to minimize the losses such as

a. Construction norms should be followed strictly for new sheds
b. Repair works should be completed immediately after the damage.
c. Windows and doors should not be built with materials which will easily shatter and injure animals.
d. Ensure that drainage channels are cleared and not covered with grasses.
e. Store the chemicals and toxic materials in a storm proof building and in secured containers.
f. Construction materials (like G.I. Sheets, Asbestos sheets etc) unsecured in the open because these materials become projectiles during high winds.

4.0 DISPOSAL OF DEATH ANIMALS

Scattered animal carcasses in open are not uncommon during flood, tsunami, and earthquake and even in prolonged drought situation. These dead bodies need to be disposed early; otherwise they can be contaminated with all kinds of micro-organisms including those which are extremely pathogenic to human and other living animals. Besides the animals which died due to direct impact of natural disaster, there may be some animals during disaster which die with the symptoms of infectious diseases with high risks. Depending upon the epidemiological situations with regards to infectious diseases, the safe disposal of carcasses can be categorized into two categories.

4.1 Disposal of High Risk Carcasses

Examination of suspected animal (live or death) and diagnosis of infectious diseases with high risk should be made by qualified veterinarian only and he should also suggest the method to be adopted for transportation and destruction of carcasses. Transportation of dead animal bodies contaminated with high risk infection should be carried out with utmost care. Following precautions need to be taken while transporting such dead animals (Kondaiah and Prasad, 2002).

a. Dead bodies scattered in different location should be collected in one place for destruction.
b. The vehicle used for transportation of such highly infectious materials must not be used for any other propose. Special care should be taken to protect people involved in handling infective materials and to prevent the spread of contamination.
c. Cleaning and disinfection of the vehicles including all possible surfaces, should be carried out after every transportation of dead animals and animal wastes.
d. Destruction of high risk carcasses are done either by
    i. ***Incineration Process*:** In most of the natural disaster, the electric power supply system generally comes to halt. Hence incineration using electric furnace may not be always feasible. Moreover in India such facilities are not common. Generally oil and heat are used for burning the dead bodies at 800°C to 1000°C. When complete incineration is not possible due to high cost, dead animals and other materials of high risk category should be burnt in underground pits and be buried as per safety observance.
    ii. If burning is not possible because of a lack of oil or wood or any other facility or when large number of dead animals must be disposed off after natural disaster or outbreaks of contagious disease, carcass may be buried. For **burial process** following points need to be considered.
        - Animals to be buried in remote places.
        - Burial places should be located on dry grounds so that spread of infectious agents to ground water is minimized. It should also be away from the stream, river or tank.

- Burial should be deep enough (at least 6 feet underground).
- Before filling the pit with earth, the dead body should be covered with quicklime or common salt which will hasten the destruction process.
- Care must be taken to prevent dogs, vultures, birds from digging the pit and carrying away any portion of the carcass.
- The infectious micro-organism may survive for long time even after burial. Hence, the burial place should be clearly indicated, registered and protected; people and animals should be kept away from this site for instance by fences.
- No lifting of under ground drinking water should be allowed in this area.

e. Disposal of low risk carcass

   i. Dead body of animal without any infectious disease is not generally destructed. The various parts of the body like skin; bones etc are used after sterilizing and processing in the Carcass Utilization Center.

   ii. However if it is not possible to utilize them, for instance if they become available in places that have difficult access (*e.g.* islands and mountain areas) or if quantities are too large to process them in short period of time or which can not be stored at proper temperature for long period (*e.g.* in the case of calamities caused by flood or earthquake or disease (not highly dangerous) outbreak which result in death of large animal population, the destruction by burial is followed.

   iii. Anaerobic method is another way of destruction of such low risk carcass. This is done by storing carcasses and offal's in closed pits or closed tanks, so that the materials are allowed to decompose in absence of oxygen. This method is not suitable for large animal. The other disadvantages of this method are that it takes a longtime before the materials disintegrate and that the microorganisms may stay viable. Anaerobic digestions may also cause very disagreeable odors.

5.0 GUIDELINES FOR DISINFECTION OF THE SURROUNDINGS

Disinfection of the surrounding from where high risk animals including other infected materials are lifted for disposal should be done very carefully. The following points should be remembered.

a. Cleaning procedure should only consist of removal of dirt with mechanical means and incinerating them afterwards.

b. No water should be used to restrict the spread of pathogens.
c. Disinfection of surroundings is done either of the following chemicals: 4% solution of caustic soda (70° – 80°C) or 2% solution of formaldehyde or chlorine solution containing 3% active chlorine.
d. In case of dead animals from anthrax like high dangerous zoonotic or animal infectious diseases, all contact surface should be disinfected with 10% caustic soda solution or 4% formaldehyde solution or chlorine solution containing 5% active chlorine or 7% hydrogen peroxide solution or 2% glutatadehyde solution.
e. Prophylactic disinfection (at least once in a month) should be carried out with either of these chemicals: 2% caustic soda solution, 2% formaldehyde solution.

6.0 REFERENCES

Emergency Management Institute (EMI) 1998. Animals in disaster – Module A: Awareness and Preparedness. Federal Emergency Management Agency (FEMA), EMI, Emmitsburg, Maryland. (www.training.fema.gov/EMIWeb/is10.htm).

Kondaiah, N. and Prasad, M. 2002. Disposal of animal carcass during natural disaster. National Seminar on 'Biological Threats and Disaster Management – Role of Veterinarians'. 9 – 11 August, 2002, Punjab Agriculture University, Ludhiana.

Mittal, J.P. 2002. Management of sheep during drought conditions in arid and semi-arid belt of the country. National Seminar on 'Biological Threats and Disaster Management – Role of Veterinarians'. 9–11 August, 2002, Punjab Agriculture University, Ludhiana.

11

PREPAREDNESS FOR UPKEEP OF LIVESTOCK DURING CALAMITY

S.K. MAHANTA, N. DAS AND BINDU MADHURI

ABSTRACT

Animals are the main source of livelihood to the poorest of the poor and to the landless, but concrete steps towards the care and management of animals are not taken during natural calamities. Many times even the loss of animals are not being considered/ projected. However, animals also play an important role in calamities through search and rescue operations. Calamity being an emergency situation need be attended with the urgency it deserves. Relevant information/data should be made use of in loss assessment and monitoring of compensation. A calamity management plan for animals shall essentially include retrospective epidemiological study of calamities (collection of data and impact assessment of past calamities etc.), action plan for the rehabilitation of rescued animals and planning of resources (like manpower, store, equipments, ambulance, medicines, logistic needs etc.). Poverty alleviation, local resource enrichment, optimal use of resource at hand, etc. should be part of calamity preparedness. Because of high tension during calamities, people get so focused on themselves that they often fail to look around and notice the other things that are going on around them. Emergency responders get better at avoiding this problem with experience and training. Usually they follow a response guideline that reminds them to take in all the other factors. Normally, the response time for a calamity is extremely limited. The accepted sequence for safety and evacuation is people first, followed by pets, livestock and property.

1.0 INTRODUCTION

Over the years it has been observed that vulnerability and impact of calamity is high on the economically and socially marginalized sections of

Indian Grassland and Fodder Research Institute, Jhansi (UP), India

the community. Land being limited, the number of marginal farmers and Landless people are increasing every year. These sections by and large depend on animals for their livelihood. Whether it is the "Masai" of Africa or "Gujjar, Bakharwal or Raikas" of India there are a large number of nomads in the world who by and large depend on animals for their livelihood. Animals have become an important cultural element, essential in maintaining man's traditional life style. For example the Yak has made it possible for humans to live in harsh areas where production of crops is virtually impossible. Animal genetic biodiversity has helped humans to survive in a wide range of environment.

Many developing countries are calamity prone. Out of 32 states and union territories (UT) of India, 22 states/UT are vulnerable to one or other form of calamity. The cyclone in the state of Andhra Pradesh, of November 14 to 20, 1977, caused an estimated loss of 230146 cattle and 344056 other animals, as against 8,515 human deaths. Similarly in Orissa cyclone (in India) during the 4th of June, 1982 there had been 11,468 cattle lost against 243 human deaths, although losses of other animals are not being projected. Earthquake has comparatively less impact on animals; but in Uttarkashi earthquake, 3100 cattle heads were lost as against 770 human lives. During 1953–1990, we lost 102905 cattle against 1532 human lives because of earthquakes. It is apparent that even though animals are the main source of livelihood to the poorest of the poor and to the landless, concrete steps towards calamity management of animals are yet to be taken.

2.0 ANIMALS' ROLE IN CALAMITIES

- It is believed that some animal species can forewarn, an incumbent calamity, particularly earthquake.
- Animals like the dogs are used for search and rescue operations. During flash floods people escape holding on to animals; all animals are natural swimmers.
- Animals and animal driven carts are the available means of transport of injured and invalid people when no other transport is possible. Animals are also used for clearance of debris.
- During calamity the people lose their property and livelihood. Losses of crops, infrastructure or machinery take time to recoup. Animals if salvaged can provide the community with food, energy, transport and the other utilities immediately.
- Unlike standing crops or infrastructure, the animals can be salvaged and brought to safer places, where a calamity can be predicted. In transit camps or in rehabilitated dwellings too the animals can be utilized for milk, manure, transport, fuel (dung cakes).

- Animals are used for fetching water or for lifting water from the water sources, where machine or man power are not available or can not be used.
- Even in their death animals serve the community; they provide some material gain through, their hide, bone, lard and carcass.
- Partially damaged crops and grains unfit for human consumption can be gainfully used as animal feed.
- Since there is major participation of women and children in the care of animals, this activity can be a major diversion to tide over the gloom and depression that fall over the victims during a calamity.
- Flooded areas where drainage is slow can be used for duck rearing and fish farming. These help in clearing pests that usually breeds during floods.
- Free living animals like the pigs play an important role in clearing garbage, waste materials etc.; they provide livelihood to economically weaker section and can be a cheap source of protein to these communities especially when other sources of food become scarce.
- Dogs (pets and community animals) keep the creeping reptiles and straying wild animals at bay or at least forewarn the community of their intrusion.
- People marooned during floods or earth quakes during winter have survived chilly nights taking refuge among herds of sheep and goats.
- War is man made calamity. Horses and mules are used for transport in difficult terrain around the border. Dogs are used in the army for reconnaissance, tracking, guarding and combat.
- Scavenging works done by living animals: Stray dogs consume discarded food, crows clear spoilt materials that are thrown around, vultures devour the dead animals, snakes control rodents, and cats do the same.

3.0 CALAMITY MANAGEMENT PLAN FOR ANIMALS

Based on the economic, political, social and environmental importance of animals and life system of the small holders who depend on them in the developing countries, one may have to consider animal calamity management for animals in that light. A calamity management plan for animals shall essentially include-

3.1 Retrospective Epidemiological Study of Calamities

This includes-

- Data collected, interpreted and analyzed (*i.e.* information), on the basis of which some prediction can be made.

- Impact assessment of the past calamities, preparedness and mitigation measures
- Calamities need not repeat every time. Experiences gained from similar hazards of nearby areas are positively helpful. This is especially useful in animal calamity management which is still in its infancy.
- Disaster vignette is a means by which mapping is done on the basis of incidence frequency, magnitude, epicenter, vulnerable areas etc.
- ***Herd Profile*:** the total animal population (herd number), vulnerable animal population as per their species, breed, age, sex etc.
- ***Community Profile*:** the total population, animal owning population, their age, sex, socio-economic status, education, cultural distribution etc.
- Animals at risk, the nature of hazard, the intensity of impact and mortality rate (immediate or delayed).
- Analysis of type of risk (identification and analyses).

3.2 Action Plan

Since animals are to be rescued following calamities and collected in relief camps, the immediate priority would be controlling and combating disease. The animal health component of disaster mitigation will include-

- Promotional herd health care such as nutrition, pregnant animal care, care of new-born and young animal etc.
- Prevention of risk through vaccination, pest/ vector, control, sanitation etc.
- Specific therapy by way of early diagnosis and treatment.
- ***Rehabilitation*:** help animals to recover from any trauma or fear. Many animals in which treatment is unlikely to be beneficial may have to be put to sleep *i.e.* euthanasia (mercy killing).
- Disposal of dead animals and carcass utilization.

3.3 Resources Planning

- Assessment of available man power *i.e.* veterinary doctors, para-veterinary staff, ancillary staff etc.
- Store and equipments include the medicine, surgical and medical appliances, diagnostics, life saving equipments etc.
- ***Logistical Needs*:** that is the need for fuels, lighting equipment, tents, sheds, grass bedding, trolleys, materials for sanitation, storage of feed and fodder and water.

- ***Ambulance and Out Reach Facility***: transporting animals is more cumbersome when they are sick, injured and non-ambulatory.
- Veterinary medical facilities such as veterinary hospital, mobile veterinary units etc.
- Grazing and watering facility.

3.4 Training Plan

- Training veterinary personnel, para-veterinarians, attendants etc.
- Training administrators like BDO, telephones and fire service personnel, civil defence personnel, sarpanch/ village headmen, administrators etc.
- Animal health awareness for trainees such as social workers, volunteers.

3.5 Allied Planning (Organization)

- Augmenting political and administrative support
- Involving NGOs, CBOs, media, Goshalas, animal welfare organizations and other volunteer groups.
- Eliciting commitment and allocation of funds
- Formation of veterinary service groups at State and Central Government level.
- Organization of district/ area level bodies, assigning specific tasks and responsibilities.
- Establishing communication channels, alternate channels like ham radios.
- Establishing alternate source of power, energy etc.
- Plan for monitoring and supervision.
- Publicity and public relation activities (Vet-PR)
- Plan for mitigation and rehabilitation of animal owners along with animals

3.6 Execution of Short and Long Term Plans

Plan implementations include estimation of animals involved, damage assessment. One may make use of media to help locating remote areas where often organizers' attention may not easily reach. Monitoring is part of implementation. Follow-up of calamity management especially that of animal management are valuable experience which may be shared with people working in different vulnerable areas. Linking with allied agencies for uniformity in all regions is part of calamity management programme, as unbiased and uniform measures help transparency which would further instill confidence of the community in the system.

4.0 CALAMITY MANAGEMENT FACILITY

4.1 Temporary Veterinary Hospitals and Control Room

The sudden increase in deaths and casualty would increase the demand for veterinary personnel. Early relief period would therefore be critical. As time passes, the threat of diseases also increases. Affected animals being under stress, inadequately managed and being herded in relief camps would be increasingly vulnerable as each day passes. Temporary camps or sheds also need facilities that are previously non-existant. In the temporary hospital the equipment and other infrastructure are provided along with stock pile of emergency equipments and drugs. As search and rescue would be over, relief and rehabilitation would acquire importance. The impact of disaster being calamitous one may come across a number of cases which are beyond mere physical injuries or contagious disease. The mental trauma due to the loss of a sibling or a partner where bullocks are used in pair, can be encountered and will have to be handled at appropriate stage. Many of the cases of extensive trauma that cannot be handled may require to put the animal sleep (mercy killing). For information exchange and co-ordination of veterinary support, control rooms are used. These units handle co-ordination of information among directorate of animal husbandry, district veterinary hospitals, rural hospitals of the affected area etc.

4.2 Control of Diseases

Immediate task after the emergency treatments are completed would be containing spread of diseases. Armed with retrospective epidemiology, authorities would identify emergency diseases and sanitation measures. This would include digging manure pits, drainage facilities (more important in floods, cyclone etc.), drinking water and water troughs. This will have to be simultaneously established in hospitals, temporary shelter camps as well as in the affected areas from time to time and as per priority. For health care operations there is a need for public co-operation. Animal welfare agencies working in the areas can help stockpiling medicines. Media can help by providing feed backs from field.

4.3 Calamity Preparedness

Poverty alleviation, local resource enrichment, optimal use of resource at hand, etc. should be part of calamity preparedness. On the basis of the earlier experience one has to improve or modify the preparedness or establish one where it is not existent. The objectives should be to

- Minimize suffering, loss and damages
- Prevent any communicable disease

- Reduce losses due to delay with prompt communication, improving and optimizing existing facilities, equipping support system and
- Assisting rehabilitation

The following characteristics of animals need to be considered during calamity management of animals-

4.3.1 *Earthquake*

Since in India animals are tied outside or in thatched sheds, chances of physical injuries are low. But when tied or caged their chances of escape are minimized.

4.3.2 *Flood*

Animals are natural swimmers; therefore can escape drowning if they are not tied or caged.

4.3.3 *Cyclone*

Though cyclone can be predicted with some accuracy, salvaging animals has never been a priority. Cyclone shelters are not designed to house animals.

4.3.4 *Drought and Famine*

Meteorological or agricultural drought, need not necessarily affect animals as they can survive without agriculture. Animal stand the stress of the famine better than humans. They can use most sources of the water which may be unclean by human standards. Dry fodder reserves, urea molasses licks, bricks made of fodder urea and molasses etc. can be part of the stock pile. Scarcity fodder can be grown as part of the preventive action.

4.3.5 *Cold Famine*

An incidence famine may occur in the alpine pastures due to unusual snow that covered grass. Although the sheep easily withstand the snow but they starved for want of pastures. As a preventive measures, feed/ forage stores can be established by storing the dry fodder in pits; the low ambient temperature prevents spoilage. Cold famines are also known in the snow deserts or part of Ladak; but Yaks do survive these.

4.3.6 *Landslides and Snow Avalanches*

Tied or caged animals stand a higher risk during these calamities. Care of animals is often a low priority in calmity management. Though animals stand better chance of survival during disasters, much less efforts have been made in this direction.

4.3.7 *Fire and Forest Fire*

Animals are more prone to fire as they are usually housed in thatched sheds. The practice of smoking the sheds to repel pests is a common practice.

Proximity of the chaff or the paddy straw makes conditions vulnerable to fire accidents. Forest fire affect wild animals and birds though many of them are more lucky than their domestic counterparts, being more athletic and free to escape from fire.

4.3.8 *Industrial and Technological*

Animals are not frequent direct victims, though indirectly they can be trapped/affected any time. During Bhopal gas tragedy many animals including livestock, were affected; but poultry is known to have escaped from the impact of the hazard.

4.4 Role of Government

Ministry of Agriculture which is the nodal agency of the country in charge of calamity management, should specifically draw its attention to calamity management in animals also. A separate cell has to be established for the purpose in the department dealing with veterinary service. Prevention measures and preparedness are to be organized by this department with the cooperation of their counterparts. Animal population profile and distribution profile should be clearly drawn and plans for salvaging and rescuing the animals in major calamities like cyclones and flood to be undertaken on the basis of this data. Agencies like NGOs and CBOs should be organized in such a way that their services can also be utilized at short notice for calamity management. Cyclone shelters are to be specifically devised for animals in cyclone prone areas of the country. Drought relief programmes like arrangement of scarcity fodder, drinking water and subsidized food be arranged as the stockpiles along with drugs and vaccines required to combat with allied calamities. Each state should organize a calamity management group in the department dealing with animal husbandry and veterinary service with specially trained staff, epidemiological data and communication facilities. The required field staff should be kept in constant readiness throughout the vulnerable months of the year. During lean period, the team should undertake preparedness and relief exercise to test their efficacy and preparedness. This will help to develop a well oiled working system. There has to be a separate control cell attached to the central control room dealing with national animal calamity management. The epidemiological and statistical information collected before, during and after calamity should be compiled and discussed among the States of each zone on a scheduled timetable. Preventive actions taken or preparedness should be constantly monitored and lacunae there of be frankly reported and corrected for future use.

Calamity being an emergency situation need be attended with the urgency it deserves. It is seen that accountability may have to be compromised to an extent to make room for faster service delivery. Relevant information/data should be made use of in loss assessment, for efficient disposal of dead animals and animal wastes, and should be used for estimating and monitoring of compensation; this process would therefore ensure transparency, impact and social justice.

4.5 Role of Society

The large amounts of stimuli and tension generated during calamities affect both humans and animals. Because people get so focused on themselves that they often fail to look around and notice the other things that are going on around them. Emergency responders get better at avoiding this problem with experience and training. Usually they follow a response guideline that reminds them to take in all the other factors. Although it's not desirable for livestock owners to face so many calamities that they also develop this broad focus, there is one key approach that helps enhance safety in tense situations is teamwork. Take help and designate one person to keep watching for additional oncoming hazards. Normally, the response time for a calamity is extremely limited. If we are not aware of pre-existing conditions, we will have even less time to respond. It's important to mentally develop a priority system for safety efforts. The accepted sequence for safety and evacuation is people first followed by pets, livestock and property. Our personal priority system may vary but we should always put the safety of people far ahead of other concerns. But our desires to address pets, livestock, and property will not precede the personal safety of everyone in the calamity-affected area.

5.0 REFERENCES

Cotton, S. and Ackerman, R. 2008. Caring for livestock during disaster. http// www.caring animals-disaster.pdf.

Heath, S.E. 2008. Mitigating the disaster affecting the livestock industry. www.animaldisasters.com/Livestock.htm

Ramakumar, V. 2008. Role of livestock and other animals in disaster management. www.vethelplineindia.com/ProfRamKumar-article.doc.

Sen, A. and Chander, M. 2003. Disaster management in India: the case of livestock and poultry. *Rev. Sci. Tech. Off. Int. Epiz.* **22**: 915–30.

12

LIVESTOCK HEALTH MANAGEMENT DURING NATURAL CALAMITIES

S.B. Maity

ABSTRACT

During disaster, disease incidences increase during both drought and flood. Increased incidences of infectious and parasitic diseases of ruminants are reported during the drought period when most of the normal water sources dried up and animals suffered from lack of clean water for consumption. During the drought, mortality rates of livestock also increases significantly particularly in arid and semi-arid region. Livestock maintained on grazing are prone to suffer most during draught and flood. During flood, lowland grass fields are generally submerged and water born disease can occur if organisms in the manure migrate to surface water or ground water sources that are used as a supply of drinking water. Outbreak of deadly contagious diseases like caprine pleuro-pneumonia (CCPP), haemorrhagic septicemia, black quarter, anthrax etc is common during flood. During draught, the main problem is feed and forage scarcity. Disease incidences are related with weakness of animal from starvation as well as scarcity of drinking water. In normal time, animal may appear healthy in spite of harboring worms, if level of nutrition is good. Due to poor nutrition during draught, internal parasite flare up and cause severe loss in production. Animals are also subjected to thermal stress which leads to silent heat, anoestrus, poor semen quality, low conception rate and embryonic deaths. In extreme heat stress, the thermoregulatory mechanism fails and the animal die due to heat stroke. The treatments of different diseases that occur during drought and flood, vaccination schedule, disposal of carcass, management of different deficiency diseases are discussed. This chapter also deals with the management of livestock during calamities.

Indian Grassland and Fodder Research Institute, Jhansi (UP), India

1.0 INTRODUCTION

During natural calamities attention usually goes primarily towards human welfare, however, welfare of animal is also of paramount importance considering their causalities from draught and flood prone diseases, epidemics and different feed poisoning. During disaster the scenario for domestic animals is by and large similar to that of humans. With sheds and enclosures damaged and destroyed by the impact of disaster, livestock moves freely, often with injuries. Some may be trapped in the debris. Animals which manage to survive need fodder, water and shelter for survivability. Disease incidences increase during both drought and flood. Increased incidences of infectious and parasitic diseases of ruminants are reported during the drought period when most of the normal water sources dried up and animals suffered from a lack of clean water for consumption. Although migration was observed throughout the year, it increased during the drought, as the search for water and forage intensified. During the drought, mortality rates of livestock also increases significantly particularly in arid and semi-arid region. During floods there is outbreak of deadly contagious diseases like caprine pleuro-pneumonia (CCPP), haemorrhagic septicemia, black quarter, anthrax etc. The detrimental effects of floods include increase incidences of parasitic and epidemic diseases among livestock and the destruction of infrastructures. There is also increase in the livestock mortality rate which is attributed mainly to diseases. Carcasses of animals are a health hazard and have to be handled with care to prevent the transmission of disease conditions that could result during natural disasters.

2. DISEASES DURING CALAMITIES

Diseases that occur in livestock after disaster may be infectious or contagious or may be of epidemic and poisonous in nature. The organisms (bacteria, virus and protozoa) flare up more in the animal body after flood or during drought when the body immunity is lowered due to low nutrition. Water born disease can occur if organisms in the manure migrate to surface water or ground water sources that are used as a supply of drinking water.

2.1. Post Flood Disease Incidences

During flood, lowland grass fields are generally submerged and livestock take shelter on uplands. Livestock maintained on grazing are prone to suffer most both during draught and flood. A minimum 30 days rest period for flooded field is recommended before livestock are reintroduced to pastures on which manure is spread. But in reality the poor farmer can not afford to follow the recommendation. The debris on the grassland are decomposed and act as predisposing factors for bacterial, viral and fungal growth and cause diseases like:

2.1.1 ***Haemorrhagic Septicemia (H.S)***

It is an acute septicemic disease of cattle, buffalo, sheep and goat caused by *Pasturella multocida*. Out breaks usually occurs in early monsoon particularly after heavy rain followed by stress due to humidity and low level of nutrition.

2.1.1.1 *Signs*

Sudden onset of fever is followed by high rise of temperature, swelling of throat and tongue with salivation and protrusion of tongue (oedematous). Acute fibrino pleura-pneumonia and respiratory distress are observed in pneumonic form whereas acute enteritis with diarrhoea is observed in enteric form.

2.1.1.2 *Remedies*

Alum precipitated H.S vaccine (sub-cutaneous S/C) may give immunity up to 6 months whereas oil adjuvant vaccine (Intra-muscular I/M) will give immunity for 1 year. In disease out break, affected animal should be treated with drugs like i) Sulphamezathine, Vesadin, Diadin, Sulphadimidin etc (100-200 ml as initial dose followed by half dose after 24 h). ii) Injection Timethoprim Sulpha- 15-30 ml by slow intra-venous (I/V) route iii) Injection Terramycin or Oxytetracycline-LA 20-30 ml I/M with Anthisan or Hostacortin 10 ml.

2.1.2. ***Johne's Disease (JD)***

It is a chronic infectious (causative agent: *Mycobacterium paratuberculosis*) disease usually affects heifers (more than one year of age) and adults. Due to high stress, during floods, sheep and goat are also affected by *Mycobacterium paratuberculosis*.

2.1.2.1 *Signs*

Chronic recurrent diarrhoea which does not respond to treatment. Intestinal mucosa becomes thick and corrugated. Loss of appetite and absorption of nutrients does not occur.

2.1.2.2 *Remedies*

Infection is resistance to treatment. Streptomycin @ 10-15 mg/kg body weight for a prolonged period may cause transitional recovery. Animal should be segregated to control further contamination. Sheep and goats are to be slaughtered at early stage to prevent further loss.

2.1.3 ***Anthrax/Splenic Fever***

It is an acute febrile disease of mammals caused by *Bacillus anthraces* and characterized by an enlarged, black, soft spleen and edema and hemorrhages of subcutaneous and sub serous tissues. Disease is transmitted through feed contaminated with soil and vegetation and biting of blood sucking insects present in over flooded areas.

2.1.3.1 *Signs*

Common signs are hyperpyrexia, tarry colored bleeding from nasal orifices and sudden death of affected animal.

2.1.3.2 *Remedies*

i) Affected animal may be treated with Crystalline Penicillin, 40-80 lakh unit, I/V or Procaine penicillin 40 lakh i.u. I/M. ii) As prophylaxis measure anthrax spore vaccine may be inoculated @ 1 ml S/C. Immunity is 1 year. Annual revaccination in enzootic areas is recommended. iii) After death, carcass of affected animal should not be opened. Deep burial of carcass and disinfection of cattle shed are recommended.

2.1.4 *Black Quarter/Black Leg*

Black Quarter is a sporadic, enzootic, acute febrile disease of cattle, sheep and occasionally in goat, swine and deer caused by *Clostridium feseri (chauvi)*. Disease is mostly found in wet weathers of rainy season and spring. Swampy low lying land remains infectious for long period.

2.1.4.1 *Signs*

i) Sudden onset with high fever (106°C), ii) lameness accompanied with inflammatory swelling of thick musculature, crepitating sound on palpation of swelling, iii) swelling femoral, lumber, gluteal or sacral region or shoulder and neck. Swelling is hot and painful at first, later colder but no pain, iv) Dysopnea, rapid and feeble pulse and v) body temperature falls but rises slightly just before death.

2.1.4.2 *Remedies*

i) No specific treatment except injection of Anthisan on swelling site after incision. Ii) As prophylaxis, alum precipitated vaccine (5 ml S/C) is used which has immunity of 6 months. Combined HS/BQ vaccine is also recommended.

2.1.5 *Foot and Mouth Disease (FMD)*

Foot and Mouth disease is an acute, febrile, highly contagious disease of cloven footed animals caused by different strains of FMD virus and it is characterize by the formation of a vesicular exanthema in the buccal mucous membrane and in the skin above the hoof and in the inter-digital spaces. FMD primarily occurs after rainy season, especially during spring months.

2.1.5.1 *Signs*

i) Loss of appetite, slight temp. ii) Retarded rumination and increased thirst. iii) Vesicles develop on buccal mucous membrane, lips and tongue which are 0.25-1.00 inch in diameter, usually circular. Vesicles rupture leaving painful erosions. iv) Large patches of epithelium may be sloughed off. v) Difficult swallowing due to spread up of infection to pharyngeal mucosa. vi) Frequently, erosions are infected by secondary bacterial invaders and this causes lameness and stiffness in gait.

2.1.5.2 *Remedies*

Only prophylactic vaccination is recommended to check the disease. The affected animal may be treated with antibiotic to check the secondary infection.

2.1.6 *Enteritis (Diarrhoea/Dysentry)*

It is generally caused by i) irritant feedstuffs in poisoning case and ii) enteric bacterial and intestinal parasitic infestations.

2.1.6.1 *Signs*

i) Shooting diarrhoea with fetid smell, ii) loss of appetite, and iii) loss of water from body followed by emaciation and weakness.

2.1.6.2 Remedies

In bacterial enteritis, antibiotic (Sulphaguanidin, Thalazole, Gastina, Pesulin, Antrima, Kaltin, Dependal-M, Dysfar-M, Lomotil, Sifezole tabs) with supplemental treatment (like dextrose I/V) are necessary. As supportive measures, liver extract may be administered. In parasitic diarrhoea, periodical deworming should be done with potent anthelmintics like Albendazole, Mebendazole or Pyrental pamoate.

2.1.7 *Calf Scours*

It is also called white diarrhoea, white scours, dysentery, three day calf disease etc. Calf scour is a disease of new born calves caused by faulty feeding of putrid milk/colostrums during group feeding. It is common after flood. Over feeding of milk, contaminated straw/water is the predisposing cause. Common causative agents are different strains of *Escherichia coli* and *Clostridium perfringens*.

2.1.7.1 *Signs*

i) High rise of temperatures, dullness, weakness, anorexia and dehydration. ii) Sunken eyes, rough hair coat, tucked up abdomen with rapid respiration. iii) In acute cases, profuse yellowish/brownish diarrhoea is common. iv) In sub acute cases, rise of temperature is followed by scouring, anorexia and increased peristalsis.

2.1.7.2 *Remedies*

Dam's blood serum -500 ml S/C within 4-5 h after birth (if possible), sulfonamides, penicillin, streptomycin, sulfacarbolates, bismuth salt is recommended. Sanitation measures are the preventive factors of diseases.

2.1.8 *Liver Disorder*

Liver disorder at post flood situation may occur primarily due to Fascioliasis, Leptospirosis, secondary bacterial infections and poisoning. Liver is the functional reserve of body. Hence, signs of liver disease are not clinically manifested, with slight disorder. Clinical signs are seen when nearly 80% of liver is damaged.

2.1.8.1 *Signs*

i) Enlargement in the size with or without pain. ii) Chronic indigestion, anorexia, constipation and diarrhoea. iii) Dullness, clay- colored feces, nervous signs. iv) Ecterus (Jaundice) needs to be ruled out for hemolytic jaundice. Van den Bergh test will confirm type of jaundice. v) Hypoproteinemia and edema can be confirmed by liver function test.

2.1.8.2 *Remedies*

Identification of specific cause like liver fluke, leptospirosis or poisoning is necessary and specific drugs like antiparasitic, anthelmintics may be administered. Liver disease is a managemental disease and may be controlled by feeding of digestible carbohydrate, sugars, fat free and protein rich diet. Dextrose by I/V may be given whenever necessary. Amino acid methionine can be used as protein supplement. Liver tonic (like Livogen, Belamyl, Liviplex, Savaplex, Nutriliv etc) and vitamin A will help the recovery of affected animal. Administration of antibiotics with corticosteroids will help to control secondary bacterial infection and prevent fibrosis in diffuse liver disease.

2.1.9 *Fascioliasis and Distomiasis*

Internal parasitic infestation occurs in certain areas of water logging and it may flare up after severe flood followed by water logging in low land areas. Here intermediate host of the parasite is snail. Infection occurs after grazing on pasture contaminated by infective metacercariea present in grass blade.

2.1.9.1 *Signs*

Chronic diarrhoea, debility, black feces, anemia, submandibular oedema (bottle jaw), razor back and pot belly (ascitis). All these are the signs of hypoproteinemia that occur in such type of diseases.

2.1.9.2 *Remedies*

Specific anthelmentics should be given like i) Avlothane or Fairmethane or Fasciola dispersible powder @ 10-15 gm as a drench in large animals and 2-4 gm/animal in sheep and goat. ii) Hexanide one bolus per 100 kg body weight. iii) Other drugs like, Distodin, Flukin or Distonex, Fasinex, Zanil, Ranide, Tolzan-F, Trodax, Albendazole etc can be given in recommended dose.

2.2 Disease Incidence during Draught

During draught, the main problem in livestock rearing arises due to feed and forage scarcity. Disease incidences are related with weakness of animal from starvation as well as scarcity of drinking water. Some protozoan disease use to flare up during this period. Endoparasitic infection is very common during natural calamities which may be due to poor nutrition and contaminated feed and fodder. During draught, due to poor nutrition, internal parasite flares up and causes severe loss in production. Worms generally utilize nutrients from host animals and leading to low production of milk, meat or wool.

2.2.1. *Parasitic Infection*

Some of the common parasitic infections observed during drought are mentioned below:

2.2.1.1 *Anaplasmosis*

It is an infectious disease of cattle, sheep and goat caused by *Anaplasma marginale* and characterized by progressive anemia due to destruction of erythrocytes. This intra-erythrocytic protozoan transmitted principally through tick bites. Indian breeds of cattle are the carriers of this disease. Disease is severe in exotic and crossbred cattle. Predisposing factors are warm climate and tick population.

- *Signs*: i) Continuous high fever, rapid and progressive anaemia and ii) icterus with presence of anaplasma bodies within RBC.
- *Remedies*: a) Terramycin, Oxysteclin, Steclin, Achromycin @ 10 mg/kg b. wt may be given for 5-7 days. b) As a supportive therapy, Belamyl or Livogen @ 5-10 ml I/M may be given twice in a week. c) As prophylactic measures, control of tick through spraying or dipping is necessary.

2.2.1.2 *Babesiosis (Tropical red water of cattle)*

It is a tick transmitted disease of bovines and recovered animal used to remain carriers.

- *Signs*: i) High fever ii) Hemolytic anaemia iii) Hemoglobinuria with coffee colored urine and some times jaundice.
- *Remedies*: Trypan blue 1% solution (80-100 ml through slow I/V), Berenil (1.6 gm for 100 kg body weight by deep I/M), Babesan (1 ml in 5% solution per 50 kg S/C).

2.2.1.3 *Theileriasis*

It is a parasitic disease of exotic and crossbred cattle, caused by intraerythrocytic pleomporphic bodies, *Th. annulata* and *purva* transmitted by ticks. Indigenous cattle remain as carrier and serves as source of infection.

- *Signs*: i) Persistent high fever, ii) capricious appetite, iii) enlargement of superficial lymph nodes, and iv) progressive anaemia followed by secondary bacterial infections.
- *Remedies*: Terramycin, Steclin, Achromycin etc @ 5-10 mg/kg b. wt. daily for 6 days, combined with Berenil 4-5 gm, I/M. Other drugs are Menoctone @ 12.5 mg/kg b. wt. in 4 divided doses I/M. Antimalarial drug like Clexon-4 may be given as single dose @ 20 mg/kg, I/M. A cell culture vaccine is developed by NDDB and successfully been introduced in the field as "RAKSHAVAC-T".

2.2.1.4 *Coccidiosis*

It is mainly observed in poultry but during draught and flood it may affect the young calves (below 6 months of age) causing heavy mortality and morbidity. Predisposing factors are deficiency of Vit-A due to lack of green fodder and proper nutrients. *Eimeria necatrix* (intestinal) and *E. tenella* (caecal) are the causative agents which spread through feces and soil.

- *Signs*: Sudden onset of severe diarrhoea, haemorrhagic enteritis and bloody diarrhoea with mucous are common. Calves become emaciated and cachectic. Anaemia and dehydration are common. Mortality is high.
- *Remedies*: Sulpher drug is the specific medicine to control coccidiosis. Sulphamezathine, Diadin, Sulmet, Vetydine etc are the oral preparation of sulpha tab./bolus or for drenching. Dextrose saline should be administered in severe dehydration during diarrhoea in calves.

2.2.1.5 *Tape Worms*

These are common in calves and difficult to eradicate as compared to round worms.

- *Signs*: i) Chronic debility, ii) Diarrhoea, iii) Stunted growth, iv) Anemia, rough body coat as well as in-coordination in movement and v) pot bellied condition.
- *Remedies*: 1.Niltape 20% powder, in calves, sheep and goat (2.5 to 5 gm/day)

  2. Dicestal- 1 tab for each 3 kg b. wt. as a single dose.
  3. Niclosan or Niclex- 2 tab daily for 2 days.
  4. Taenil: 12-15 gm for calves, sheep and goat.

2.2.1.6 Round Worms

Round worm (Ascaris, Neo ascaris, and Toxocara sp.) are very common in cattle and buffalo. Buffalo calves are often suffers from Neo-ascaris infections. Vitamin A deficiency and hypoproteinemia are also associated with worm infestation.

- *Signs*: i) General unthrifitness, ii) Diarrhoea, iii) Stunted growth, iv) Pot belly, v) Starring body coat, vi) Grinding of teeth, and vii) In severe infections, convulsions and fits may be observed.
- *Remedies*: (Any of the followings)
  i. Phenovis for growing calves, 4-10 gm above 6 months of age as drench and repeat after 3 months.
  ii Piparizine adipate or citrate powder, calves- 8-10 gm repeat monthly interval for up to 6 months.
  iii. Piparizine liquid 45% soln. @ 4 ml/10kg b. wt. in calves.
  iv. Vermex @ 5 ml per 10 kg in calves.
  v. Banminth forte as broad spectrum anthelmintic, 200 mg tab. for calves @ one for 40 kg b. wt.
  vi. Panacure @ 5-10 mg per kg b. wt.
  vii. Wopell (a herbal product) @ 25 gm for each calve.

2.2.1.7 *Ectoparasites*

During hot and humid climate, ectoparasitic (ticks, lice and fleas) infestation will increase.

- *Signs*: i) Apart from constant irritation and itching, there is loss of hair, ii) ticks cause anemia, iii) fleas produce allergic dermatitis.
- *Remedies*: i) Dusting of affected animal can be done with 5% Gammexine powder, ii) Butox 1% solution. Malathion 50% w/v, 0.1% soln. can be sprayed over animal body and 2.0% over animal houses.

2.2.2 *Heat Stress*

To maintain constant body temperature, animal has to lose extra heat by adopting various physiological processes like vasodilatation, sweating, panting, evaporation etc. When the animal fails to do so, it suffers from hyperthermia and may die. Comfort zone is the temperature range within which no demand is made on the thermo-regulating mechanism of body; it varies between 30-60°F at low humidity. As the air temperature increases from 60°F, there is relatively little effect on the animal system until the critical temperature is reached. Animal will experience stress whenever the body temperature exceeds the critical temperature. The *critical temperature* for different categories of dairy animal is given below:

- Europian Cattle 80°F
- Zebu cattle 95°F
- Crossbred cattle 85-90°F
- Buffaloes 100°F

With the rise in environmental temperature, many of the normal body functions starts impairing. Hot air coupled with humidity aggravates stress by increasing heat load on the animal body. To maintain constant body temperature, body heat must loss the additional heat; this can be achieved by various ways like breathing at higher rate, increasing peripheral blood flow that helps heat loss from skin, eating less food, adopting different postural changes, avoiding sun radiation, reducing body movement and by sweating. Milk production decreases during heat stress especially when environmental temperature increases more than 35 degree Celsius. Growth of young animals is also reduced because of appetite for feed during summer. High temperature has also effect on reproduction system which leads to silent heat, anoestrus, low conception rate and embryonic deaths. Semen quality in male is affected due to high ambient temperature and libido is reduced. In extreme heat stress, the thermoregulatory mechanism fails and the animal die due to heat stroke.

2.2.3 *Malnutrition and Starvation*

Malnutrition is defined as the inadequate intake and / or mal-absorption of required nutrients whereas starvation is characterized by a lengthy and continuous deprivation of food. Malnutrition can also occur in an animal, which is getting food, but it is not able to ingest, digest, absorb, or utilize nutrients. Impact of inadequate food intake can be exacerbated during extreme climate like drought or flood when animal is subjected to physiological stress. At the time of malnutrition, animal uses energy sources (carbohydrate, protein, fat) which are stored in the body. At the initial stage of malnutrition, glycogen (carbohydrate) stores is rapidly exhausted and after that stored fat is utilized. Bone marrow is the last body store of fat to be used. During starvation, when glycogen and fat stores have been depleted, the only sources available are protein, the catabolism of which results in the development of ketosis. If this negative energy balance is not corrected, the animal will die.

2.2.3.1 *Signs*

The clinical signs related to malnutrition and starvation includes i) loss of skin texture ii) dull hair coat, iii) sunken eyes, iv) tucked-up abdomen, v) prominence of the bones of shoulders, ribs, vertebra and pelvis, vi) atrophy of muscle, and vii) decrease or absence of subcutaneous, perineal, pericardial and bone marrow fat.

2.2.3.2 *Remedies*

During natural calamities, malnutrition and starvation in a section of livestock is inevitable but adequate management practices should be taken to decrease its impact. Following administrative and relief care can be taken up.

- Supply of sufficient feed and fodder to the affected animal.
- Utilization of locally available non-conventional food materials to check hunger death.
- Utilization of other energy sources like molasses and urea-molasses-mineral blocks.
- Treatment of crop residues with alkali, urea, molasses, salt to improve their nutritional quality.
- Supplementation of salt and mineral mixture.
- Rotational grazing in available pasture or grazing field.

2.3 Toxicity (Poisoning)

Diseases may occur from various toxic agents that enter into the body through contaminated feed and fodder or through infectious organisms. Some feed and fodder may contain substances which have harmful effect on livestock. During draught and flood, animals are compelled to eat excess amount of those feed or fodder which can be life threatening.

2.3.1 *Cyanogenetic Glycosides (HCN Poisoning)*

Glycosides which yield hydrocyanic acid (HCN) upon hydrolysis are termed as cyanogenetic. The toxicity of the compound is caused solely by its HCN components acting as free molecules after hydrolysis. In natural condition, hydrolysis is brought by enzymatic action in plant or in animal body. HCN poisoning is common in ruminants due to ingestion of sorghum (Jowar) plant at certain stage. During draught, plant growth used to become stunted and leaf becomes wilted. Those plants contain higher concentration of HCN. New sprout which appears after recently harvested field of jowar cause poisoning when such field grazed by animal. Linseed and certain other cyanogenic plant (*Hiwar pods*) can also cause poisoning.

2.3.1.1 *Signs*

i) Severe dyspnea, ii) Restlessness, iii) Bloat, iv) Convulsions, v) Bright red eye with dilated pupils, vi) Death within 1-2 h of poisoning.

2.3.1.2. *Remedies*

Sod. Nitrate	3 gm
Sod. Thiosulphate	15 gm
Dist. Water	200 ml

Injection should be given through I/V route and repeat after one h. Sodium thiosulphate may be drenched @ 30-60 gm at hourly interval for 4-5 times. Calborogluconate and dextrose may be given as intensive supportive therapy.

2.3.2 *Mycotoxins*

Food that has been molded may contain toxic substances also. The most common toxins is aflatoxin, produced by *Aspergillus flavus*. Toxins are also produced by other fungus like *A. clavatus, Penicillin citrinum, P. rubrum* and *Stackybotrys atra*. The fungus grows on grains, groundnut, maize, jowar when they become moldy during continuous rain or flood.

2.3.3.1 *Signs*

Symptoms appear slowly if the toxic feed is consumed in small quantities over a long period. The animal suffers from a chronic indigestion and diarrhoea and shows nervous signs like in-co-ordination, blindness and grinding of teeth. Milk production decreases.

2.3.2.1 *Remedies*

i) Rumen evacuation if possible, ii) Administration of intravenous dextrose saline, calcium and liver tonics, iii) Removal of affected feed from storage.

2.3.3 *Bacterial Toxins*

When putrefied feed/foods are fed to animals, it causes poisoning. The common toxin producing organisms are bacteria like clostridia, salmonella, shigellea or staphylococci. Animal foods like boiled gluten, rice hulk or moistened concentrate generally putrefy quickly at high temperature during summer or during draught and feeding of such feed/food may cause poisoning in animals. During flood, bacterial toxin is produced in stored feed. Scarcity of food leads to competitive feeding of contaminated feed.

2.3.3.1 *Signs*

Sudden indigestion followed by salivation and diarrhoea. Animal becomes off fed and isolated from group. Muscular tremor and death occur in acute cases.

2.3.3.2 *Remedies*

Dextrose saline (I/V) and injection of antihistaminic should be given immediately followed by specific antibiotic therapy.

2.3.4 *Insecticides, Pesticides or Fungicides*

With the increased use of potent chemical additives, the possibilities of their misuse became high. This may occur in different ways:

i. Toxic materials if spilled on feedstuffs.
ii. Pesticides if left exposed near feed store.
iii. Indiscriminate use of insecticides during post flood situation to control flies and mosquitoes or use of other fungi and pesticides spray in post calamity period.

iv. Treated seed/grains when used as animal feed.
v. Errors may occur in the intentional addition of drugs etc.

2.3.4.1 *Signs*

i) Salivation, ii) Dyspnoea, iii) Diarrhoea, iv) Muscular stiffness and staggering, v) Constriction of pupils, tremors of head, and vi) Bloat, collapse and death.

2.3.4.2 *Remedies*

Atropine sulphate @ 0.25 mg/kg for cattle and 1.0 mg/kg body weight for sheep and goat. Dose should not be more than 50 mg. Half the dose should be given by slow I/V route and half by I/M route, repeat after 5-6 h and give supportive treatment by injecting calcium borogluconate through I/V route.

2.3.5 *Urea Poisoning*

Urea is the most common fertilizer extensively used by farmer. Accidental ingestion by animal may occur particularly during scarcity period. Urea is also used as cheap sources of protein supplement to animal. The concentration of urea in the straw/bhoosa should not be more than 2-3%. Feeding of urea treated straw may also lead to accidental higher intake if urea concentration in bhoosa exceeds the recommended level. During urea feeding, adequate proportion of carbohydrates like maize, molasses etc must be available in animal diet. Overdose of urea usually results in poisoning. 100 gm of urea can be poisonous for average size cattle if the animals are not accustomed.

2.3.5.1 *Signs*

i) Severe colicky pain, ii) In-coordination, tremors, iii) Bloat, dyspnoea, violent struggling and bellowing. The animal may die within 3-4 h. There is a strong smell of ammonia in the breathing air and death is due to respiratory arrest.

2.3.5.2 *Remedies*

It is usually grave. Orally weak acid drenching may be effective in early stage. Large volume of continuous I/V dextrose saline drip may help to recovery.

2.3.6 *Oxalate Poisoning*

Oxalate in the form of potassium oxalate is present in some plants and grasses. During flood or draught animal compelled to graze on unwanted grassland or whatever is available nearby. *Aspergillus niger* (black fungus) are capable of producing oxalates. If fodders infected with this fungus is stored under humid condition, it will act as potential sources of oxalates to livestock. Although there is fairly high limit of tolerance in animal but continuous feeding of infected fodder over a prolonged period can cause poisoning to animal. Sudden ingestion of highly infected fodder may produce acute poisoning in stressed animal.

2.3.6.1 *Signs*

i) Acute poisoning leads to muscular tremors, ii) Staggering, recumbancy and coma, iii) Sub-acute poisoning leads to muscular weakness, ruminal atony, bloat, incoordination of gait, hematuria, edematous swelling on perineum and around genitals. For confirmation, suspected fodder or grasses should be analyzed for oxalates.

2.3.6.2 *Remedies*

Intravenous infusion of 25% calcium solution may reduce the problem. Feed may be mixed with lime water. The fungus infected fodder, hay should be sun dried.

2.3.7 *Arsenic Poisoning*

After flood, mass dipping or spraying of animals in commonly done for control of ectoparasites. Further spraying of insecticides/weedicide having arsenic compound is also done in the field where animals go for grazing.

2.3.7.1 *Signs*

i) Severe gastroenteritis manifested by acute abdominal colic, ii) Grinding of teeth, iii) Vomiting even in cattle, fetid diarrhoea, iv) Cyanosis, weak pulse, shock and collapse.

2.3.8 *Nitrate and Nitrite Poisoning*

If plant grows on soils heavily fertilized with nitrates, the concentration of nitrate in the leaves and stems will be high. Feeding of such plant will cause poisoning. Some times animal drinks water from the field which is fertilized with ammonium nitrates.

2.3.8.1 *Signs*

i) Salivation, ii) Abdominal pain and anoxia, iii) Muscle tremors, iv) Staggering, vi) Subnormal temperature, vii) Acute gastroenteritis, viii) Secretion of dark red or coffee colored blood from natural orifice.

2.3.8.2 *Remedies*

The specific anti-dote is methylene blue (20-40 ml of 1% solution I/V). The drug may be repeated if required followed by intensive supportive therapy like injection of dextrose solution (I/V) and liver tonics (I/M).

2.3.9 *Strychnine Poisoning*

Strychnine based preparation is commonly used for killing rats in the store where relief materials are kept during calamities. Accidental mixing of the poison with animal feed may cause illness in animals.

2.3.9.1 *Signs*

i) Dyspnoea, ii) Cyanosis, iii) Severe tonic and colonic spasm with opisthotonus and iv) Death due to respiratory paralysis.

2.3.9.2 *Remedies*

Stomach evacuation needs to be done using emetics like Apomorphine 20 mg I/V. Give sedative to the effected animal (pentobarbital or intraval sodium 0.5 gm in 20 ml solution @ 30 mg/kg b. wt. I/V).

2.3.10 *Snake Bite*

During flood, poisonous snake bite is a common cause of livestock death. Animal used to take shelter on the upland areas where different poisonous snake also migrate from low land. Chances of accidental tramping on snake is very high in such situation.

2.3.10.1 *Signs*

Symptoms of snake biting depend on type and species of snakes. There will be i) fang marks on lower extremities, head and muzzle, ii) local swelling and pain, iii) excitement, salivation, hyperaesthesia, tetany, iv) recumbancy and paralysis. Death due to asphyxia occurs in 1-48 h in cattle.

2.3.10.2 *Remedies*

Polyvalent anti-venoum serum should always be in stock during calamities to save animals from snake bite.

2.4 Anti-Metabolic Components in Agro-Industrial Byproducts

Large numbers of non-conventional byproducts are usually utilized during scarcity period for animal feeding. These include rubber seed cake, salseed cake, prospis juliflora pods, tapioca starch waste, tamarind seed powder, tea waste, mango seed kernels etc. These are extensively used by feed industry @ 10-15% level and used to supply during scarcity. The total contribution of such non-conventional materials is 30-35% in the concentrate mixture of cattle and 15-20% in poultry meshes. While using agro-industrial byproducts on large scale, we should remember that many of them contain anti-metabolic factors that not only lower down the nutritive value but are also harmful for the health and may adversely affect the productive and reproductive efficiency. Anti-metabolic factors may be of natural inhibitory agents, artificial inhibitory agents or mycotoxins. Natural inhibitory agents are mainly alkaloids, glycosides and polyphenolic compounds. Some of these byproducts and their effects on animal body have been depicted in Table 1.

Table 1 Agro-industrial byproducts and their toxic effects

Name	Toxic factors	Actions
Neem seed cake (*Azadirachta indica*)	Alkaloid nimbin	Lowered protein digestibility thereby growth
Mahua seed cake (*Madhuca indica*)	Mowrin & Tanin	Reduce protein digestion in body
Datura seed (*Datura stromonium*) in feed	Alkaloid atropine	Poisonous and may cause death.
Sorghum fodder & seeds, Sudan grass, Johnson grass, Linseed cake, Rubber seed	Cyanide units is converted into glycoside which is further hydrolysed in the cake or Kusum cake rumen and converted into HCN.	Cyanide poisoning which may cause death of animals.
Castor cake (*Ricinus communis*)	Ricin	Poisonous
Mustard and Toria cake	Glucosinolates	Create goiter and some times affect growth in calves
Sweet clovers, Lucerne	Coumarin glycosides Phytoestrogen	Reduce the production of prothrombin in liver
Horse bean	Vicine glycoside	Damage RBC
Soybean cake, Barley, Guar seed, Lucerne	Tripsin inhibitors (Protease inhibitors)	Inhibit protein digestion, deficiency of essential amino acids
Sunflower seed cake	Lipoxidase	Oxidizing agents
Subabool leaves	Mimosine	Loss of hair/wool, reduce appetite, affects thyroid function.
Paddy straw	Selenium	Selenocystine & seleno methionine cause degnala
Khesari dal (*Lathyrus sativa*)	Lathyrogens	Lathyrism, neural damage
Linseed cake	Linamarin	Pyridoxine deficiency
Dry paddy (*Oryza sativa*) straw, water hyacinth	Oxalates	Disturb calcium metabolism and utilization
All types of babool seed and pods, non-conventional seed cakes, tea wastes etc.	Tannins	Hampers protein utilizations
Cotton seed cake	Gossipol	Reduce hemoglobin in blood, reduce absorption of Fe.
Ground nut cakes	Aflatoxins B_1, B_2, G_1 and G_2.	Anti-metabolic activities

2.4.1 *Remedies*

Over dose of alkaloids are poisonous to animal. The affected animal may be given specific anti dote. Common anti dotes are dextrose solution infusion, atropine sulphate, laxative, and supportive therapy of liver tonics. The feed and fodder contaminated with alkaloids should be withdrawn from animal diet.

2.5 Vitamin Deficiency

Vitamins are the essential substances necessary for normal body growth and functions. Pre-formed Vitamins or their precursors are present in feed and fodder. During natural calamities like draught and flood, scarcity of feed and fodder as well as damages of feed resources leads to deficiency of vitamins. There are no as such clinical symptoms of avitaminosis. The deficiency diseases are therefore called as “Responsive Diseases”.

2.5.1 *Vitamin-A Deficiency*

Hypovitaminosis-A occurs due to dietary deficiency of vitamin A or its precursor carotene. Feeding green only in advance pregnancy does not increase vitamin A levels in foetus but it is increased in colostrums which act as a source of vitamin A for new born calves. Vitamin A deficiency may cause chronic liver and intestinal diseases during draught or hot summer.

2.5.1.1 *Signs*

i) Night blindness, ii) Thick and cloudy cornea, iii) Stunted growth, iv) Reproductive problems, infertility, abortions and retained placenta.

2.5.1.2 *Remedies*

Injection of Prepalin forte or Vitacept may be given I/M. Vitablend AD_3 may be used orally. Daily dietary requirements are 30-40 i.u/kg body weight in calves and 70-80 i.u./kg body weight in pregnant and milking animals.

2.5.2 *Vitamin-D Deficiency*

Vitamin D, together with hormones calcitonin and parathormone helps in regulating calcium absorption and its metabolism. It is also known as “antirachitic factors” because of its role in prevention of rickets in small animals. Vitamin D is formed when ultraviolet solar radiation falls on the skin of animal or on the plant. During natural calamities proper sun dried hay may not be available that may cause vitamin D deficiency.

2.5.2.1 *Signs*

Weakness in growing animal, rickets in dog, parturient paresis in milking animals.

2.5.2.2 *Remedies*

Parental administration of calcium and phosphorus should be given along with administration of Vitamin D.

2.5.3 *Vitamin-E Deficiency*

Vitamin E is a fat soluble vitamin and exerts its main biological function in association with selenium. It prevents necrosis of liver and muscular dystrophy in animals. Vitamin E also enhances the reproductive efficiency in farm animals.

2.5.3.1 *Signs*

i) Nutritional muscular dystrophy in calves characterized by stiffness, weakness, trembling and inability to stand, ii) Muscle becomes hard, rubbery and swollen and iii) Animal may die suddenly.

2.5.3.2 *Remedies*

Injection of Vitamin E is given. Germinated wheat may be included in the feed of animals.

2.5.4 *Vitamin-B Deficiency*

In general, vitamin B deficiency does not occur in ruminants as rumen microbes are able to synthesize it. But during scarcity period, when green fodder are scarce, some type of vitamin (like Vitamin $B_{1,}$ thiamine) becomes deficient in animal body.

2.5.4.1 *Signs*

Thiamine deficiency causes neuromuscular symptoms such as ataxia, circling, convulsions, head pressings etc. Diarrhoea may also be present.

2.5.4.2 *Remedies*

Injection of Berin or Neurobion or Neuroxine is recommended.

3.0 LIVESTOCK MANAGEMENT AFTER FLOOD

- During rapid rise of water, animals should be driven out immediately. Animal generally resist leaving the shed, so care should be taken while moving them. Animals kept in loose can take their care during floods. Do not let them be trapped in low lying pens. Animal should be evacuated before flood water enters barns or in enclosed area. They should be kept on plains where flood water is seldom deeper than 3 or 4 ft.
- Hog houses, barns and chicken houses should be cleaned regularly. Buildings should be sprayed with a good disinfectant before animals occupy them again. Airy buildings will help quick drying. Debris from

animal sheds should be removed by scrubbing and disinfecting the walls, ceilings, floors, stanchions and other equipment. The milk house and equipments should be scrubbed with detergent and hot water. Dairy sanitizer may be used to sanitize milk equipments.

- If the farm building or grazing field is flooded, livestock needs extra care in shed as well as in grazing field. Clean the grazing field of harmful debris and cleans animal building as soon as possible. Block off the narrow passageways in the shed so that animal can not enter. A few heavy animals in a narrow passageway with dead end can be dangerous both for the animals and building. Storing of farm implements or machinery in vicinity of animal should be avoided.
- Care should be taken on flood related diseases, such as common injuries, fever, difficult breathing, muscle tremors, muscle swelling, swelling on neck, throat, chest or back. Animals must be kept in watch for any indication of infectious pneumonia, sneezing, oozing from nostrils or eyes, any lameness, injuries or foot rot, swelling on any body parts, isolating behavior of animals etc. Lameness and body swelling usually occur when cattle are crowded on wet ground where house fly or horn flies are abundant.
- The incidence of any disease needs to be reported to nearby veterinarian who will arrange for vaccination of the remaining healthy animals.
- Clean and uncontaminated water should be provided in the temporary shed. Animals should not be fed damaged or moldy hay unless it has been tested for mycotoxins.
- Feed or forages contaminated by chemicals or pesticides should not be used. Care should be taken that herbicide, pesticide and treated seeds are not accessible to livestock. Feed or forages should be stored in a high area where floodwater will not reach.
- Management of Pasture and Grazing Land: Water logging may ruin pasture or grazing field. Lack of adequate forage could force animals to eat poisonous plants. After flood, debris and other foreign objects like barb-wire or metallic pieces should be removed.
- Management of Dairy Herd: Milking animal especially higher yielder should be milked at regular interval to avoid body stress and damage in the mammary gland. If animals from different herd maintained separately, they should also be milked separately. If the feed supplies are limited, the largest portion of available feed should be given to the high producing animals and those in advance pregnancy. The parlor, animal shed and equipments in use should be clean. Milch animals should be watched for the sign of mastitis, udder injuries and insect problems. Mosquitoes and other pests may be abundant after flood and they not only annoy the animals but also carry diseases.

- Livestock damage due to electrocutions can be avoided by turning off the main electric switches during flood.

Table 2 Vaccination schedule in post flood and draught situation

Animal	Disease	Vaccine	Dose/route	Immunity
Cattle and buffalo	FMD	Tetravalent IVRI, BAIF and Heechst	-10 ml S/C	6 months. Booster dose during calamities.
		BAIF VAC.	5 ml, S/C	
		RAKSHA	3 ml, S/C	
	HS+BQ	Combined vaccine by BAIF	5 ml S/C	1 year. Booster dose during calamities.
	Anthrax	Spore vaccine	1 ml S/C	1 year
	Theileriasis	RAKSHA-T	3 ml S/C	1 year
Sheep	Enterotoxemia	BAIF, IVRI, Bio Med Vac	2.5 ml S/C, 2 dose in 3 week apart	1 year
	Sheep pox	Tissue culture vac	0.5 ml s/c	2 year
Goat	PPR	TCRP Vac	1 ml, S/C	1 year
	CCCP	IVRI Vac	0.2 ml, S/C	1 year
Swine	Swine fever	Freeze dried Tissue culture	1 ml, S/C	1 year

4.0 LIVESTOCK MANAGEMENT DURING DRAUGHT

Ruminants, owing to their larger body size need huge quantity of feed. Generally one adult ruminant needs feed (dry matter, DM) equivalent to 2-4% of its body weight. For meeting this requirements animal needs more fibrous roughage feed. It is also essential to give minimum quantity of concentrates along with roughages. However, roughage feeds are usually not available in adequate quantity during scarcity. Following practices help to mitigate the fodder shortage problem.

- Crop residues like wheat straw, paddy straw, pearl millet husk, mustard gotar should not be burnt in field and can be collected as scarcity ration.
- Village common grazing land should be developed by water harvesting technology and plantation programme.
- Boundary trees may be planted around field with sesbania, subabool and neem.
- Fodder bank may be set up near forest area to collect forest grasses and tree by products.
- India is rich in alternative forage resources that need proper tapping.
- Plantation in river beds to get lopping materials during draught.
- Drain water diversion for fodder production in semi urban areas.
- Community silage bank for silage making with surplus fodder.

- Grouping of animal as per their production, feed requirements and strategic culling thereafter.
- Strategic supplementation of minerals, vitamins and salt.
- Grazing at morning and evening to save animal metabolic energy.
- Provision of fodder enrichment with molasses and urea.
- Chaffing of fodder to reduce wastages.
- Rational use of available feed resources.

5.0 DISPOSAL OF CARCASSES

Natural calamities bring lot of damages in terms of losses of crops, national wealth and human as well as animal life. In severe calamities, intensity of losses and casualities increases when disposal of carcasses also becomes a national relief issues. During that times proper disposal of dead carcasses (the body or a part of a domestic animal or fowl that has died or has been killed otherwise than by being slaughtered for human or animal consumption) are necessary to maintain hygiene and sanitation in the surrounding and to reduce the chance of disease out break. There is different process of hygienic carcass disposal. It may vary with the number of bodies to be disposed. In individual cases, process may be simple burial but when animal losses are large, mass burial or burning is needed. General rules of carcass disposal are:

- Carcass must be disposed of as soon as reasonably possible *i.e.* 48-72h.
- Burying a carcass requires that the carcass be 5 ft above the water levels and covered with 3 ft of dirt.
- Incineration must be in an incinerator that is approved by the local municipalities.
- Transportation of carcass or discarded animal parts must be in the vehicle or containers that are leak proof and covered. The vehicle should have permits /license from local bodies.
- If poultry or rabbit carcass is received by fur farms, that need a permits.
- If carcass is to be used as pet food, it must pass an inspection by a veterinarian and must be processed under clean and sanitary conditions.
- Carcass piled up for further transportation must be away (at least 200 yards) from human dwellings. Carcass must be picked up within 72 h.

5.1 Disposal Through Burning/Incineration

When death is due to an infectious disease, burning of the carcass can prevent the spread of disease. When large number of carcasses are to be disposed particularly after severe flood, mass burning may be arranged in an isolated place away from localities and livestock rearing areas. In mass burning, care should be taken for the procurement of sufficient burning materials like dry wood, grasses, and firing agents. Burning pyre should be arranged on a higher platform or places. Wooden blocks should be arranged crosswise considering the direction and velocity of wind. Firing woods is stacked first with a height of 2 ft or more on which dead bodies may be arranged in mass or in individual which should again be covered with burning agents and dry wooden blocks and dry grasses. After setting on fire, complete burning of bodies are necessary, therefore, sufficient time should be given. After complete burning, washing of burning places is necessary with disinfectant solution like phenyl or sprinkling of limes.

5.2 Burial Method of Disposal

Burial is the obvious choice of carcass disposal during calamities considering low cost and traditional process. Topographical considerations are important, especially the dead animal's proximity to wells, streams and other places which could be contaminated due to leaching or run off. The diameter of the grave depends on the body size of animal. For adult cattle, it should be 6 ft × 6 ft and depth is also 6 ft. For mass grave, size should be calculated accordingly. Before burial, a layer of lime should be laid at the bottom of grave and on the top of carcass as well. This hastens decomposition and reduces the chance that predators will open the grave site. Some times thorny bushes may be used to cover the carcass inside to keep away the predators.

5.3. Composting

The heat and bacterial action of composting is a good way to break down carcass. Like incineration, this is a good way to destroy transmissible disease. The carcass can be buried in a big manure or bedding pile for one year. Another method is to cover the carcass with thick layer of straw and lime. Sawdust can be a substitute for the straw. Whatever may be the process, water proof topping should be provided to avoid any entry of rain water in the compost pit. Make sure that composting is done in an area that would not be flooded. Water from compost pit should not leach into wells or other water supply system. There may be some legal ramification to composting, so local legal process needs to be followed.

6.0 VETERINARY DISASTER MANAGEMENT PLAN

In many Indian rural families, animals have a position of provider of additional family income as well as companion and their conditions during disasters an important consideration because their "value" (not wholly economic) is recognized. However, the government institutions engaged in disaster management do not give much consideration to animals during times of disasters. This attitude has to change in this country, just as it is beginning to change worldwide. For instance, a joint emergency management programme has been initiated by the American Veterinary Medical Association in association with the American Red Cross because they have recognized that the fate of people and animals in a disaster is intertwined. They found that often people were reluctant to leave their animals in times of disasters, or landed up at human shelters with animals, often creating public health concerns. Considering India's large population of livestock and our dependence on them, we should think of a veterinary disaster management plan that will include the synchronization of the available human and material resources for addressing appropriately the impacts on animals during disasters. Some of the basic interventions that could take place are mentioned below:

A. Management During Disaster
   - Search and rescue of animal.
   - Provision of community based temporary shelter.
   - Emergency veterinary care.
   - Hygienic disposal of carcasses.
   - Vaccination of surviving animals against epidemic outbreaks.

B. Post-disaster Management
   - Monetary assistance to the poor farmers whose animals are lost or died.
   - Financial assistance to reconstruct the animal shed.
   - Rehabilitation of sick and abandoned animals.

Great deals of resources, both material and human are needed for the above works. Professional veterinary assistance could form part of the force, while the rest may be made up of para-veterinary teams and volunteers depending on the skill and technical expertise required. These supporting staff can help in the search and rescue operation and transportation of animals. They may require some training in animal handling and first aid and vaccination. Veterinarians and animal scientists could be directly involved in the following operations.

- Veterinary assistance/intervention (possibly operating at the first aid clinic or at the makeshift hospital at base camp).
- Management of rescued animals (shelter, fodder, and water provided, along with a record of animals rescued, identification).
- Advice on public health issues (disposal of carcasses, immunisation of humans and animals), in coordination with medical teams.

7.0 REFERENCES

Blood, D.C., Henderson, J.A and Radostits, O.M. Veterinary Medicine. The English Language Book Society and Bailliere Tindall. Red Lion Sqare, London

Merchant, I.A. and Packer, R.A. Veterinary Bacteriology and Virology (7th edn.).The Iowa State University Press, Ames, Iowa, USA.

Soulsby, E.J.L. Helminths, Arthropods and Protozoa of Domesticated Animals (7th edn.). Bailliere Tindall and Cassel Ltd, London.

http://www.ndmindia.nic.in/documents/document.html

http://www.ag.ndsu.edu/diaster/flood/livestockhealthafterflood.html

http://www.ag.ndsu.edu/disaster/flood.html

http://www.uwex.edu/ces/newsinfo/drought.pdf

13

DISASTER MANAGEMENT OF POULTRY IN FLOOD

R.P. Singh and A.B. Mandal

ABSTRACT

Like domestic animals, poultry also deserves important consideration in disasters including flood. The adverse impact of flood on poultry includes decrease in productivity and output of poultry products, income erosion, unemployment and relocation of poultry following migration of poultry farmers/keepers. Focus areas suggested include construction of poultry farms away from water-logging area, maintaining updated records of poultry pockets in flood prone areas, cage identification mark, flood-related kits for poultry, pre-monsoon mass vaccination as per endemicity of poultry diseases in such area, response plan for timely and effective evacuation of stranded chickens and their proper care at the emergency camp, re-unification with the owner and short-term rehabilitation measures for affected poultry farmers in terms of immediate relief. The affected farmers/entrepreneurs may be encouraged for backyard (free-range) or small rural (semi-intensive) systems of poultry production using dual purpose cross-breed chickens (like CARI Nirbheek, Hitcari and Upcari) and some selected native fowl- based improved chicken varieties (like Giriraja from UAS, Bangalore, Krishna-J from JNKVV, Jabalpur, Kuroiler from M/s. Kegg Farm, Gurgoan and also by KAU, Mannuthy, Kerala) which require less cost of rearing and reasonably good return on the investment. Restoration and strengthening of small scale rural poultry farming activities, utilizing cross-bred chicken varieties and rearing model especially developed for this purpose, exploring contract broiler farming, promoting duck farming and deriving benefits under Govt's Poultry Venture Capital Fund scheme to be implemented by the NABARD have been suggested for diversified and sustainable poultry production as long-term rehabilitation measures.

Central Avian Research Institute, Izatnagar (UP), India

1.0 INTRODUCTION

A disaster is an extreme incident that causes extensive damage, economic destruction, loss of human, livestock including poultry lives, other fauna and flora on a scale requiring extraordinary emergency response from agencies outside the affected community to deal with it. Like domestic animals, poultry also deserves important consideration in disasters including flood as the poultry production has emerged as the fastest growing sector of our livestock economy. With current annual output of about 56 billion eggs and 2.5 million chicken meat, with annual growth rate of 5 to 7% in layers and 10 to 12% in broilers over the last decade, India has emerged as the 3rd and 5th largest egg and poultry meat producer in the world, respectively. The spectacular growth of poultry production in India over almost last four decades is shown in Table 1. Poultry sub-sector contributes to nearly 0.8 % out of approximately 4% contribution of livestock sector to India's GDP. India has a mix of small, medium and large poultry farms. Genetically improved hybrid layers and broilers are exclusively used for commercial production whereas hardy but less productive indigenous (*desi*) fowls are reared in small number usually under scavenging or semi-intensive system by small and marginal farmers and agricultural labourers. Although the latter category of small rural poultry production units/ backyard poultry keeping units still contribute to nearly 25% of total output of eggs and poultry meat but these scattered units in villages suffer more during flood, causing adverse impact on the livelihood security, supplementary income and nutritious food to family of socially unprivileged sections of society.

Poultry includes all domesticated avian species of commercial importance like chicken, duck, turkey, quail, guinea fowl, ostrich, emu, geese and pheasant. Among them, chicken farming is much more popular and accounts for nearly 92% of total egg and poultry meat output in India,

Table 1 Selected indicators of India's poultry development (1970 -2020)

Indicators	1970	1980	1990	2000	2007	2020 (Projected)
Egg production (billion, no)	5	12	23	33	48	120
Egg availability (no.)/head/year	10	18	28	33	45	90
Broiler production (million, no)	4	30	190	700	1800	4000
Poultry meat output (million tons)	0.12	0.18	0.41	0.75	2.20	4.20
Poultry meat availability (kg) /head/year	0.22	0.27	0.50	0.70	2.00	3.10
Human population (million)	548	683	846	1000	1080	1340

followed by duck (7%) and the rest by other poultry species. Ducks are more popular in coastal states and NE region of our country. In view of more popularity of chicken farming, an attempt has been made here to present an account of different phases of flood disaster management practices *viz.* mitigation, preparedness, response and recovery including post-flood rehabilitation of affected people engaged in this venture for sustainable poultry production.

2.0 IMPACT OF FLOOD ON POULTRY

In times of crisis like flood, people worry not only about their family but also about the animals and / or poultry they rear. This calls for focus on the integration of livestock including poultry issues in the disaster planning processes by the State/Central Government's Disaster Management Division/Department in association with Veterinary / Animal Husbandry Department to address public concern for the safety of their animals and poultry in the time of crisis. The scenario prevailing in the wake of disaster for human and animals including poultry is by and large the same. With barns and fences destroyed, the direct impact of flood makes animals/poultry confused, restless, shelter less and they often get injured. Those who manage to survive need shelter, feed, water and health care immediately to protect them from further adverse impact of disaster including flood. Unlike water fowl (duck) and to some extent non-descript (*desi*) chicken, commercial hybrid chickens inherently lack the ability to swim for a long time or climb/ fly up nearby short trees during flood and get perished, unless evacuated in time. Livestock / poultry carcasses pose health hazard and have to be disposed off promptly to prevent epidemics and zoonosis. Rescued animals / poultry need to be treated for ailments like physical injury, vector-borne/respiratory and other diseases, gastro-intestinal parasitic infestation, etc. at composite livestock-cum-poultry camp set up for this purpose. Damage to veterinary hospital/ dispensaries makes it difficult to attend to ailing livestock/ poultry.

The indirect adverse impact of flood on poultry includes decrease in productivity and output of poultry products, income erosion, unemployment and relocation of poultry following migration of poultry farmers/keepers.

3.0 MANAGEMENT OF POULTRY DURING FLOOD

In view of the growing economic importance of poultry farming as a vehicle of employment generation, upliftment of rural economy and nutritious food, the four recognized phases of disaster management *viz.* mitigation, preparedness, response and recovery also apply to it and post-flood rehabilitation of their owners.

3.1 Mitigation

It is preventive measure of evolving disaster-specific hazard reduction strategy and developing awareness and understanding of the elements of hazards among stakeholders to minimize vulnerability of people to disastrous situation like floods. Avoiding water logging sites for construction of poultry houses/sheds is an important step in this direction. Poultry houses should have strong foundation and high plinth level to withstand flood among other considerations for hot and humid flood prone zones. Increasing the height of slatted floor serves twin purposes of reducing the vulnerability of chicken to flood water and their contact with their faeces is also reduced. Even chicken huts made up of wood and thatched roof or wire mesh fenced sheds shall be raised on elevated platform for backyard poultry farming for obvious reason. Planting deep-rooted small trees nearby backyard poultry sheds will help domestic (*desi*) fowls save themselves till rescued. In addition, poultry litter pits need to be emptied before the onset of monsoon to avoid surrounding water contamination. Taking appropriate engineering measures for prevention of breach of river embankment and/or regulating its water flow during monsoon is itself an institutional mechanism to mitigate flood.

3.2 Preparedness

Strategies for disaster preparedness include awareness of disaster that is most likely to occur at a particular time and location, identification of human and material resources, risk assessment, fixing roles and responsibility, coordination, communication, capacity building of response team so as to reach to a level for responding timely and effectively to specific emergency situation. Disaster-specific action plans must be prepared before a disaster strikes and should be flexible to actual disastrous situation. The community likely to be affected should be included, volunteers among them identified and trained, apart from involvement of NGO and other stakeholders as per need. Poultry should form a component of disaster preparedness plan for livestock by involving poultry specialists of the State Animal Husbandry Department, R & D institutions and other stakeholders. Following suggestions need consideration for a flood like situation.

- The state disaster management department shall maintain a district-wise database of poultry pockets and poultry population to assess evacuation needs in flood-prone areas.
- Poultry farmers in such area must ensure that in case of emergency their chickens have access to high ground to perch by keeping extra galvanized iron/wooden/bamboo cages, feed and feeding trough, clean drinking water from hand pump as wells often get contaminated during flooding. They should also keep durable and visible chicken

cage identification mark/ tags bearing their address with telephone number for restoration of rescued chickens to them.

- Depending on the endemicity, commercial poultry stocks should be immunized against Ranikhet, Marek's, infectious bursal and fowl pox diseases prior to the onset of monsoon in accordance with disease and vaccination calendar developed for this purpose. Domestic (*desi*) fowls are generally resistant to these diseases and may not be vaccinated.
- Poultry owners should be advised to evacuate along with their chicken stock during flood, unless their own life is threatened. Emergency response team would provide necessary assistance for the evacuation of stranded chickens, as per the actual situation.
- Flood related kits for poultry shall be prepared well in advance. The major items of the kit include updated list of poultry farms, flock size, location and telephone number of farmers. Cages, mash/ pellet feeds, feeding troughs, waterers, basic first aid kit, tools, detergent / sanitizer for multi-purpose livestock-cum-poultry camp, poultry vaccines, veterinary drugs, carcass disposal kits, worker's personal protection uniform kits, communication tools and vehicle for transport of rescued chicken in crates, etc. are the other important requirements.

3.3 Response and Recovery

The response and recovery action plan for deployment of personnel and equipment/ materials must be kept in readiness to address immediate priority needs resulting from a disaster. The accepted sequence of emergency evacuation is people first, then livestock including poultry and property. The following action plan will help ensure a safe, effective and timely response to search, rescue and recovery of marooned chickens during flood. It must be emphasized here that mere evacuation of stranded chickens from certain death is not enough but they need to be gently handled, taken to safer places in cages/ crates or to composite livestock camp to be kept separately away from other farm animals from bio-security viewpoint.

- The response time for the evacuation of commercial hybrid chickens is very short due to their inherent inability to swim for a long time, unlike duck. Hence, once they get submerged in flood water, they must be evacuated preferably within first 10 minutes of submergence as they can not survive in such situation for more than 15 minutes even in stagnant flood water. Likewise, indigenous (*desi*) chickens need to be evacuated within first 20 min of submergence in water.
- The rescued chicken showing signs of lethargy, depression, loss of appetite or injury need to be examined by the Veterinarian, treated, sheltered, and fed until re-united with their owner for which owner

will have the responsibility for fixing proper identification mark on cage prior to flooding.

- In case of shortage or non-availability of supplies like compounded balanced poultry feed at the composite livestock camp, they may be fed survival (holding) diet, using locally available feedstuffs like paddy grain, rice *kani* (broken rice), rice bran, beaten rice or crushed maize in combination with 5 to 7% mustard cake or some other locally available vegetable protein source at the emergency camp @ 90 to 100 g /layer or broiler/day. Clean water, preferably chlorinated (2 to 3 drops of chlorine bleach / lit) be provided *ad libitum*. As a general guideline, 20 lit drinking water / 100 chickens /day will be required. Floor space of about 1.5 sq. ft/ head shall be provided.
- Poultry, being more sensitive to fumes / smoke than most other animals, need protection against it at rescue and relief camp.
- Prompt disposal of dead/ decomposed carcasses by deep burial away from water body or by incineration must be carried out for restoring general hygiene and cleaning of environment by the Veterinary/ Animal Husbandry Department. Likewise, for proper disinfection of composite livestock-cum-poultry camp, bleaching powder, phenol and carbolic acid, etc. are obvious requirements.
- Arranging post-flood camps in flood affected areas is necessary to restore the normal health and production potential of poultry stock.

3.4 Rehabilitation

Disaster and development should go hand in hand. For short-term rehabilitation of affected poultry farmers / scavenging poultry keepers, they need to be compensated/ assisted promptly to revive their lost stock. Government norms for financial assistance exist for small and marginal farmers / agricultural labourers @ Rs.30/- per poultry bird subject to a ceiling of Rs.300/- per beneficiary household in case of notified natural calamity including flood. However, this relief will not be available to those availing such relief from any other government's scheme.

In the aftermath of flood, it is necessary to collect information on the number of poultry farmers affected, type and flock size maintained by them, physical and financial damage including percent mortality of chicken attributable to flood and cost involved in the restoration of farms, etc. to work out appropriate rehabilitation programme and associated action plan for implementation on priority basis.

In order to provide quick source of income from poultry farming, the farmers should be encouraged to rear broiler chickens. This apart, arrangement be made for exploring contract broiler farming wherein farmers get needed inputs like chicks, feed and medicine/ vaccination, etc.

from the poultry companies and the farmers are required to rear them as per company's guidelines under buy-back arrangement for which they get Rs.3 to 5 / broiler as return for rearing chicks. The major broiler integrators operating in India include M/s. Suguna, M/s. Pioneer and M/s. VHL in South India, M/s. Arambagh in West Bengal, M/s. Godrej Agro-vet in Maharashtra and M/s. Skylark in Haryana, etc.

For long-term rehabilitation, as already embodied in our national disaster management policy framework for linking it to sustainable development, the affected farmers/ entrepreneurs may be encouraged for backyard (free-range) or small rural (semi-intensive) systems of poultry production which require less cost of rearing and reasonably good return on the investment. For this, dual purpose cross-breed chickens like CARI *Nirbheek, Hitcari* and *Upcari* have been developed which resemble phenotypically similar to those of original native breeds *viz.* Aseel, Naked Neck and Frizzle fowl respectively. These cross-breds have double or triple egg production potential, increased egg weight and ability to retain desirable features like tropical adaptability, considerable resistance to common poultry diseases and to perform under sub-optimum conditions of feeding and management. These improved stocks have been field tested (Table 2) and supplied to a large number of beneficiaries in public and private sectors.

Table 2 CARI* developed chicken varieties for rural poultry production

Traits	**Nirbheek**		**Upcari**		**Hitcari**	
	Farm	**Field***	**Farm**	**Field***	**Farm**	**Field****
Body weight at 20 weeks (g)	1350	1290	1285	1184	1320	1280
Age at sexual maturity (days)	176	174	165	165	178	172
Annual egg production (no.)	198	166	220	182	200	174
Egg weight at 40 weeks (g)	54	53	60	58	61	59
Fertility (%)	88	–	90	–	92	–
Hatchability, FES (%)	81	–	84	–	81	–

* CARI – Central Avian Research Institute, Izatnagar, Bareilly (UP) –243122
** Tested on a large scale at Hisar, Bikaner, Anand, Hazaribagh

Similarly, two other chicken varieties, *Vanaraja* and *Grampriya* have also been developed at Project Directorate on Poultry (PDP), Hyderabad, field tested and propagated for rural poultry production. In addition, some other native fowl- based improved chicken varieties developed for promoting rural poultry production include *Giriraja* from UAS, Bangalore, *Krishna-J* from JNKVV, Jabalpur, *Kuroiler* from M/s. Kegg Farm, Gurgaon and also by KAU, Mannuthy, Kerala. These improved varieties are also well adapted to tropical/ sub-tropical climatic conditions and are adopted by a large number of poultry keepers in rural areas.

The cost-benefit analysis of rearing CARI, Izatnagar developed improved chicken varieties under backyard (free-range scavenging) system is presented in Table 3.

Table 3 Economics of backyard poultry rearing per dozen dual purpose improved *Desi* Chicken

Rearing Cost #:	Rate (Rs.)	Amount (Rs.)
Non-recurring:		
12 Straight-run (6-week old) chicks*	25/chick	300
Recurring:		
Supplementary feed**		
for 6 male chicks for 14 weeks	0.25/chick/day	147
for 6 female chicks for 66 weeks	-do-	693
Medication	3/chick	36
Interest (@ 7%)		82
a) Total rearing cost		1,258
Income:		
Sale of 5 cocks at 20 weeks of age	200/cock	1000
Sale of 850 eggs (170 eggs/hen/year)	4/egg	3,400
Sale of 5 culled hens	150/hen	750
b) Total income		5,150
Cost : Benefit ratio		1: 4.1
Net income (a-b)		3,892
Average monthly income		324

* 10% mortality (5% each due to disease & predation),

** Imputed value for 30 g broken rice, coarse grain, kitchen waste/chicken/day,

\# Cost of kachha shed/labour cost ignored.

Of the total input cost, supplementary feed alone accounts for about 2/3rd of the rearing cost. Cocks are reared and sold at 20 weeks of age with an average live weight of 1.5 kg, whereas hens are reared up to one year of egg production. Since these chicken varieties scavenge during day time, they need only night shelter made up of mud/thatch or bamboo baskets. Labour cost is almost negligible, since family members tend them during morning and evening. The cost-benefit ratio worked to 1:4 with a monthly supplementary income of Rs.325/dozen chicken. They can lay 120 to 130 eggs in a year even without supplementary feed. It is advisable to keep not more than 25 chickens per household family under scavenging system of rearing.

Fig. 1 depicts a suggested model for popularizing backyard poultry rearing in rural areas. The R&D institution's responsibility is two-fold *i.e.* to develop improved germplasm and location-specific supplementary feed formulae to be supplied to district level Govt's poultry farms/NGOs/KVKs which will maintain the parent stock to hatch chicks. This intermediate level unit will have backward linkage with chick rearers at Block/Tehsil / Panchayat levels who will rear the chicks up to 6-8 weeks of age prior to selling them to a small scale poultry farmers, apart from providing tips on management and health care of the flock.

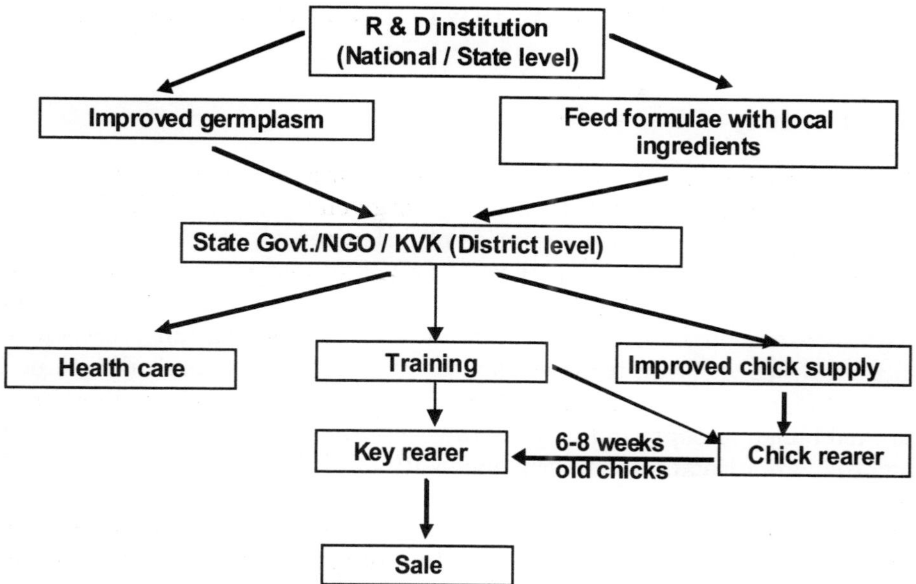

Fig. 1 Suggested model for backyard poultry rearing

3.5 Poultry Venture Capital Fund Scheme

This is a centrally sponsored scheme launched by the Department of Animal Husbandry, Dairying and Fisheries, Min. of Agriculture, G.O.I. to be implemented by the NABARD for promoting small scale (low input) rural poultry production. Under this scheme, interest-free loan up to 50% of the project cost is provided by the commercial banks for setting up small poultry farms in rural areas, apart from mini feed-mill, mini poultry dressing (300 chickens/day) unit and for purchasing egg –cum- chicken cart by the retail vendors.

Apart from chicken, ducks can be reared with or without swimming water facility and they are as good layers (>300 eggs/year) as commercial hybrid layer chickens, while meat type ducks become market-ready by 7 to 8 weeks of age with 2 to 2.5 kg live weight. Above all, being a water fowl they possess inherently good swimming ability and, thus, can be popularized in flood prone areas as a supplement / alternative to chicken farming. The CARI, Izatnagar and Central Poultry Development Organization (CPDO), Bangalore may be contacted for technical inputs, training and source of supply of farming inputs for duck rearing. The measures suggested above and taking advantage of incentives provided by the NABARD under its Poultry Venture Capital Fund scheme for promoting rural poultry production will go a long way in rehabilitating disaster affected poultry farmers and entrepreneurs.

4.0 FEEDING MANAGEMENT UNDER DISASTER OR SCARCITY

4.1 Feeding of Backyard Poultry

Native fowls and improved native fowls are reared mostly under backward or free range system, thus little attention is given to their feed requirement. The feeds under backyard comprise of fallen cereal grains, grass seeds, oilseeds, insects, earthworms, green tender leaves/grasses, small fishes, snails, aquatic weeds, etc. After returning back at evening, they are given with some amount of home grown concentrates like cereal grains, rice polish, rice bran, wheat bran, thrashed fishes/snails, kitchen waste/garbage, etc. depending upon the availability. During early age, the chicks or ducklings are fed boiled rice, ground rice, earthworms etc. After one to two weeks of age they are also allowed to scavenge with adult birds. The availability of feeds however is variable, and depends upon season, number of birds in an area and cropping pattern of that area. Generally, feed availability is more during rainy season and after harvesting of crops. Though, extra care is not required in feeding and nutrition when few birds are being reared under conventional backyard rearing system, but for intensified farming with sizable number of birds under scavenging/backyard rearing system, definitely require supplementary feeding to the extent of 30% of *ad lib* feeding level to augment their productivity. The nutritional requirements for optimum production performance of native birds or their crosses are given as below (Table 4). However, for holding purposes, the requirements can be reduced by 30%.

Table 4 Nutrient requirements of native birds and their crosses at different ages (week)

	Frizzle or Naked Neck X CARI–Red (HitCARI and UpCARI)				Aseel or kadaknath			
Age	**0–4**	**4–12**	**12–20**	**Laying**	**0–4**	**4–12**	**12–20**	**Laying**
CP, %	20.00	16.00	12.00	16.00	18.00	18.00	14.00	16.00
ME, (kcal/kg)	2600.00	2600.00	2600.00	2600.00	2700.00	2700.00	2500.00	2600.00
Lysine, %	0.92	0.75	0.56	0.85	0.85	0.85	0.65	0.85
Methionine, %	0.42	0.34	0.26	0.31	0.38	0.38	0.30	0.31
Threonine, %	0.75	0.63	0.47	0.63	0.70	0.70	0.55	0.63
Calcium, %	1.0	0.90	0.80	3.20	1.0	0.90	0.80	3.00
Av. Phos, %	0.40	0.35	0.32	0.30	0.40	0.35	0.32	0.30

4.2 Feeding Management of Commercial Broilers

The basic objective of broiler feeding is to achieve good growth and feed conversion efficiency, good skeletal health, survivability and quality meat at shorter period of time. Broiler grow at a rapid rate and reaches to about

2.0 to 2.2 kg at 6 weeks of age with an average feed conversion (FCR) of 1.8 kg on whole mash and 1.65 kg on pellet. The broilers are generally reared under three or four phases, *i.e.* 0 to 10/14 days (pre-starter phase), 10/14 to 21/24days (starter phase) and 21/24 days to 42/49 days of age (finishing period) or even up to 8 weeks of age to produce larger broilers (roasters). The finisher period may further be divided into 21/24 to 35 days (developer) and 35 days to 42/49/56 days (finisher). The large broilers are suitable for production of different value added products. The basic purpose of phasing out is to provide nutrients as per need depending upon the growth rate and type of growth, and thus to reduce feed cost. As the broilers grow at rapid rate, the concentration of all the nutrients in their diets is more. Increasing the dietary energy level causes increased gain and improved Food Conversion Ratio (FCR). The energy levels in broiler diet range from 2800 to 3200 kcal ME per kg diet. At low level of dietary energy, the broilers are unable to meet requirement even by increasing feed intake to compensate energy intake resulting poor growth and feed conversion. High-energy diet also leads to fat deposition in abdominal cavity, but the effect is more pronounced when energy to protein or limiting amino acids is wider. High-density diet adequately balanced in energy and protein improves gain and FCR but may not increase abdominal fat pad.

The concentrations of essential amino acids are more important than protein as such. The dietary protein levels for broilers should be in between 21.5 to 23% during pre-starter, 20-21% during starting and 18-20% during finishing period at energy level of 3000 to 3200 kcal ME per kg in both the phases. Calorie-protein ratio must be maintained for effective utilization of protein as well as amino acids. Ambient temperature and age at marketing of broilers determine calorie-protein ratio. The concentrations of amino acids especially of limiting amino acids are important. The first, second and third limiting amino acids in diets based on maize- soybean meal are methionine (or methionine plus cysteine), lysine and threonine, respectively; but in the groundnut meal-fish meal- maize based diets these are lysine, methionine, arginine and tryptophan in order of deficiency.

The requirements of vitamins and minerals of broiler chickens are also higher to support rapid growth including mineral growth, and metabolism of nutrients. Vitamin A, riboflavin and vitamin D3 are the most limiting vitamins in practical broiler diets. Other water soluble vitamins are also added as per requirements, however, their dietary supply may be reduced to half after 3 weeks of age when reared on litter (broilers are generally reared on litter). The requirements of calcium is 0.9 to 1.2% of total ration, while available phosphorus requirements are 0.45-0.50% during starting period, 0.40-0.45% during 21 to 42 days of age and 0.35% to 0.40% thereafter. The cost of vitamins and minerals are about 5 to 8% of total cost.

Certain feed additives or growth promoters *viz.* antibiotics, coccidiostats, enzymes, probiotics, antioxidants etc. can be added in diets as per recommended dose depending upon need. Under commercial environment, the birds are generally exposed to infection. Antibiotic feed additives (zinc bacitracin, verginiamycin, aureomycin, etc. which are not used for therapeutic/treament purpose) not only prevent birds from subclinical infection, but also promotes growth and reduce stress. These additives are beneficial only when the broilers are kept under poor hygeinic conditions and with diets that are marginally deficient in micronutrients. The antibiotic feed additives must be withdrawn from diet one week prior to sacrifice of the birds. A coccidiostat (amprolium, bifuran, neftin, madhuramycin, monensin and many more anticoccidial feed additives are available in market) is must in diet especially when the litter is moist. Enzymes may be added in high fibre diets but their effect is always not consistent. Table 5 presents the nutrient requirements of commercial broilers under Indian conditions. Under scarcity, the requirements can be reduced by 30% for holding.

Table 5 The nutrient requirements of commercial white and coloured broilers (on dry matter basis)

Nutrients	Unit	BIS, 1992		Suggested for tropical condition*		
		Starter	Finisher	0–3wk	3–6wk	6–9wk
ME	kcal/kg	2800	2900	2950	3050	3100
Crude protein	%	23.0	20.0	22.0	19.5	18.7
Lysine, min.	%	1.20	1.00	1.20	1.05	0.98
Methionine, min	%	0.50	0.35	0.48	0.42	0.39
Meth.+Cystine	%	0.9	0.7	0.86	0.76	0.70
Calcium, min.	%	1.2	1.2	1.0	0.90	0.80
Av. Phos. , min	%	0.5	0.5	0.45	0.40	0.35
Manganese	mg	90.0	90.0	50	50	50
Zinc	mg	60.0	60.0	50	50	50
Copper	mg	12.0	12.0	10	10	10
Vitamin A	IU	6000	6000	5000	5000	5000
Vitamin D^3	IU	600	600	400	400	400
Riboflavin	mg	6.0	6.0	6.0	6.0	6.0
Pyridoxine	mg	5.0	5.0	5.0	5.0	5.0
Pantothenic acid	mg	15	15	15	15	15
Nicotinic acid	mg	40	40	40	40	40
Choline	mg	1400	1000	1400	1000	800

* Suggested based on reports from CARI, Izatnagar and elsewhere in India.

4.3 Feeding Management of Layers

The laying hens for production of table eggs (for consumption of human being) are called layers; while for production of settable eggs (for hatching) are called breeders. The body weight and nutrient requirements are almost similar for both the groups except certain micronutrients. The replacement pullets are generally reared in three phases *viz.* starter (0-8 weeks of age), grower (8-20 weeks of age) and layer (20 weeks or above). In certain countries a five phases feeding system has been recommended *i.e.* starter (0-6 weeks), grower-I (6-12 weeks), grower-II (12-18 weeks), grower-III (18 to 1st egg) and layer (20 weeks of age and above). The major objective is to raise them at good health to attain sexual maturity at a given weight depending upon genetic make up of the birds with minimum nutritional input. Modern layers reach to the sexual maturity (*i.e.* the point of laying egg) at little earlier age than ever before and thus the improved modern birds have been able to produce 300-310 eggs per year. However, reaching to sexual maturity at an earlier age is always associated with smaller size of eggs. A low protein and/or low energy diet, or quantitative restriction of a balanced feed delays sexual maturity. A larger egg size is obtained from delayed sexually matured birds. The requirements of nutrients of layer birds are given in Table 6. In laying hens the dietary energy concentration ranges from 2550 to 2700 kcal ME/kg. On an average one laying hen producing 90% of egg laying capacity requires 16-18 gram of protein and 285 to 290 kcal ME per day. The protein and amino acid profile in diet are important for optimum growth, feed conversion, egg size and total egg production.

Meeting calcium requirement (0.9 to 0.7% of total ration) is important during overall growth and laying phases, but it becomes most crucial during

Table 6 Nutrient requirements (% or unit/kg feed, DM basis) of egg type pullets (BIS, 1992)

Nutrients	Unit	Growth phases of Layers			Breeder
		Chicks 0–8wk	Grower 8–20wk	Layer 20wk>	20wk>
ME	kcal	2600	2500	2600	2600
Crude protein	%	20.0	16.0	18.0	18.0
Lysine	%	0.9	0.6	0.65	0.65
Methionine	%	0.3	0.25	0.3	0.3
Calcium	%	1.0	1.0	3.0	3.0
Available P	%	0.5	0.5	0.5	0.5
Manganese	mg	90.0	50.0	55.0	90.0
Zinc	mg	60.0	50.0	75.0	100.0
Copper	mg	12.0	9.0	9.0	12.0
Vitamin A	IU	6000	6000	8000	8000
Vitamin D_3	IU	600	600	1200	1200
Thiamine	mg	5.0	3.0	3.0	3.0
Riboflavin	mg	6.0	5.0	5.0	8.0
Pyridoxine	mg	5.0	5.0	5.0	8.0
Pantothenic acid	mg	15	15	15	15
Nicotinic acid	mg	40	15	15	15
Vitamin B_{12}	mg	0.015	0.01	0.01	0.01
Choline	mg	1300	900	800	800

laying stage. The requirement of calcium during laying is about 3.0 to 4.5g/bird/day in diet. Just prior to initiation of egg production, huge amount of calcium is stored in bones, which is sufficient for 6 to 30 eggs. Therefore, calcium concentration is increased to about 2% of diet a week before onset of egg production. Once egg lying is started the birds adjust calcium requirements from dietary intake and also drawing from body reserve. The specific gravity of eggs shell increases with the increase of calcium concentration in diets. The White Leghorn hens producing 90% of egg laying capacity require about 4.5 g of calcium daily. Half of the calcium in diet should be supplied through coarsely ground powder and the remaining half as grit so that palatability of feed is maintained and hens may select and consume calcium as per their need. Therefore, in addition of dietary calcium (3 to 3.5%), there should be continuous access to shell grit to the birds. Shell grits also help in grinding of feed materials in gizzard and thus improves digestion and utilization of other nutrients. Addition of grit of appropriate sizes in mash at an interval of every two to four weeks may be beneficial. The retention of calcium during first 40 weeks is about 55% and decreased thereafter with age. Thus, requirement of calcium is increased with age. Heavy birds consume more feed, thus dietary concentration of calcium should be lesser. Any factor that affects feed intake like ME content of diet, environmental temperature, etc. will affect calcium concentration in diets. The available phosphorus requirements are 0.4% of total diet in starting, 0.35% during growing and 0.25 to 0.3% in laying phase.

Feed intake is an important factor in the economic production of eggs. Generally a White Leghorn hen consumes 8 kg of feed up to 20 weeks of age and 39 kg during 360 days (about 110g/bird/day or 130g/egg and producing 300 eggs per year) of laying phase. However, feed intake is

Table 7 Tentative feed intake of starting and growing pullets

Age (wk.)	Summer		Winter	
	Weekly	**Cumulative**	**Weekly**	**Cumulative**
1st	85	85	85	85
2nd	115	205	125	210
3rd	150	360	180	390
4th	190	550	230	620
5th	220	770	280	900
6th	280	1050	320	1220
7th	320	1370	360	1580
8th	370	1740	410	1990
9th	400	2140	425	2325
10th	420	2560	460	2785
11th –12th	900	3460	950	3735
13th –14th	950	4410	1050	4885
15th –16th	980	5390	1100	5985
17th –18th	1050	6440	1150	7085
19th –20th	1260	7700	1300	8385

governed by many factors such as dietary energy and protein concentration, environmental temperature and humidity, stress, vaccination, beak trimming and diseases etc. The tentative feed intake of starting and growing pullets is given at Table 7.

4.4 Selection of Feed Ingredients During Disasters

During disaster, selected feed ingredients are generally fed as a mixture to achieve the desired level of nutrients for holding. The inclusion level of certain commonly used for energy as well as protein feedstuffs / supplements are given at Tables 8 & 9 respectively. The inclusion rate of few miscellaneous alternate feed resources which can be used during scarcity period is mentioned at Table 10. If birds are to be hold for longer period say for more than a week, then must be supplemented with minerals (about 1 kg limestone/marble powder, 0.5 kg to 1 di-calcium phosphate, 0.3 kg common salt, 8 g of zinc oxide, 8 g of manganese sulphate and 3 g of copper sulphate) and the most critical vitamins (vitamin A, B_2, D_3, vitamin B-complex and choline chloride). Addition of a coccidiostat will be helpful in diets of birds below 10-12 weeks of age. The water consumption is about 2 to 2.5 times of feed intake. Due consideration must be given for water quality also. Water sanitizers (hydrogen peroxide-6%, 1-2 ml per every 10 litres of water) may also be used.

Table 8 Inclusion level (% w/w in diets) of certain energy feedstuffs/supplements

Ingredient	Inclusion rate*	Remarks
Maize, yellow	70	It is used as sole source of cereal
Maize, white	70	Not preferred due to lack of pigment but can be used during scarcity of yellow maize
Maize bran	10	High fibre, but can be included to increase the volume and for maintenance.
Hominy feed	10–20	Can also be fed as sole ration supplemented with minerals and vitamins during scarcity.
Maize grit	10	Can be fed as sole ration supplemented with minerals and vitamins during scarcity.
Wheat, grain	50	Contains arabinoxylans, new wheat should be avoided otherwise should be coarsely ground and mixed with bran to prevent choking,
Wheat, damaged	25	The inclusion rate or its sole feeding during scarcity depends on degree of damage.
Wheat bran	10	High fibre & costly in terms of nutrient availability as it is very low in available energy value.
Pearl millet (*bajra*)	30	Contains polyphenolic substances, generally not ground and may be used even upto 50% of diet.
Sorghum , red	15	Contains 1.8–2.2% tannin, low in oil, should be coarsely ground and may constitute 25% of diet

(Contd...)

(*Contd...*)

Ingredient	Inclusion rate*	Remarks
		during scarcity.
Sorghum, brown	30	Contains 0.5–0.6% tannin, low in oil, should be coarsely ground and may constitute 50% of diet during scarcity.
Sorghum, white	40	Contains 0.25–0.35% tannin, low in oil, coarsely ground, and may constitute 75% of diet during scarcity.
Finger millet	15	Grain is discharged as such through excreta, hence grinding is recommended.
Kodo millet	15	High fibre, generally can replace 25% of maize but can be used even at higher level in holding diets for various classes of poultry.
Fox–tail millet	30	Generally can replace 25% of maize but can be used even at higher level in holding diets for various classes of poultry.
Paddy, crushed	25	It is high in fibre and AIA. Generally can replace 25% of maize but can be used even at higher level in holding diets for various classes of poultry.
Rice, broken	40	Generally it contains less protein and thiamin, and quality of protein is also poor. It can form major portion of holding diet if it is supplemented with protein ingredients.
Rice kanki	20	Generally it contains less protein, quality of protein is also poor and deficient in thiamine. It can form major portion of holding diet if it is supplemented with protein ingredients.
Rice bran or polish	20	High fibre but contains good amount of oil and protein. It may be used in holding ration once it is supplemented with some cereals and protein supplements.
Rice bran, deoiled	20	High fibre, less palatable and increases dustiness of feed. . It may be used in holding ration once it is supplemented with some cereals and protein supplements.
Barley	20	High fibre, beta–glucan present, fibre injures mucosa
Oat	15	High fibre, beta–glucan present, fibre injures mucosa
Tapioca flour	5–10	Contains prussic acid
Molasses	10	High potassium, problem of diarrhoea
Animal fat	5	–
Vegetable oil	8	Including of dietary oil
Biscuits & bakery wastes	10	Should be free from mould growth
Bitten rice (*poha*)	10	–

Table 9 Inclusion level (% w/w in diets) of certain protein feedstuffs/supplements.

Ingredient	Inclusion rate	Remarks
Soybean meal, solv. ext.	40	Can be used as sole source of protein supplement, deficient in methionine
Soya seed	10	Can be added in ration after proper roasting and grinding.
Ground nut, deoiled	20–25	Aflatoxin problem especially during rainy seasons or on wetting. Otherwise it can be added as sole protein supplement in holding ration.
Ground nut, exp.	30	Aflatoxin problem especially during rainy seasons or on wetting. It can be added as sole protein supplement in holding ration.
Rapeseed/mustard meal, exp.	5–10	Contains tannins, erucic acid and goitrogen, however can be used with cereal mixture to have maintenance or holding ration.
Rapeseed/mustard meal, solv.ext.	10–15	Contains tannins and goitrogen, however can be used with cereal mixture to have maintenance or holding ration.
Sunflower meal, undec., solv.ext.	10	High fibre, presence of chlorogenic acid but may be added even upto 20% in holding ration.
Sunflower meal, dec., solv.ext.	20	High fibre, presence of chlorogenic acid but may be added even upto 20% in holding ration.
Safflower oil meal, dec.	20	High fibre, but may be added in holding ration.
Safflower oil meal, un–dec.	10	High fibre, but may be added in holding ration.
Sesame/*til* cake	15	With soybean meal form good protein and also available in many places. It can also form sole source of protein supplents.
Karanja cake (solv. extracted)	6–10	Karanjin
Kusum cake (un–decorticated)	10	High fibre
Linseed meal	3	Anti–nutritional factors
Cottonseed meal, delinted	10	Meal with low free gossypol (less than 0.04%)
Cottonseed meal, linted	5	Meal with low free gossypol (less than 0.04%)
Maize gluten meal	5	Rich in methionine, deficient in lysine and arginine
Maize germ meal	10	Rich in methionine, deficient in lysine and arginine
Mahua seed meal	3	Mowrin
Neem seed meal	3	Anti–nutritional factors
Coconut meal	5	High fibre
Ramtil/niger meal	8–10	
Guar meal, raw	5	Contains gum, trypsin inhibitor
Guar meal, toasted	10	Contains gum but may be used even upto 15%

(Contd...)

(Contd...)

Ingredient	Inclusion rate	Remarks
		in holding ration.
Peas, undersized	10	–
Palm kernal cake	5	High fibre
Fish meal	10	Problems of salt, urea and silica
Jawala fish	10	–
Fish scrap	4	–
Fish soluble	4	–
Meat meal	10	Must be sterilized, may be restricted up to 5%
Meat–cum–bone meal	5	Must be sterilized
Blood meal	3	Must be sterilized
Poultry byproduct meal	2–5	–
Poultry hatchery byproduct meal	3	Must be sterilized
Feather meal	2	Less digestible
Feather meal, hydrolyzed	5	–
Silk worm pupa meal	3	–
Shrimp meal	5	–
Skim milk powder	3	–
Tarula yeast	5	About 20% nitrogen is in form of nucleic acid
Yeast dried	4	About 20% nitrogen is in form of nucleic acid
Distillers grain, dried	10	–
Distillers soluble, dried	10	–
Penicillin mycelium waste	15	–
Earth worms	–	Free access with cereals or cereal mixture

Table 10 The inclusion rate (% w/w in diets) of miscellaneous alternate feed resources

Ingredient	Inclusion Rate	Remarks
Alfalfa meal	5–10	High fibre, can be given after chopping or in the form dried meal.
Berseem meal	5	High fibre, can be given after chopping or in the form dried meal.
Subabul leaf meal	3–5	Dried ground meal @3% in broiler and 5% in layer diet, mimosine is the toxic principle present in it.
Groundnut haulm meal	5	High fibre
Cassava meal	5	Prussic acid is present
Animal fat (tallow, lard)	3	–
Fish oil	2	Fishy smell in products
Vegetable oil	3	–
Dried poultry droppings (untreated)	5	Presence of pathogens
Dried poultry droppings (heat treated)	10	Uric acid is present

5.0 CONCLUSIONS

India, owing to its diverse geo-climatic conditions, is prone to different types of disaster including flood depending on the vulnerability of the area. It has experienced some devastating floods in the past including the Kosi deluge of Aug. 18, 2008 in Bihar. Being predominantly an agrarian society, India's livestock including poultry sector not only contributes about 6% to national GDP but also as a valuable source of livelihood security and supplementary income to the socially unprivileged section of society. There is, thus, growing realization of the need for integration of disaster management plans to address livestock including poultry related issues. An attempt has been made in this paper to present direct and indirect adverse impacts of flood on poultry, particularly chickens, flood related pro-active mitigation efforts, preparedness plan and response approach to minimize loss of poultry as well as short- and long- term strategies for rehabilitation of affected poultry farmers in the aftermath of flood. Focus areas suggested include construction of poultry farms away from water-logging area, maintaining updated records of poultry pockets in flood prone areas, cage identification mark, flood-related kits for poultry, pre-monsoon mass vaccination as per endemicity of poultry diseases in such area, response plan for timely and effective evacuation of stranded chickens and their proper care at the emergency camp, re-unification with the owner and short-term rehabilitation measures for affected poultry farmers in terms of immediate relief. Restoration and strengthening of small scale rural poultry farming activities, utilizing cross-bred chicken varieties and rearing model especially developed for this purpose, exploring contract broiler farming, promoting duck farming and deriving benefits under Govt's Poultry Venture Capital Fund scheme to be implemented by the NABARD have been suggested for diversified and sustainable poultry production as long-term rehabilitation measures.

6.0 REFERENCES

Banerjee, G.C.1991. A Text Book of Animal Husbandry. Oxford &IBH publishing Co.Pvt.Ltd., New Delhi.

Gopalakrishnan, C.A and Mohanlal, G.M.1985. Livestock and poultry enterprises for rural development. Vikas Publishing house Pvt., Pahari Dhiraj, Delhi-110006.

ICAR .1997. Hand Book of Animal Husbandry. Krishi Anusandhan Bhavan, New Delhi-110012.

Singh, R.A.1997. Poultry Production. Kalyani Publisher, New Delhi-110002.

http://statespcaorissa.com/website/activities/disaster-management.htm

http://www.ndmindia.nic.in/management/Revised norm2009.pdf

http://www.vitcdmm.org/journel_on_disasters.pdf

http://www.vmat.org/

14

POTENTIAL AND PROBLEMS OF NON-CONVENTIONAL FEED RESOURCES AND THE STRATEGIES FOR THEIR EFFICIENT UTILIZATION

K.K. Singhal

ABSTRACT

Three main categories of underutilized feeds are crop residues, agro-industrial by-products (like molasses, rice bran, pineapple waste, conventional oilcakes, guar meal, etc) and non-conventional feed resources (like bagasse, cassava, sweet potato, poultry litter, coconut cake etc.). Underutilization of non-conventional feed resources may be attributed to scattered production and low quality of collected materials, high cost of collection, difficult to process, poor palatability due to the presence of anti-nutritional factor (such as tannins, saponins, cynogens, alkaloids). Expanding the use of agro-industrial by-products and non-conventional feed resources in India should be given priority, not only as a feed resource, but as a raw material to convert it in to valuable product either through animal or through industrial application. However, urgency relates to continuing low animal performance, inadequate utilization of the available feed ingredients and poor efficiency of existing animal production systems. The strategy to correct this situation calls for more innovative technical application of the information on hand, backed by strong industrial support that would focus on use of more agro-industrial by-products and non-conventional feed resources innovatively in the feeding systems and for this purpose institutional, and resource support is required besides generating precise information about the regional imbalances in availability of feed resources. It will facilitate the management system of feed resources and provide a strong base for improving the dairying in different regions. An infra-structure may be created for the collection and processing of non conventional feed resources, which may be helpful in managing the natural

National Dairy Research institute, Karnal (Haryana), India

calamities as being experienced these days. Critical analysis and review of policy for export of protein supplements such as oil cakes and fish meal needs to be done in view of local demands.

1.0 INTRODUCTION

India is bestowed with large geographical area, diverse climatic conditions and huge livestock population of different species and breeds. In spite of being the largest milk producer and sizable production of meat and other animal byproducts, the productivity of Indian livestock is poor and the sole reason of this is their low level of nutrition because of dependence on the crop residues and agro-industrial byproducts as a source of nutrients. Natural and nutritious feeds such as concentrate ingredients, fodder and grasses are in short supply. Realizing these facts animal nutritionists are working from last several decades with a most challenging objective to enhance the utilization of available feed resources including the agro-industrial byproducts. The situation is extremely compelling due to the serious feed deficits that do not appear to be contained in the face of increasing human and animal population. Nutritional deficiency is affecting the animal production adversely and causing the hindrance in meeting the national targets for milk production. Therefore, there is need to think about the efficiency of existing animal production systems and how to change it by the efficient utilization of available feed resources.

While the gravity of the problems of utilization of agro-industrial byproducts have been reviewed extensively in the past, yet the situation has not changed significantly. The intent in this paper is to focus on the availability of feeds, possibilities of augmenting the new feed resources in view of their limitations and examine possible strategies that can encourage their expanded utilization. The discussion will include the likely impact of these strategies on potential increased contribution from the animal resources.

1.1. The Animal Resources

Animal resources in India and the average annual growth rate specific to each species are depicted in Table 1. Population of buffaloes and goats is increasing followed by those of sheep and cattle. It is significant to note that 465 million livestock heads distributed in approximately 6 million villages in the country provide livelihood security to over 100 million households of rural areas. The livestock are renewable resources and their function varied from food production (meat and milk) to provide draught power, transportation and manure besides fulfilling the religious and social needs. Unlike dairying, poultry production has taken a shape of industry in peri-urban areas of the country, managed by the educated and resource

Table 1 Current animal resources of India

Species	Population (Million)	As % of total world population	Annual growth rate (1997-2002) (%)
Cattle	176.60	13.00	0.5
Buffaloes	98.70	55.58	1.9
Goats	125.46	15.10	3.6
Sheep	64.27	5.91	0.2

FAO, 2007

rich entrepreneurs. Transfer of nutritional management technologies for dairying is not easy, which is still in the hands of landless and marginal farmers having poor resources and low education. In view of these facts, the incorporation of scientific innovations for improving the productivity of livestock is becoming the Herculean task.

1.2 Feed Resources

Three main categories of underutilized feeds are crop residues, agro-industrial by-products (AIBP) and non-conventional feed resources (NCFR). The crop residues are mainly fibrous materials that are the byproducts of crop cultivation, which form the main constituent of animal ration in all the animal production systems being followed in the country.

Agro-industrial byproducts (AIBP) refer more to byproducts derived in the industry dealing with the processing of the main products. These are comparatively less fibrous, and have higher nutritional status than crop residues. Good examples of AIBP are molasses, rice bran, pineapple waste, conventional oilcakes, guar meal etc.

Non-conventional feed resources (NCFR) have been identified separately from crop residues and AIBP although they can be components of both these feed categories. NCFR are those feeds that have not been traditionally used in animal feeding in spite of their sizable availability and nutritional attributes. These are generally having one or more anti-nutritional factors, which restrict their use in animal ration. Some NCFR can not be utilized as such, but these can be utilized for the production of nutritious feeds such as wastes from fruit and vegetable units and fermentation industry, single cell protein etc. Others can be obtained by the processing of slaughter house waste, and processing of forests products including the top feeds. These may be designated as new feed resource. Usually the byproducts from food processing industry are the organic end products of production and consumption of food items in the solid, liquid or slurry form that have not been used, recycled or salvaged and their economic value is often less than the cost of their collection and transportation. Some NCFR such as cassava and sweet potato are good sources of fermentable carbohydrates and have the potential for utilizing the inorganic nitrogen in the ruminant's diet. Forest's wastes, generally have deleterious effects on animals, and not enough is known about the nature of the active principles and ways of alleviating theirs effects.

Efforts have been made to work out the feed availability, its requirements and the deficit (Jain *et al.*, 1996; Ramachandra *et al.,* 2001). A data base on feed requirements and availability for livestock has been prepared by the scientists of National Institute of Animal Nutrition and Physiology, Bangalore in 2005. It was reported that dry fodder, green fodder and concentrate availability in the country was 526 million tons and the overall deficiency of feed was to the tune of 19%, which will continue to persist till 2020.

Recently, an exercise was made at the National Dairy Research Institute, Karnal to project the total availability and requirements of DM, protein and energy for Indian livestock in view the productivity trend of different crops and the ratio of agricultural products and their crop residues, current population of different categories of livestock and their growth rate, cattle feed production and the availability of industrial products. DM availability appeared to be adequate even the wastage of about 10 percent in handling; storage and transportation is taken in to account (Table 2). However, availability of protein and energy appeared to be deficient due to the poor nutritional value of feedstuffs. It appears that deficiency of energy and protein will persist in years to come but magnitude will be reduced in view of declining trend of livestock population with increasing agricultural production. In a broader perspective it may appear good but concern have

Table 2 Current status and projections of availability and requirement of nutrients (DM, CP and TDN) in India (× 10^6 million tons)

Feed resources	2005	2010	2015	2020
Green fodder & grasses	104.780	111.720	118.660	125.590
Crop residues	467.500	517.650	574.060	637.520
Oil cakes	19.290	21.650	24.320	27.310
Brans	7.320	8.240	9.270	10.430
Cereal grains	5.150	5.670	6.260	6.920
Top feeds	5.000	5.640	6.340	7.130
Vegetable & fruits waste	5.240	5.490	5.770	6.060
Other agro byproducts	31.260	32.830	34.460	36.190
Total DM availability	645.570	708.900	779.150	857.160
Total DM requirements	622.370	626.740	642.570	670.920
Surplus/deficiency (%)	+ 3.730	+13.110	+21.250	+27.760
Total CP availability	34.946	38.083	41.563	45.435
Total CP requirement	59.606	60.671	62.832	66.211
Surplus/deficiency (%)	– 41.370	– 37.230	– 33.850	– 31.380
Total TDN availability	268.360	298.560	332.590	370.960
Total TDN requirement	354.580	358.860	369.610	387.500
Surplus/deficiency (%)	– 24.320	– 16.800	– 10.020	– 4.320

Jain and Singhal (unpublished data)

been expressed about the regional imbalances, unpredictable weather conditions as being experienced this year and the control of private sector on the trading of animal feed resources, which may result in price hike of the feed resources including crop residues. Such phenomena is being experienced with no respite to the milk producers as the price of milk and animal products have not registered proportional increase, resulting a direct economic loss to the dairy farmers.

Crop residues (637 million tons), green fodder (125 million tons) and conventional and non-conventional concentrate ingredients (82 million tons) worth Rs. 2,34,800 crores at current prices are expected to be available in the country by the year 2020 and the contribution of agro-industrial byproducts continue to be at about 10%. To bridge the gap between supply and demand of nutrients, we have to increase the use of agro-industrial byproducts and NCFRs, whose potential, limitations and the methods of improvement have been discussed in details by NRC (1983), and Mudgal and Singhal (1993). They have considerable potential as feed materials, and for some, their value can be increased if there were economically justifiable technological means for converting them into some usable products in the areas of their availability. Recycling, reprocessing and utilization of NCFR (Table 3), offers the possibility of economic use.

Table 3 Nutritional value of some NCFR

Feeds	Moisture (%)	Crude protein (%)	Crude fiber (%)	Organic matter digestibility
Bagasse3.9–4.7	2.9–6.9	10.3–39.3	–	49.0
Brewer's grain	9.8–10.8	24.0–27.4	15.9–17.1	60.0
Coconut cake	10.0	18.0	12.0	78.0
Molasses	24.5	1.6	–	108
Pal kernel cake	5.7	14.2	20.2	66.8
Palm press fiber	78.0–89.0	9.6	11.5	30.8
Pineapple waste	13.8	4.0	36.4	76.0
Poultry litter	6.8	4.9	20.8	54.2
Rice bran	6.4	40.4–45.7	18.0–21.2	62.0
Wheat middlings	6.7–9.7	1.5–2.8	14.3–41.4	69.0–71.4

1.2.1 *Food Processing Wastes*

Currently, food processing industry is experiencing a fast growth in view of the consumers demand and changing life style of the people. Food processing wastes include the residual materials available after processing of foods which are unfit for human consumption. Such byproducts include conventional oil cakes, brans and fruits and vegetables residues, acid oil, spent grains, molasses, bakery waste etc. Conventional byproducts do not have any problem as they have ready market, but the unconventional byproducts, having the problem of bulk, moisture and/or palatability, which are not only causing the economic losses to the food processors, but also creating the pollution hazards. For example, fruits and vegetable wastes, having high moisture content and sizable sugar levels, are disposed off from the processing units to the secluded places, where they emit obnoxious smell as a result of putrefaction and ultimately polluting the rivers or water bodies. These wastes are usually high in energy and low in protein. Wastes from fruit and vegetable processing industry, starch industry, egg, fish, meat and meat processing industry and breweries have a sizable potential for animal feeding.

As per a recent estimate, the total availability of fruits and vegetable wastes in India is about 5.24 million tons on DM basis, which can be utilized as a feed resource in view of their nutritional value (Table 4) without any problem of palatability and anti-nutritional factor. In addition to the high moisture content, their seasonal availability is a constraint for their utilization for animal feeding. In view of their perishable nature, ensiling and/or drying can be employed to preserve their nutritional value. Alternatively, these may be preserved alone or in combination of crop residues employing the technology of ammoniation.

Table 4 Chemical composition and nutrient digestibility of wastes of different vegetable (% DM basis)

Waste	DM	OM	CP	NDF	Ca	P	NBDMD
Cabbage*	7.1	90.95	21.66	35.00	0.68	0.34	54.55
Cauliflower*	7.1	85.35	22.40	37.00	0.74	0.4‘	50.60
Carrots*	10.41	91.90	6.78	24.50	0.38	0.35	58.52
Pea pods*	13.34	95.40	15.40	45.50	1.35	0.28	50.98
Pumpkin*	8.20	96.10	14.20	47.00	0.66	0.16	51.22
Okra*	11.30	92.40	20.02	38.00	0.98	0.39	51.98
Potato peel	87.00	87.45	17.52	60.29	–	–	72.05
Maize bran	85.50	98.38	9.81	82.79	–	–	66.28
Spent grains	86.20	98.32	19.18	66.89	–	–	49.52

* Khan, 2003

Ensiling of such wastes alone or in combination with dry roughages is a good technique for their preservation and utilization for animal feeding. Ration containing silage of carrot waste consisting crown and tail portion with small, split and woody carrots (treated with 0.25% urea and 0.5% common salt) and pea pods consisting empty shells and vines can replace berseem hay in a total mixed ration of growing and lactating animals without sacrificing their productive efficiency.

Apples are processed for manufacturing many products for human consumption and the residual material, known as apple pomace (25% to 35% of fresh apple) contains 5.5% CP and 16.4% high quality fiber besides sizable level of minerals including calcium and phosphorus. Due to high palatability it can be fed as fresh or after drying. Its nutritive value in terms of DCP and TDN is 1.62, and 64.84%. Urea treatment of fresh apple pomace under air tight condition for 2 days followed by its drying increases its nitrogen content. On replacing the maize grain with urea treated apple pomace in the concentrate mixture, the growth and milk production performance of dairy animals remained uninfluenced but cost of feeding reduced significantly (Table 5).

Table 5 Performance of growing and lactating cows on feeding urea treated apple pomace (AP) in place of maize grain in concentrate mixture

Parameter	Control	Urea treated AP
Growth		
DM intake (kg)	483.60	483.40
Live wt gain (kg)	53.00	51.90
Daily wt gain (g)	541.60	529.80
Cost/kg gain (Rs)	13.80	10.75
Lactation		
DM Intake (kg)	6154.00	6268.00
Milk production (kg)	5137.00	5248.00
Total fat production (kg)	233.70	232.00
DM intake/kg milk yield (kg)	1.20	1.20
Feed cost/kg milk (Rs.)	1.25	1.04

Sharma *et al.*, 1989

Treating the mixture of apple pomace and wheat straw (1:1) along with 4% urea (on DM basis) for 30 days under air tight conditions not only preserves the apple pomace but also improves the nutritive value of wheat straw. Similar approach can be employed for the utilization of other high moisture containing food processing wastes such as maize bran (from starch industry), spent grain (from breweries).

Whey, a waste from milk processing industry, contains 0.7% to 0.9% CP, and 0.5% to 0.6% fat. It can also be used for feeding ruminants and swine as such or after drying, however, cost of drying is the major constraint in its utilization. Dried whey can be fed to young calves successfully; however, its use in poultry is restricted due to their lower lactose tolerance. Ghee residue is another byproduct from dairy industry, having high crude protein and energy value and it can be used by compounded feed industry.

1.2.2 *Other Industrial Wastes*

1.2.2.1 *Fermentation Industry*

Antibiotics and other therapeutic compounds are being produced using the fermentation process. India has made considerable progress in the area of antibiotic production. Currently the country consumes about 30 bulk antibiotics and produces 13 numbers by fermentation. The major basic manufacturing plants set up for the production of penicillin, ampicillin, amoxycillin, streptomycin, erythromycin, tetracyclin, oxytetracyclin, etc. Large quantities of certain antibiotics like rifampicin and chloramphenicol are being synthesized by fermentation with an investment of about Rs 1.20 billion. The installed capacity of penicillin crystals is likely to increase to about 6000-7000 MMU. There are large quantities of diverse nature of wastes from fermentation are available and currently these wastes are primarily burned, treated in waste treatment system or used as fertilizer. Most of the waste from antibiotic manufacturing is the fungal mycelia that are removed from the fermentation broth by filtration. The solid content of the waste is usually around 15%. Mycelia contains 20-50% protein and about 10-30% ash, which reveals its potential for livestock feeding. The palatability of such wastes may be low, but it can be improved by mixing it with other palatable ingredients. However, presence of residual antibiotic in such wastes may need an attention.

1.2.2.2 *Petroleum Refining Units*

Keeping in view the unique inherent capability of ruminants to utilize NPN sources as protein supplements, various NPN compounds *e.g.* urea, biuret were investigated in great details but their utilization have certain limitations such as non availability of energy sources like molasses, microbial adaptation period, and chances of toxicity. Biotechnology has offered yet another attractive alternative *i.e.* cultured biomass which is commonly referred to as single cell protein (SCP). Microorganisms including yeast have received much attention as a source of high quality protein because of their high rate of multiplication and their ability to utilize variety of renewable (bagasse, molasses, animal manure, corn steep liquor, sugar starch, cellulose, ethyl alcohol etc.) and non-renewable (methane, methanol, gas oil, n-paraffins etc) substrates. Besides carbon, some nitrogen source is also required for the growth and multiplication of microbes. Various nitrogen sources can be used for this purpose including the animal manure. Crude protein content of SCP depends upon the organism and substrate employed for its production; however, it varies from 55% to 60% in which sizable part is in the form of nucleic acids. Candida species grown on petroleum waste contains about 57% CP and 8% fat. The amino acid and mineral profile of SCP is comparable to the conventional protein supplements indicated its potential as protein supplement for livestock feeding.

Doubts had been raised about its feeding value in view of its higher nucleic acids and lipid fractions (practically negligible amount of triglycerides and more than 50% odd chain fatty acids, mainly of the 17 carbon atom series). In view of high nucleic acid content, safety evaluation of yeast grown on n-paraffins, indicated that feed conversion ratio in broilers and egg production efficiency in laying hens were comparable to the conventional diet. SCP can replaced milk powder or fish meal in milk replacer and high quality vegetable protein supplement (groundnut cake or soybean meal) in the ration of lactating animals without any adverse effect on the performance of animals in terms of growth or quantity and quality of milk as nucleic acids were found to be metabolized because of higher (5-50 times) pancreatic ribonuclease in ruminants. SCP was not only found highly digestible but also highly assimilable in ruminants. Economical production of SCP using renewable and non renewable sources is highly desirable in the present context of protein shortage for livestock.

1.2.2.3 *Waste from Edible Oil Refining Industry*

Acid oil, a byproduct of edible oil refining industry, has an immense potential for animal feeding as a source of energy as the crop residues based rations are generally deficient in energy, however, this valuable byproduct is being used for soap making industry. The acid oil is having high level of free fatty acids but the ruminants have the capacity to utilize them to fulfill their energy requirement. Care must be taken about the palatability while fixing its level in animal ration.

In view of increasing demand of oil for human consumption, cottonseed is being extracted. The oil extraction from decorticated cottonseed is efficient than from undecorticated seeds. Therefore, decortication technology is being adopted in the country; which is resulting increased availability of cottonseed hulls. Cottonseed hulls have low bulk density and high lignin content, but its palatability is not a problem. Several studies demonstrated its use as roughage in the ration of ruminants. Recently, it was demonstrated that cottonseed hulls can be used as roughage in complete feed having roughage to concentrate ratio 60:40. Such complete feed in mash form can be turned into flakes by passing it through an oil expeller. Such complete feed was found palatable and the performance of animals fed on cottonseed hulls based ration was similar to those fed on wheat straw based ration (Table 6). Higher bulk density than that in mash form make it easy and economical for its handling and transportation. Sunflower hulls and sunflower heads can also be used in similar fashion for the feeding of livestock.

Table 6 Effect of differently processed cottonseed hull (CSH) based complete feed on growth of crossbreed calves

Parameter	Wheat straw based complete feed	CSH based complete feed	
		Flaked	Mash
Initial body wt (kg)	126.80	112.90	115.95
Final body wt (kg)	156.25	183.201	179.95
Live wt. gain during 4 months	$29.45^{a} \pm 1.43$	$70.30^{b} \pm 3.36$	$64.00^{b} \pm 3.33$
DMI (kg/day)	$3.99^{a} \pm 0.38$	$5.20^{b} \pm 0.03$	$5.19^{b} \pm 0.02$
DMI/100 kg body wt	2.78^{a}	3.55^{b}	3.55^{b}
Feed eff. (DMI/kg body wt. gain)	15.61^{a}	8.49^{b}	9.24^{b}

a, b Figures bearing different superscripts in row differ significantly ($P<0.05$)

1.2.2.4 *Paper Industry Waste*

Various types of fines and screen rejects are generated from pulp mills and paper mills, which consist of about 60 percent water content. The *in vitro* DM digestibility varied 44 to 95 depending upon the fiber size and type of raw material used for making the pulp. Fines and screen rejects are low in protein and high in ADF with a sizable level of minerals due to their exposure to chemical processing during pulp making. Therefore, care must be taken for the presence of toxic minerals while feeding such waste to the animals.

1.2.2.5 *Shelling Industry*

Being one of the major producers of groundnut, cotton, rice, soybean etc., a large amount of hulls is available from their processing as shells/hulls constitute about one third of main produce. All these byproducts are rich in complex carbohydrates but the presence of lignin not only makes them poorly digestible but also unpalatable, particularly in case of rice hulls. These are generally being used as fuel and poultry litter. The chemical composition of these wastes indicates their poor nutritional status (Table 7).

Table 7 Nutritive value of shells of various products

NCFR	CP	CF	Hemi-cellulose	Lignin	Organic matter digestibility
Groundnut shell	6–7	66–70	7–8	11–30	–
Cottonseed hulls	3.0–12.4	37–48	16–23	16–27	–
Rice hulls	11.4–17.4	10.4–20.0	–	–	37.0
Soybean hulls	12.0	40.0	–	2.1	–
Coffee seed hulls	6.9	45.6	–	–	31.0
Cocoa pod husk	6.0	31.5	–	–	45.0

Being dry and fragile, groundnut shell can be easily ground and mixed with other ingredients. Several experiments have been indicated that 5–15% groundnut shell can be incorporated in the concentrate mixture without any adverse effect on the animal performance, provided their nutritional requirements are fulfilled with other dietary sources.

Different wastes are emanated from rice shelling such as rice polish, rice husk and broken rice. Fresh rice polish is used for oil extraction by solvent extraction industry and the deoiled rice bran is an important feed ingredient for the cattle and poultry feeds; however, rice hulls have very limited potential for animal feeding as its palatability is extremely low due to thorny surface and very high silica content (<20%). Alkali (NaOH) treatment of rice hulls improved its fiber utilization but it appears to be uneconomical.

Unlike groundnut shell and rice hulls, soybean hulls have low lignin content and its cell wall constituents vary in the range of 80–87%, however, its palatability is quite low. It can be ground and used as an ingredient of the concentrate mixture of dry or low producing cows.

1.2.2.6 *Sugar Industry*

Annual availability of sugarcane bagasse, a byproduct from sugar industry, is about 30.72 million tones annually (assuming 12% of sugarcane crushed by sugar mills) and it is mainly used as a source of fuel in sugar industry itself while partly it is utilized by paper and board industry. Sugarcane bagasse is a soft, white and moist (40–60%) byproduct and consists particles of outermost hard covering of the sugarcane and the pith portion. The composite bagasse in dry form is usually softer than wheat straw. It is not fed to the animals because of its low palatability. Palatability problem of sugarcane bagasse can be overcome by using it at lower level along with the supplementation of concentrate mixture. Experiments indicated that 40 percent of wheat straw in completes feed (R: C 60: 40) can be replaced by sugarcane bagasse. To facilitate the handling and transportation, complete feed was turned in to block form and feeding of both types of feed blocks resulted similar growth performance in crossbred calves (Table 8), indicating the scope of incorporation of bagasse in the ration of ruminants economically.

Table 8 Average body weight, DM intake, average daily gain, feed efficiency and cost of feeding of crossbred calves during experimental period (120 days)

Attributes	Straw based feed block	Feed block with sugarcane bagasse + wheat straw
Initial body weight (kg)	203.24 ± 8.07	200.54 ± 7.69
Final body weight (kg)	261.50 ± 9.40	262.07 ± 10.99
Total gain (kg), 120 days	58.26 ± 3.07	61.53 ± 5.80
Average daily gain (g/d)	485.46 ± 43.51	512.76 ± 35.81
DMI (feed block, kg/d)	5.82 ± 0.30	5.90 ± 0.31
Total DMI (kg/d)	6.21 ± 0.30	6.29 ± 0.31
Feed efficiency	12.79 ± 0.83	12.27 ± 0.80
Cost of feed block (Rs./kg)	3.85	3.54
Feed cost/day/animal (Rs.)	23.22 ± 1.01	21.69 ± 0.91
Feed cost/kg DMI (Rs.)**	3.74 ± 0.00	3.45 ± 0.00
Feed cost/kg gain (Rs.)*	47.82 ± 1.28	42.29 ± 2.09

* $P<0.05$; ** $P<0.01$

This brief account of possibilities of various NCFR from various industries for the feeding of animals indicate that such feed resources have tremendous potential, and we have to take proper initiatives for their utilization, incorporating the scientific innovation involving the compounded feed industry.

1.2.2.7 *Aquatic Plants*

India, having about 5700 km long coastal line, has a vast potential for aquatic feed resources. Red, black and brown algae are available to the tune of 10, 30 and 60 million tons in west coast of Gujarat alone. Aquatic plants are usually considered as a problem rather than an asset as they create hindrance in navigation and prove the breeding grounds for mosquitoes. Aquatic plants have sizable nutritional value but their harvesting and processing are the major constraints in their utilization. Water hyacinth is high in fiber and low in protein while algae are higher in minerals with sizable protein content (CP 7–11%). Use of algae in animal ration was envisaged in view of its high mineral content. Incorporation of dried brown algae (*Sargassum* sp) at 10% level in the ration of lactating Sahiwal cows proved economical as the milk production of such cows was found to be similar to those supplemented with mineral mixture. Another widely available aquatic plant water hyacinth has very high growth (88 kg DM/ha/day). Therefore, concerted efforts may be made for the harvesting and processing of aquatic plants for their utilization even as a source of minerals, as mineral deficiency is one of the impediments in improving the animal productivity.

1.2.2.8 *Forest Residues*

Forests are the source of large number of unconventional feed ingredients such as tree leaves, oilseeds (neem, mahua, karanj, kosum, sal, etc.) and grasses. These ingredients have sizable level of nutrients, but remain under utilized because of the presence of one or other incriminating factor. Seasonal availability, collection and processing of these feed resources are other constraints.

Tree foliage, easily available and accessible to the farmers, are the main feed resources for sheep and goats. Tree foliage can be utilized for dairy animals as these are highly digestibility biomass to provide deficient nutrients including minerals and vitamins and enhance microbial growth and digestion of cellulosic biomass in the rumen besides supplying the undegradable protein. Other advantages are the laxative effect and reduction on the dependence of concentrate mixture, thereby reducing the cost of feeding. Though their availability is seasonal, but it is recommended to use the tree foliage wherever and whenever these are available for the animal feeding. The more important tree forages are *Acacia* spp., *Ficus* spp., *Cassava*, *Erythrina* spp., *Gliricidia* spp., *Leucaena* spp., *Sesbania* spp., *Cajanus cajan*, *Grewia optiva*, *Oendrocalamus* spp., *Robinia pseudoacacia*, *Morus alba*, *Prosopis* spp. etc. Optimum dietary level on DM basis is 30–50% and as % of live weight is 0.9–1.5.

1.2.2.9 *Animal Waste*

Animals yield huge amount of different kinds of wastes and their utilization for the feeding of animals has been explored in details. The wastes include excreta from farm houses as well as slaughter house waste, which include rumen contents (partially digested fibrous feed of slaughtered animals), blood, meat, bone and offal. Animal excreta is low in energy, protein and mineral contents as the Indian livestock is already on low plane of nutrition and its utilization is limited not only due to the presence of pathogens but also considered unethical in Indian context as the animals are the source of milk. Poultry excreta generally have sizable nitrogen content besides energy and minerals and technologies have been developed to control the presence of pathogens by ensiling or drying the poultry waste prior to its mixing with feed. Ample evidences are available to show that properly processed poultry excreta can be incorporated in the ration of ruminants without affecting their productivity and quality of milk or meat. Utilization of poultry excreta as a feed resource for animals is not being adopted in India, in spite of shortage of nutrients, just because of ethical reasons. Feather meal, which can be manufactured by hydrolyzing the poultry feathers, can be good source of bypass protein, however, collection and processing of feathers are the major impediments.

Slaughter house wastes *i.e.* bone, meat, blood etc, are converted into meals for animal feeding, but as per Govt. of India circular in 2002, the byproducts of animal origin are banned in the ration of ruminants to control the 'Mad cow disease'. Therefore, these are not currently being used in cattle feeding. In view of high protein content in blood and meat meals and high calcium and phosphorous content in bone meal, these are being used in the ration of poultry, pigs and pet animals, but these should be decontaminated and processed thoroughly to make them free from pathogens, parasites and other harmful substances.

1.3 Current Constraints in Utilization of NCFR

Underutilization of NCFR may be attributed to the following reasons:

- Scattered production and the quality of collected material is not good enough for processing
- Collection of some NCFR is costly and not remunerative as compared to the conventional feed ingredients.
- Tendency to dispose off rather then the utilization
- Processing is difficult and in many cases it is problematic.
- Poor palatability due to the presence of anti-nutritional factor(s)

1.3.1 *Anti-nutritional Factors of NCFR*

Though NCFR's of forest origin, oil cakes of neem, mahua, karanj, sal, ambadi and rubber, and other wastes such as tapioca waste, tea waste, babool seed and pods, mango kernels, cassia tora seeds etc. are being utilized by compounded feed industry, mainly in their unprocessed form at low levels along with conventional feed ingredients including molasses, which masks not only masks the bitter taste but reduce the dustiness of NCFR.

Presence of anti-nutritional factors (ANF) such as tannins, saponins, cynogens, alkaloids etc in NCFR (Table 9), affects the animal performance due to the following reasons:

- Depression in food intake
- Interaction and formation of complex with dietary components, digestive enzymes and endogenous protein resulting in a drain on the nitrogen supply
- Injuring the alimentary tract
- Toxic effect after absorption

Table 9 Agro–industrial by products, their ava'lability (× 1000 million tons), nutritional value (%) and incriminating factors

Byproducts	Availability	DCP	TDN	Incriminating factor(s)
Ambadi cake*	0.02	–	–	Cycloprene acid, tannins
Babul pods	NA	5.82	62.63	Tannins
Castor cake*	737.00			Ricin (protein) and ricinine (alkaloid)
Cassia tora seeds	0.40	16.60	59.40	Crysophanic acid and tannins
Karanj seed cake*	8.24	26.00	71.40	Karanjin, gallabrin, tannins
Kosum seed cake*	0.32	10.00	50.00	Prussic acid, tannins
Mango kernal*	7.60	6.10	50.00	Tannins
Neem cake*	52.90	8.80	57.80	Nimbin
Prosopis juliflora pods	–	–	–	Tripsin inhibitor, haemaglutinine
Rubber seed cake*	0.17	18.60	54.00	Prussic acid, tannins
Subabul leaves/ seeds				Mimosine (amino acid)
Salseed meal*	50.50	–	–	Tannins
Tapioca waste	65.00	–	71.10	Prussic acid
Tamarind seed*	22.00	5.34	60.10	Tannins
Thumba seed cake*	79.00	–	–	unidentified

*Mehta, 2000

Though there are several anti-nutritional factors, present in animals feeds, but tannins, saponins are the major ones. Therefore, an important feature of these factors is presented here along with the mimosine, as the availability and scope of utilization of subabool (*Leucaena lucocephala*) is immense in the country.

1.3.1.1 *Tannins*

Majority of NCFR contain tannins in varied concentration as the main ANF. Tannins, the polyphenolic compounds, are categorized as hydrolysable and condensed tannins on the basis of their structure and reactivity. Hydrolysable tannins have low molecular weight and are cleaved under enzymatic or acid conditions to a monosaccharide. Condensed tannins have no carbohydrate core and are polymers of flavanoid units of varying composition and molecular weight. The tannin content of plants is affected by plant species, genotype and stage of growth, and may vary with plant part (leaf, stem, inflorescence, seed), season of growth and other specific environmental factors such as temperature, rainfall, cutting and defoliation by grazing herbivores including insects. Tannins not only bind strongly with proteins but also with many other compounds like polysaccharides, nucleic acids, steroids, alkaloids and saponins.

Because of protein-binding properties, tannins are known to be strongly astringent, which is the main reason of poor palatability of many NCFR. Binding the tannins with protein in an animal's gut, is believed to be responsible not only for growth depression, but also for low protein digestibility and increased fecal nitrogen concentrations. Condensed tannins are known to inhibit several digestive enzymes, including proteases, pectinases, amylases, cellulases, and lipases.

A major benefit of tannins in feed has been thought to be the protection of plant proteins from digestion in the rumen and their subsequent release as protein available for digestion and utilization by the ruminant. However, a review could not find any evidence which supports a view that tannins, even at low concentrations, improve the nutritive value of tropical legume forages and fodder trees. It thus seems feasible that tannins and other polyphenolics may also affect micro-organisms by adversely influencing their physiological functions, interfering with the uptake of minerals and other nutrients and with secretion, as has been demonstrated in animal and *in vitro*. The sequestration of minerals such as Cu, Co and Fe may influence the metabolic capability of micro-organisms as well as their ability to attach to their various substrates and secrete appropriately active enzyme complexes. The ability of PEG to reduce and alleviate the adverse effects on microbial function may be due to PEG: protein: tannin interactions as well as to reversal of the chelation of essential mineral elements.

1.3.1.2 *Saponins*

Saponins are glycosidic compounds composed of a steroid or triterpenoid sapogenin nucleus (aglycone) with one or more carbohydrate branches. Saponins are widely distributed in plants and are particularly important in temperate legume forages and are generally not a problem in tropical legumes. Saponins are bitter compounds and affect the palatability and feed intake, thereby depress growth in poultry and swine. Differences in biological effects of the saponins are associated with their chemical structures and the digestive process of the ingesting animal.

Saponins may have significant effects on all phases of animal metabolism, from the ingestion of feed to excretion of waste. The saponin rich extracts of *Yucca schildigera* have been of extensive interest in animal and poultry production, because of their ammonia binding properties. Saponins affect the protozoa adversely, however, it was demonstrated that the saponins from *Y. schidigera* were most effective than other sources of saponins. *Y. schidigera* extract, have antimicrobial properties, particularly in suppressing ciliate protozoa, which may prove beneficial to rumen fermentation and may lead indirectly to lower rumen ammonia concentrations as well as methane emission. Saponins present in lucerne fodder also suppressed methane emission by 20 percent in cattle and buffaloes maintained on wheat straw based total mixed ration. Therefore, incorporation of saponins into ruminant diets has a potential as additive to manipulate rumen fermentation for higher animal productivity and reduce environmental pollution.

1.3.1.3 *Mimosine*

This aromatic non-protein amino acid occurs in the foliage and seeds of the tropical legume subabool. Mimosine is widely regarded as an analogue of the physiologically important amino acid, tyrosine, and its neurotransmitter derivatives, dopamine and norepinephrine, found in the brain. Mimosine acts in animal tissues by interfering with cellular mitosis and the symptoms of toxicity are alopecia, reduced appetite and finally death. Metabolism of mimosine by ruminants depends upon geographical differences in rumen microbial ecology. During metabolism, 3-hydroxy-4(1H)-pyridone (3, 4-DHP) is synthesized by normal rumen organisms, which is further degraded by another microbe in adapted animals. This microbe is not present in ruminants that have not been exposed to leucaena.

Mimosine content of subabool leaves declined with maturity and it also decreased as the temperature of drying was increased. The induction of the ability of rumen organisms that totally detoxified leucaena by trans-inoculation of rumen fluid of an adapted to a non-adapted animal has raised the possibility of overcoming other problems of feed quality through specific microorganisms for dealing with tannins and phytotoxins.

1.4 Level of NCFR in the Feed

Most of the NCFRs contain some toxins and/or incriminating factors; therefore, their incorporation in animal feed results in poor palatability thereby poor productivity. Such byproducts may be included up to the level of 35-40% of the concentrate for low producing animals. To overcome the palatability problem, flavouring agents and sweeteners such as molasses may be added. It will be better to use the low levels of different NCFR in the ration along with molasses in the concentrate mixture followed by pelleting to avoid the selectivity and to check the palatability problem. To achieve the optimum performance of animals, it is recommended that dietary nitrogen level must be maintained at 30 g/kg digestible organic matter by the supplementation of additional nutrients so that N requirements of ruminal bacteria can be fulfilled. To raise the N status to the desirable level, haulms of potato and groundnut, tops of sugarcane and cassava as well as tree leaves (such as subabool) may be incorporated after their processing such as chopping. Alternatively, commercially available urea-molasses-mineral block lick may be used for this purpose. Whatever may be the approach, the acceptability of the ration should not be compromised. Additionally efforts should be made for the synchronized release of energy and N from the ration in the rumen to maximize the protein synthesis.

1.4.1 *Strategies for Utilizing the NCFR*

NCFR including agro-industrial byproducts usually have problems of either low palatability and/or the low shelf life and a common farmer have a difficulty in their optimum utilization. Though cattle feed industry is utilizing such feed ingredients by blending them with the conventional feeds but large number of ingredients remains underutilized because of their low, scattered and seasonal availability. Scientific data have been generated during last several decades about the nutritional value of available NCFR and processing methods for their better utilization. Therefore, it is desirable to use the information for the benefits of animals and the overall economy of the dairy farmers. In addition to this, National Dairy Development Board has recently launched a country wide programme for the ration balancing. In this programme, field workers will be educated to use the computer for balancing the ration, based on locally available feed resources. It is assumed that this strategy, appeared to gigantic, will succeed to improve the productivity by utilizing the NCFR to use them in a nutritionally balanced ration as majority of animals are fed on unbalanced ration. Another approach for enhancing the utilization of agro-industrial byproducts is to utilize them in urea-molasses-mineral block licks or in complete feed blocks. Of course, it needs the cooperation of feed industry as processing of the feeds is the only way to enhance their utilization in ruminants. This approach will not only reduce the cost of transportation and handling of feeds but will also be helpful in tackling the calamities such as drought and flood, which strike annually one place or other in the country.

1.4.1.1 *Urea Molasses Mineral Block (UMMB) Lick*

Urea and molasses can be fed as UMMB licks to supply the deficient nutrients. UMMB generally contains urea (10% to 15%), molasses (40% to 45%), mineral mixture (4% to 6%) along with binder (calcium oxide, bentonite or cement), and fillers to facilitate the dough making of desired shape and texture as well as to supply the nutrients such as preformed protein, bypass protein etc. to improve the rumen fermentation and increase the microbial protein yield, so that nutritional requirements al least up to the maintenance level can be fulfilled by licking the block @ 300 to 500 g/ day depending the size of the animal. The lick block generally supply 80-90% of total mineral requirement of animal and supplement the straw based diet with soluble sugars and performed proteins to improve (15-20%) the intake and utilization of straws. The UMMB, manufactured by hot or cold process, provides an excellent avenue for the incorporation of NCFRs as fillers. Normally, it is generally perceived that farmers don't have the know how to utilize the NCFRs in animal ration and they are unpalatable. These problems can be overcome by the use of UMMB as presence of molasses may take care of their palatability, presence of urea may reduce the effect of tannins and anthelmentics problem, which is widely prevalent and adversely affecting the animal performance. Additionally, plant secondary metabolites of NCFR's such as saponins and tannins may reduce the methane emission from the animals.

1.4.1.2 *Use of Complete Feed Blocks*

In the last decade, concept of complete feed block has been introduced at commercial level after through investigation. The complete feed block, containing roughages, tree foliages, molasses, and concentrate ingredients including urea and mineral mixture, not only provide the balanced nutrients to the animals but reduce the cost of transportation, handling and storage, avoid selective feeding thereby reduce the precious feed resources and above all improve the intake and utilization of low-grade cellulosic roughages. These blocks can be prepared in the surplus area and transported to the deficit regions to tackle the natural calamities, which is a regular feature in the country affecting the productivity of animals heavily. Technology of complete feed block also offers a scope to incorporate the NCFR's under the guidance of experts and the problems associated with these feed resources can be overcome in a manner mentioned under UMMB lick. Sugarcane bagasse and cottonseed hulls have been incorporated in complete feeds successfully.

2.0 FUTURE THRUST AREAS

One of the greatest opportunities arising from a better understanding of how ANFs limit nutritive value is the prospect that we may be able to make better use of NCFRs, which are now having limited value. It is possible that better understanding of the tannin structure may help explain how some low CT -containing feeds increase milk yield and have anthelminthic effects in animals. There may also be significant opportunities for developing microbial inoculates which might modify, inhibit or destroy ANF action in the rumen. The selection of fodder trees for low ANF content and high nutritive value is another obvious course for future action.

Technologies have been developed for the detoxification of some NFFR's such as oil cakes of manhua, neem, karanj, sal etc. However, these are not being utilized at the commercial level due to one reason or other.

Therefore, applicability and economic aspects of detoxification process of the technology must be looked in to. For example, process of gum extraction from guar seeds, makes it free from anti-trypsin activity and the cost of processing is recovered from the guar gum, which have industrial application. The resulting guar meal is free from its anti- nutritional factor. Similar approach may be applied for NCFRs to improve the availability of safe feeds for livestock. The ANFs present in unconventional oil cakes may prove a potential insecticide/nematodicide/anti-helmentic, etc. It will also be a good idea to incorporate neem cake or its active principle in urea-molasses-mineral block lick. Presence of molasses will mask its bitter taste, urea will take care the tannins and active principle may control the parasitic infection in animals.

Recent reports suggest that tannins of babool pods (*Acacia nilotica*) at the level of 3% of the total diet did not affect the intake and utilization of nutrients adversely on one hand and on the other hand cows fed on such diet excreted tannins metabolites in their milk, which have health benefits for human beings. In other words, dietary incorporation of tannins enhanced the nutraceutical value of milk and thus results in to value addition of the milk. Reports are appearing to show that plant extracts of various herbs have a potential to reduce the methane emission from ruminants, for which India is being blamed for global warming in view of its large animal population. Plants extracts contain some plant secondary metabolite(s), which control the protozoa population having a symbiotic relationship with methanogens in the rumen. If we could pinpoint the natural compound to reduce the methane emission from ruminants, it will be a great achievement.

Collection and processing of byproducts of forests origin, especially the non-conventional oil seeds have to be organized with an emphasis on their quality. During the process of their detoxification, pharmaceutical(s) should be isolated to recover the cost of processing.

The potential of NCFRs is enormous and to take best possible advantage of this enormous genetic resource, future research should be focused on:

- Identification of secondary plant compounds in top feeds.
- Studying the mechanisms of anti-nutritional factors to find out the ways of neutralizing their effects.
- Finding the application of secondary plant compounds as insecticide, pesticide, anti-helmentics, pharmaceutical or medicine
- Identification of microbes able to degrade ANFs under anaerobic conditions, better methods for analysis so as to correlate ANFs with biological activities.

3.0 CONCLUSIONS

Expanding the use of agro-industrial by-products and non-conventional feed resources in India should be given priority, not only as a feed resource, but as a raw material to convert it in to valuable product either through animal or through industrial application. However, urgency relates to continuing low animal performance, inadequate utilization of the available feed ingredients and poor efficiency of existing animal production systems. The strategy to correct this situation calls for more innovative technical application of the information on hand, backed by strong industrial support that can focus more particularly on the primary agro-industrial by products and non-conventional feed resources innovatively in the feeding systems and for this purpose institutional, and resource support is required besides generating precise information about the regional imbalances in availability of feed resources. It will facilitate the management system of feed resources and provide a strong base for improving the dairying in different regions. A infra-structure may be created for the collection and processing of non-conventional feed resources, which may be helpful in managing the natural calamities as being experienced these days. Research work has to be taken up to exploit the potential of wastes from food industry whose availability is likely to show a quantum jump. Efforts have to be made to recover pharmaceuticals from non-conventional feeds by devising appropriate technologies to obtain the raw material free from incriminating factors, and pharmaceuticals may be utilized for industrial and agricultural application as insecticide/pesticide/anti-helmentics and or mitigation of methane from ruminants. Critical analysis and review of policy for export of protein supplements such as oil cakes and fish meal needs to be done in view of local demands.

4.0 REFERENCES

FAO, 2007. FAO Year book. Food and Agricultural Organization, Rome.

Jain, D.K., Sharma, K.N.S., Walli, T.K. and Rai, S.N. 1996. Estimates of nutrients requirement and availability for bovine population across major states in India. NDRI Publication No 281, National Dairy Research Institute, Karnal.

Khan, I. 2003. Utilization of vegetable residues from agro processing plants in dairy rations, Ph.D. thesis, National Dairy Research Institute, Karnal, India.

Mehta, B.V. 2000. SEA Millennium Handbook on Indian Vegetable Oil Industry and Trade. The Solvent Extractors Association of India, Mumbai.

Mudgal, V.D. and Singhal, K.K. 1993. Pashu Poshan main Krishi avam Udyogon ke Upodpad, Rajasthan Hindi Granth Academy, Jaipur.

NRC, 1983. Underutilized Resources as Animal Feedstuffs. National Research Council, Washington, D.C.

Ramachandra, K.S., Anandan, S. and Raju, S.S. 2001. Livestock feed resources availability in India and need for a national database. Proc. X Animal Nutrition Conference, National dairy Research Institute, Karnal. pp. 134-144.

Sharma, D.D., Bakalkar, R.K. and Singhal, K.K. 1989. Apple pomace: a new feed resource for animals, Publ No. 149, National Dairy Research Institute, Karnal, India.

15

PLANT SECONDARY METABOLITES AS FEED ADDITIVES FOR DAIRY ANIMALS

S.K. TOMAR AND BHUPINDER SINGH

ABSTRACT

The use of antibiotics as feed additives in ruminant's diet has been banned in many countries and is likely to be banned in others due to risk of antibiotic residues in milk and meat, therefore attention has recently been shifted to natural herbal feed additives as rumen fermentation modulators. Saponins, tannins and essential oils are the group of secondary metabolites that have been widely investigated for their pro-nutritional activities in ruminants. In appropriate dose, saponins or saponin containing plants suppress protozoal population, increase bacteria and fungi population, propionate production, yield and efficiency of microbial protein synthesis, decrease methanogenesis and dietary protein degradation and improve growth performance in ruminants. Numerous studies have demonstrated that saponins and saponins containing plant extracts have toxic effects on protozoa. Some of the plant species showing this effect have been horsetail, Sesbania sesban, Acacia angustissima, Sapindus sp., Terminalia chebula, Populus tremuloides, Syzygium zromaticum, Psidium guayaba and extracts of fennel, clove and garlic. Essential oils have attracted attention for their potential as alternatives to feed antibiotics and growth promoters in livestock. Essential oils from a variety of sources have been shown to alter growth and metabolism of several types of bacteria, including rumen bacteria. Thymol is one of the major compounds of essential oil of thyme (Thymus vulgaris) and oregano (Origanum vulgaris). Limonene is the most abundant monocyclic monoterpene in lemons (Citrus limonum), oranges (Citrus aurantium), grapefruit (Citrus paradisi), peppermint (Mentha piperita) and spearmint (Mentha spicata). Adverse effects of tannins include inhibition of nutritionally and therapeutically relevant gastrointestinal

National Dairy Research Institute, Karnal (Haryana), India

microflora of the host through various mechanisms. Condensed tannins reduce the digestibility of proteins and carbohydrates in the rumen as the tannin-protein complex (TPC) formed is stable over the pH range of 3.5-7.0 but dissociates at pH <3.0 and >8.0. Tannin–protein interactions are pH dependent and the TPC formed in the rumen are hydrolyzed in the abomasums.

1.0 INTRODUCTION

Plant secondary metabolites (PSM) also known as phytochemicals, represent a diverse group of natural products some of which may be nutritionally valuable but many of which have no nutritional value. Secondary metabolites were originally considered as peripheral to the essential metabolism of the cell and often as merely waste products of metabolism. The role of secondary metabolites includes a) Protection against attack or the environment, b) Protection against UV radiation, c) Participation in allelopathy, and, d) Other plant-animal interactions. World Health Organization estimates show that about 70% of the world population use plants for medicinal purposes, with high usage mainly in Asia, South America and Africa. In the last few years, there has been an increasing interest in exploiting natural products as feed additives to solve problems in animal nutrition and livestock production. Bioactive constituents of plants are mostly secondary metabolites generated from the shikimate, acetate-malonate and acetate-mevalonate pathways. The chemically distinct classes of constituents are mainly terpenoids (sesquiterpenes, saponins, iridoids, carotenoids, steroids), phenolics (tannins, quinones, salicylates, lignins), and their glycosides (flavonoids, glucosinolates, cyanogens), alkaloids, polysaccharides (gums, mucilages), peptides and essential oils and resins, which often contain several of the above constituent classes.

Use of medicinal plants and their extracts in livestock production has a long history, as they were traditionally applied to promote animal health—a strategy now called “phytotherapy”. The increased interest in phytochemicals in animal diets has been prompted by the disapproval and decline in the use of ‘in feed’ antibiotics, the removal of animal proteins from the diet and thus the increased variety and inclusion levels of vegetable protein sources. The use of antibiotics as feed additives in dairy cows has been banned in many countries and is likely to be banned in others due to risk of antibiotic residues in milk and meat, therefore attention has recently been shifted to natural herbal feed additives as rumen fermentation modulators. Saponins, tannins and essential oils (EO) are the group of secondary metabolites that have been widely investigated for their pro-nutritional effects in ruminants. Plant extracts have been shown to stimulate the immune system, thereby enhancing the resistance of animals to inflammatory and infectious diseases.

2.0 FUNCTION OF PLANT SECONDARY METABOLITES (PSM)

The structures of many PSM have been shaped to interact with many different molecular and cellular targets, including enzymes, hormone receptors, neurotransmitter receptors and transmembrane transporters, and can thus mimic a response at the corresponding molecular target. There is hardly any cellular target that some PSM cannot modulate. In many cases PSM, *e.g.* the terpenoid essential oils, can be more effective than chemically-synthesised pure compounds because they are a complex mixture of components. Their complexity enables the PSM to interact with multiple molecular targets and, thus, it is more difficult for target microorganisms or herbivores to develop any effective response because resistance at different targets would be required.

There is considerable interaction between ingested PSM and tissues, enzymes and other compounds within the animal. The interaction during absorption, deposition and metabolism, and excretion are highly dependent on the physico-chemical attributes of the compounds involved and their susceptibility to transformation. When ingested the relevant PSM can pass through the animal unchanged or combine with bile salts and be excreted in the faeces. They can also interact with tissue and other compounds within the GIT, pass through and be excreted. The original compound can also be either absorbed directly or transformed within the GIT and then absorbed and deposited. If absorbed, the compounds can undergo further transformation, usually to increase hydrophilicity, and then excretion via the kidneys in the urine.

Due to the presence of anaerobic microbial fermentation in the rumen, the organic matter of feed is converted into volatile fatty acids, microbial mass and gases. However, the process is not very efficient as only 10% to 35% of energy intake through fibrous forages is captured as net energy because of poor digestion of structural carbohydrate. One of the major losses of energy by the ruminants is through the methane production amounting to 5% to 15% of ingested gross energy. Methane is one of the important green house gases that contribute to global warming. The generation of methane can be decreased by promoting shift in fermentation towards propionate production.

Protozoa can represent up to 50% of total rumen microbial biomass. It has been shown that defaunation reduces methanogenesis by 30% to 45% since protozoa are active hydrogen producers and are closely associated with methanogens. Plant secondary metabolites (PSM) at lower concentrations could be used to manipulate rumen fermentation favorably. In appropriate dose, saponins or saponin containing plants suppress protozoal population, increase bacteria and fungi population, propionate production, yield and efficiency of microbial protein synthesis (MPS), decrease methanogenesis and dietary protein degradation and improve growth performance in ruminants.

2.1 Saponins

Saponins are high MW glycosides in which sugars are linked to a triterpene or steroidal aglycone (sapogenin) moiety. The different classes of saponins are the triterpenes, the steroids and the steroid alkaloids. A number of factors, such as physiological age, environmental and organic factors, affect the saponins content of plants. Saponins have detergent or surfactant properties because they contain both water-soluble and fat-soluble components. They consist of a fat-soluble nucleus, having either a steroid or triterpenoid structure, with one or more side chains of water soluble carbohydrates. Saponins may have potential as natural ruminal defaunating agents. However, a complicating factor is that saponins are hydrolyzed by ruminal bacteria that remove the carbohydrate side chains. Because there may be an adaptation of ruminal bacteria for metabolism of saponins, one approach for retaining antiprotozoal activity would be to feed saponins intermittently.

The influence of saponins or saponins containing plants on DM, OM, nitrogen and fibre digestion has been shown to vary depending upon the nature of basal diet, dose, and source of saponins reported that dietary incorporation of *Yucca* extent (Sarsaponin, 44 ppm) increased ruminal degradation of OM, while degradation of ADF was not affected when added to diets with medium and low concentrate proportion. Nitrogen degradability decreased in presence of alfalfa saponins, ruminal digestibility of OM and NDF were inhibited. Supplementation of *Sapindus saponaria* had no effect on OM degradation but decreased the degradation of fiber fractions. Lower *in vitro* disappearance (24 h) of DM of hay plus concentrate in response to *Yucca* extract addition at high levels (1.2 to 3.2 g/l).

The sensitivity of ciliate protozoa towards saponins may be explained by the presence of steroids in protozoa, but not bacterial members. Thus

Fig. 1 *Yucca plant*

Fig. 2 *Quillaja saponaria*

the sterol binding capability most likely causes the destruction of cell membranes. Sarsaponin (0 to 77 ppm) from *Yucca schidigera* decreased protozoal numbers but not bacterial numbers in an *in vitro* ruminal system. Numerous studies have demonstrated that saponins and saponins containing plants have toxic effects on protozoa.

Saponins from *Quillaja* and *Acacia aurculoformis* and foliage from *Sesbania sesban* were anti-protozoal *in vitro. In vivo*, powdered *Y. schidigera* decreased rumen protozoal numbers in heifers. Methanolic extracts of alfalfa root (containing 28% saponins) reduced 92% protozoal numbers in sheep fed grass hay. The antiprotozoal effect was highest with hay plus concentrate as substrate followed by corn starch (32%) and potato starch (30%) at higher level of saponin.

Rumen protozoa ingest and digest bacteria and fungi degrading their cellular protein to ammonia. Hence, compensatory increase in bacterial and fungal zoospores population could be possible if saponins containing plants or extracts selectively suppress rumen protozoa. Addition of sarsaponin (*Yucca* extract) at 77 ppm to an *in vitro* ruminal system increased bacterial numbers significantly. Supplementation of *S. sesban* leaves (250 g/d) to sheep also increased the total and cellulolytic bacterial population by 70 and 80%, respectively after 2 days of supplementation confirming predation of bacteria by protozoa.

Reduced ammonia nitrogen (NH_3-N) concentration and increased molar proportion of propionate are typical in rumen when rumen protozoa are inhibited. Decreased NH_3-N in the rumen may be due to higher uptake of NH_3 for the synthesis of bacterial cells or decreased dietary protein degradation. Increased propionate is desirable since it acts as alternate hydrogen sink in the rumen thereby reducing methane (CH_4) formation. Decrease in branched chain fatty acids can occur due to inhibition of deamination of amino acids. Total volatile fatty acids (TVFAs), in general is considered as an index of extent of OM digestion.

Effect of intra ruminal administration of alfalfa root saponins in sheep at concentration of 0, 4 and 8% of DMI on the VFA concentrations was studied. On day 2 of saponins administration, there were linear increases in total acetic, propionic, butyric and valeric acid concentrations with increasing level of saponins. The branched chain acid showed linear decrease with increasing levels of saponins. Total, acetic and propionic acid concentrations did not differ on day 14, but butyric and valeric acid continued to show a linear increase in concentration with saponins dose. The concentration of isobutyric and isovaleric acid were linearly decreased in response to saponins.

The NH_3-N concentration significantly decreased and TVFA concentration increased as concentration of sarsaponin increased from 0 to 3.2 g/l. The molar proportion of propionate and butyrate significantly increased and acetate decreased in a dose dependent manner with corresponding decrease in A: P ratio. Dietary incorporation of some saponins containing extracts enhanced propionate production in rumen.

Some plant secondary compounds may be of interest to decrease methane produced in the rumen. Some of the plant species showing this effect have been horsetail, sage, *Sesbania sesban*, *Acacia angustissima*, *Sapindus* sp., *Terminalia chebula*, *Populus tremuloides*, *Syzygium zromaticum*, *Psidium guayaba* and extracts of fennel, clove and garlic. Some of the phytochemicals identified as responsible for this effect are thymol, eugenol, carvacrol, tannins or tea saponins.

As the concentration of sarsaponin increased in *in vitro* system the CH_4 production decreased, in a dose dependent manner. *S. saponaria* fruit supplementation decreased daily methane production by 20% in RUSITEC. Raw fenugreek seeds supplementation at levels 0.5, 1.0, 2.0 and 3.0% of DM intake reduced the methane production in crossbred steers. 0.2 and 0.4 mg/ml tea saponin decreased the *in vitro* methane emission by 12.7% and 14%, respectively.

Several factors influence microbial protein synthesis in rumen and microbial protein supply at intestinal level. The major factors include the extent of synchronization of available fermentable energy and degradable nitrogen during fermentation process. Duodenal microbial nitrogen flow was 34% higher due to intra ruminal supplementation of *S. saponaria* at 10% level. Microbial efficiency also increased significantly. Apparent total tract digestibilities of OM, nitrogen and NDF were not affected while ADF digestibility was significantly reduced by supplementation of *S. saponaria* at 10% of the gross hay based diet.

In ruminants the source and dose of saponins, nature of basal diet (concentrate or roughage) and sex of animal affects the efficacy of saponins on growth performance. Incorporation of sarsaponin (150 mg/h/d) in steers receiving corn based diets with urea resulted in significant improvement in daily gain. However, no increase in ADG was observed when urea in basal diet was replaced by soybean meal. Dietary incorporation of saponins rich *S. rarak* (12 g/d) in sheep increased ADG significantly (37g vs. 53g).

Blood biochemical profile has been used to identify both beneficial and deleterious effect of PSM on animal system. Supplementation of *S. saponaria* to sheep decreased the concentration of plasma urea nitrogen concentration by 12% while activities of glutamyl transferase and aspartate transferase were not affected. Significant reduction in blood urea nitrogen was recorded in sheep fed plant extract of *Asparagus adscendens.* Feeding of fenugreek seed @ 20% as replacement diet improved profile of functional fatty acids in milk and reduced cholesterol concentration by 4% units, as compared to control.

Growth of steers has been reported to improve when supplementing sarsaponin, a steroidal saponin extracted from *Yucca schidigera*. Saponins, contained in fenugreek, have antimicrobial properties, and suppress the growth of ciliate protozoa, peptidase-producing bacteria, and cellulolytic bacteria in the rumen which are normally acetate, but not propionate-producers.

The antiprotozoal effect of saponins is due to their capacity to form irreversible complexes with the cholesterol present in the protozoal cell membranes, causing cell lysis. Cholesterol and other sterols are components of the cell membranes of all organisms except prokaryotes (bacteria). Thus, reductions in ruminal protozoa numbers observed when saponins are fed and within *in vitro* ruminal fermentation systems are caused by reaction of saponins with cholesterol in the protozoal cell membrane, causing breakdown of the membrane, cell lysis, and death. The antibacterial properties were most pronounced against gram positive bacteria, similar to the action of ionophores. Ruminal protozoa are unable to adapt to or detoxify saponins. The mode of action of antibacterial effects of saponins seems to involve membranolytic properties, rather than simply altering the surface tension of the extracellular medium. Thus, their inhibitory activity is associated with adsorption to microbes and is, therefore, influenced by microbial population density. Interactions of saponins with cholesterol and other sterols account for many of the biological effects of saponins, particularly those involving membrane activities.

2.2 Essential Oils

Essential oils are complex mixtures of secondary metabolites consisting of low boiling point terpenes and phenyl propenes found in limited plant families. They comprise of mainly cyclic hydrocarbons and their alcohol, aldehyde or ester derivatives. Some EO contains degradation products of non-volatile constituents. Essential oils have terpenoid structures and their effect is the result of the combination of all their constituents, which in some oils may number >100 compounds. Major components can constitute up to 95% of the essential oil, whereas other components are present only as traces. Essential oils may also include a variety of low-molecular-weight aliphatic hydrocarbons, acids, alcohols, aldehydes, acyclic esters, or lactones and, exceptionally, nitrogen- and sulfur-containing compounds, coumarins, and homologs of phenylpropanoids.

One of the well-established properties of plant essential oils is their antimicrobial activity. Phenolic components such as eugenol (present in clove bud), or thymol and carvacrol (present in oregano), are responsible for the antibacterial properties of many essential oils. Some constituents in themselves are bioactive, while others may affect physical variables such as absorption rates or bioavailability. In addition, the enantiomeric composition of various terpenes in different plant species can complicate the biological activity of particular oils. More recently, essential oils have attracted attention for their potential as alternatives to feed antibiotics and growth promoters in livestock. Essential oils from a variety of sources have been shown to alter growth and metabolism of several types of bacteria, including rumen bacteria. Most essential oils are classified as Generally Recognized as Safe (GRAS), and have been approved for food and beverage consumption.

Thymol is one of the major compounds of essential oil of thyme (*Thymus vulgaris*) and oregano (*Origanum vulgaris).* Thymol has wide-spectrum antimicrobial activity, inhibiting gram-positive and gram-negative bacteria due to its capacity to act as membrane permeabilizer. It is a potent inhibitor of L-lactate production by *Streptococcus bovis* and *Selenomonas ruminantium*. Thymol at 500 mg/L reduced total VFA concentration, the proportion of propionate, ammonia N concentration, and increased the proportion of acetate, acetate to propionate ratio and rumen pH. Limonene is the most abundant monocyclic monoterpene in lemons (*Citrus limonum*), oranges (*Citrus aurantium*), grapefruit (*Citrus paradisi*), peppermint (*Mentha piperita*), and spearmint (*Mentha spicata*).

Fig. 3 *Origanum vulgaris*

Fig. 4 *Thymus vulgaris*

The influence of essential oils on rumen micro-organisms has been determined *in vitro*. Among the predominant species of rumen bacteria, only *Prevotella ruminicola*, *Clostridium sticklandii* and *Peptostreptococcus anaerobius* are prevented from growing at 40 mg essential oils/l. The last two species are so-called 'hyper-NH_3-producing' (HAP) bacteria. Furthermore, many species could adapt to grow at higher concentrations. Rumen ciliate protozoa are unaffected, and *Methanobrevibacter smithii*, a methanogenic archaeon related to those found in the rumen, is similarly unaffected. *Ruminobacter amylophilus* grows in the presence of 40 mg essential oils/l, but the oils greatly enhance its lysis in the stationary phase; thus, it may also be considered to be sensitive to essential oils. The effects of essential oils on the degradation of protein supplements may be mediated via *R. amylophilus*, because *R. amylophilus* is amylolytic as well as proteo-lytic, it may be the loss of the amylolytic activity from the consortium digesting the supplement that affects the colonising microbial consortium more than the loss of the contribution of *R. amylophilus* to proteolysis. This factor would explain why the degradation of rapidly-degraded starchy protein meals is affected by essential oils, but less-rapidly-degraded starchy meals or non-starchy protein supplements are unaffected. Reduction in protein degradation suggests that at least part of the effects of essential oils can be attributed to decreased proteolysis.

Rumen anaerobic fungi are sensitive to essential oils at concentrations similar to those for the most sensitive bacteria. Thus, fibre digestion may be decreased by essential oils under conditions in which the fungi are prevalent, such as high-fibre low-quality rations.

Many essential oils have dose-dependent effects on bacteria, protozoa, and fungi. In general, gram-positive bacteria appeared to be more susceptible to inhibition by plant essential oil compounds than did gram-negative bacteria. This effect has been related to the presence of an outer membrane on gram-negative organisms, which endows them with a hydrophilic surface that acts as a strong impermeability barrier. The activity of lemon grass oil affects electron transport, ion gradients, protein translocation, phosphorylation steps, and other enzyme-dependent reactions, causing the affected bacteria to lose chemiosmotic control. The effects of plant extracts on ruminal fermentation may differ according to ruminal pH.

Most of the investigations conducted to date on essential oils have been laboratory-based (*i.e.*, *in vitro* incubations) and of a short-term nature. A mixture of EO (thymol, eugenol, vanillin and limonene in an organic carrier) reduced the rate of deamination of amino acids. Inhibition of deamination of amino acids was found in the presence of commercial blend of EO. Effects of commercial blend of EO compounds (110 mg/d) on *in sacco* nitrogen degradation showed inhibitory effect on ruminal degradation of soybean meal nitrogen but not a heat-treated rapeseed meal or hay.

The influence of EO on the end products of rumen fermentation depends on nature and extent of microbial species affected by EO. Feeding peppermint to Holstein steers as feed additive significantly decreased ruminal NH_3 concentration. However, supplementation of commercial EO to sheep had no influence on NH_3 production, VFA concentration and microbial protein supply as estimated by excretion of purine derivatives.

Supplementation of medicinal plants (5% *Cymbopogan* herbage plus 3% *Eucalyptus* leaves) in concentrate based diet significantly increased live weight gain and feed conversion efficiency in ram lambs.

Among the plant extracts from spices namely, garlic, cinnamon, anise, oregano or pepper, garlic (7.5 ppm) decreased NH_3-N by 22.5% in a dual flow continuous culture fermenter. Garlic (31.2 mg/L) and cinnamaldehyde (312 mg/L) increased the proportion of propionate and butyrate and decreased the proportion of acetate compared with control. Garlic oil (300 and 3,000 mg/L) reduced acetate and increased propionate and butyrate proportions. At 3,000 mg/L, capsicum oil, carvacrol, carvone, cinnamaldehyde, cinnamon oil, clove bud oil, eugenol, fenugreek, and oregano oil resulted in a 30% to 50% reduction in ammonia N concentration. Feeding of aqueous extract of *Zingiber officinale* (0.625 mg/ml of rumen liquor) increased microbial protein and efficiency of microbial protein synthesis.

2.3 Tannins

Tannins are naturally occurring water soluble polyphenols of varying molecular weights and are the most abundant polyphenolic compounds with the ability to precipitate proteins from solutions. Chemically there are two main classes of tannins widely distributed in vascular and woody plants. Hydrolyzable tannins (HTs) are esters of polyol or hexahydroxydiphenic acid (ellagitannins). In plant cells, the tannins are located separately from the proteins and the enzymes of cytoplasm, but when tissue is damaged, for example, when animals feed, tannins may react with proteins, making them less accessible to the gastric juices of the animals. Tannins could intervene by binding to either digestive enzymes (*e.g.*, trypsin) or the substrate (*e.g.*, dietary proteins) or to both, thus rendering them unavailable to the host. Condensed tannins (CTs) or proanthocyanidins, the polymers of flavanol units, are probably the most extensively studied PSMs with reference to their physiological and nutritional consequences. Their adverse effects include inhibition of nutritionally and therapeutically relevant gastrointestinal (GI) microflora of the host through various mechanisms. CTs reduce the digestibility of proteins and carbohydrates in the rumen as the tannin-protein complex (TPC) formed is stable over the pH range of 3.5–7.0 but dissociates at pH <3.0 and >8.0. Tannin–protein interactions are pH dependent and it has been proposed that tannin–protein complexes formed in the rumen are hydrolysed in the abomasums.

HTs and flavonols undergo extensive microbial metabolism in the GI ecosystem. CTs prevent bloat formation in ruminants and protect dietary protein in the rumen, enhancing amino acid absorption and utilization by the ruminants. Forages containing moderate concentrations of CTs can have positive effects on sustainability and production in intensive grazing systems through increasing the efficiency of animal production by decreasing urinary nitrogen excretion and reducing chemical inputs such as detergents to control rumen bloat in cattle and anthelmintics.

CT, when present in forages in moderate concentrations (20–40 g/kg DM) can exert beneficial effects on protein metabolism in ruminants, by

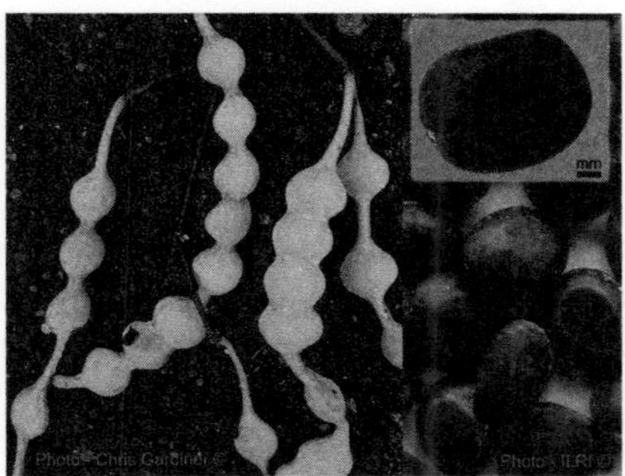

Fig. 5 *Acacia nilotica*

slowing the rapid microbial degradation of dietary protein and increasing protein outflow from the rumen, thus increasing the absorption of amino acids in the small intestine of the animal. Dietary CT may also contribute to animal health by reducing the detrimental effects of internal parasites in sheep and the risk of bloat in cattle.

Tannins decrease methane production. Total phenol and total tannins were good predictors of methane production reduction potential. As chemical analysis of these parameters is relatively simple, they could be used to screen large numbers of samples in a relatively short time and, with commonly available laboratory facilities, for their antimethanogenic potential in the quest to obtain plants and plant products as alternatives to conventional antibiotic growth promoters. The seeds of *Terminalia chebula,* leaves of *Rheum undulatum*, *Vaccinium vitisidaea*, *Bergenia crassifolia*, *Rhus typhina* and *Peltiphyllum peltatum*, and roots of *B. crassifolia* have considerable potential (*i.e.*, >25%) to decrease enteric methane production from ruminants.

It is possible to use PSM to manipulate ruminal fermentation by selective suppression of certain microbial species. A particular chemical extracted from a medicinal herb might in pure form have only a fraction of the biological activity that it has in the plant matrix. The major components of essential oils from lemon grass geranial and neral individually elicited antibacterial action, the third main component, myrcene, did not exhibit any activity. However, myrcene enhanced activities when mixed with either of the other two main components. Sennoside A and sennoside C from senna have similar laxative activities in mice. However, a mixture of these compounds in the ratio 7:3 (which somewhat reflects the relative levels found in senna leaf) has almost two-fold laxative activity.

Plant bioactives (PSM) are still an under-explored area of research and in many cases although biological activity has been observed, the natural phytochemicals responsible for the activity have not been identified. Delivery of bioactives is an important consideration. The form the bioactive presented to the animal will affect not only bioavailability but also cost of delivery.

The use of plant bioactive for animal health is an area of increasing research importance. Research is still required to identify the active constituents of plant extracts and understand their mode of action. A collaborative approach between animal nutritionists, microbiologists, biochemists with studies taking into account the effects of manufacturing processes are required. Such collaboration would also seem to extend to plant breeders and agronomists as herbs become more widely cultivated. Maintaining bioactivity in plants must be considered even more important than increasing crop yield. Moreover, little information is available on the effective dose of essential oils that can be used in animals without inducing toxic effects or imparting unwanted taints to meat or milk products. More research is needed in this area. Issues such as toxicity, photosensitivity, residues, taint, allergenicity and cost effectiveness still need to be addressed before these compounds will gain widespread acceptance in the agricultural industries.

3.0 REFERENCES

Alexander, G. 2005. Effect of plant extract on rumen fermentation and nutrient utilization in sheep. PhD Thesis, Indian Veterinary Research Institute, Izatnagar, India.

Baruneton, J. 1999. Pharmacognosy: Phytochemistry of Medical Plants. Hampshire Intercept, U.K.

Busquet, M., Calasmaiglia, S., Ferret, A. and Kamel, C. 2006. Plant extracts affect *in vitro* Rumen Microbial Fermentation. *J. Dairy Sci.* **89**: 761–771.

Cheeke, P.R. 1998. Natural Toxicants in Feed Forages and Poisonous Plants. (2nd edn.) Interstate publishers Inc., Danville, USA. pp. 281–284.

Cowan, M.M. 1999. Plant products as anti microbial agents. *Clinical Microbiol. Rev.,* **12**: 564–582.

Evans, J.D. and Martin, S.A. 2000. Effects of thymol on ruminal microorganisms. *Current Microbiol.,* **41**: 336–340.

Hess, H.D., Kreuzer, M., Diaz, T.E., Lascano, C.E., Carulla, J.E., Solvia, C.R. and MacMuller, A. 2003. Saponin rich tropical fruits affect fermentation and methanogenesis in faunated and defaunated rumen fluid. *Anim. Feed Sci. and Technol.,* **109**: 79–94.

Hosettmann, K. and Matson, A. 1995. Saponins. Cambridge University Press, Cambridge, U.K.

Hristov, A.N., Ivan, M., Neil, L. and McAllister, T.A. 2003. Evaluation of several potential bioactive agents for reducing protozoal activity *in vitro*. *Anim. Feed Sci. and Technol.,* **105**: 163–184.

Jounay, J.P. and Ushida, K. 1999. The role of protozoa in feed digestion. *Asian-Australian J. Anim. Sci.,* **12**: 113–128.

Lila, Z.A., Mohammed, N., Kanada, S., Kamada, T. and Itabashi, H. 2003. Effect of sarsaponin on ruminal fermentation with particular reference to methane production *in vitro*. *J. Dairy Sci.,* **86**: 3330–3336.

Makkar, H.P.S., Sen, S., Blummel, M. and Becker, K.1998. Effects of gractions containing saponin from *Yucca schidigera, Quillaja saponaria* and *Acacia auriculoformis* on rumen fermentation. *Journal of Agric. and Food Chem.*, **46**: 4324–4328.

McIntosh, F.M., Williams, P., Losa, R., Wallace, R.J., Beever, D.A. and Newbold, C.J. 2003. Effects of essential oils on ruminal microorganisms and their protein metabolism. *Appl. Environ. Microbiol.,* **69**: 5011–5014.

Newbold, C.J., McIntosh, F.M., Williams, P., Losa, R. and Wallace, C.J. 2004. Effects of specific blend of essential oil compounds on rumen fermentation. *Anim. Feed Sci. Technol.,* **114**: 105–112.

Rejil, M.C. 2005. Mitigation of methane emission from ruminants by dietary supplementation of fenugreek seeds. M.V.Sc Thesis, National Dairy Research Institute, Karnal, India.

Wink, M. 1999. Introduction: Biochemistry, Role and Biotechnology of secondary products. *In*: Biochemistry of secondary product metabolism (*Ed.* Wink, M.) CRC Press, Boca Raton, FL., USA. pp. 1–16.

16

IMPACT OF THERMAL STRESS ON PHYSIOLOGICAL ADAPTABILITY AND REPRODUCTIVE EFFICIENCY OF ANIMALS

V.P. Maurya, V. Sejian and S.M.K. Naqvi

ABSTRACT

Thermal stress challenges to an animal's homeostasis or equilibrium result in a stress response. The animals reared in arid and semi-arid region have to adapt by special physiological means to maintain their thermal equilibrium. The changes in the biological functions of animal due to exposure to heat stress include depression in feed intake and utilization, disturbance in metabolism of water, protein, energy and mineral balances, enzymatic reactions and hormonal secretions. In livestock mammals, heat stress may cause delayed puberty, delayed onset of sexual season, irregularity in cycle length, long duration of estrus, low ovulation rate, high frequency of anovulatory estrus, morphological abnormalities in ova, embryonic mortality, fetal abnormalities, abortion and undesirable semen characteristics. However the effects of extremes of temperatures on reproduction vary with the species, breed, age, stage of reproductive cycle, period of temperature exposure, nature of temperature fluctuation and altitude. Exposure of the pregnant ewe during mid and late gestation to warm ambient temperature greatly reduces the total embryo cell number and placenta size. In addition, lamb birth weight, live body weight gain or growth rate, as well as, total body weight and daily weight gain are impaired by exposure to elevated temperatures. Cows with elevated body temperature exhibit lower dry matter intake (DMI) and milk yield and produce milk with lower efficiency, reducing profitability for dairy farms in hot, humid climates. Although adequate cooling systems are available but their efficiency in humid climates is less than in arid climates and these systems often lack the ability to maintain normal body temperature. There is genetic variation in cattle for cooling capability, which suggests that more

Central Sheep and Wool Research Institute, Avikanagar, Via-Jaipur, Rajasthan - 304501, India *heat tolerant cattle can be selected genetically, and cross-breeding may also offer opportunities. Developing nutritional strategies which support yield but which also address metabolic and physiologic disturbances induced by heat stress will help the cow to maintain normal metabolism that should enhance performance.*

1.0 INTRODUCTION

Thermal stress challenges to an animal's homeostasis or equilibrium result in a stress response. While a degree of stress is beneficial for adaptation and survival, chronic or long-term stress is detrimental. The impact on the endocrine, immune and nervous systems compromise animal performance and prevents realization of optimum reproduction. High ambient temperature, relative humidity, and radiant energy compromise the ability of the lactating animals to dissipate heat, and coupled with metabolic heat makes it difficult to maintain thermal balance. Elevated body temperatures initiate compensatory and adaptive mechanisms to reestablish homeothermy and homeostasis. Stott (1981) stated that these readjustments to maintain homeostasis by homoeostatic mechanisms are referred to as adaptations and may be favorable or unfavorable to the economic interests of humans. Nevertheless, they are essential for survival of the animal. The impact of heat stress on livestock is broad in geographic terms. Environmental factors, animal factors and thermoregulatory mechanism have a significant impact on the energy exchange between the animal and the environment (Nienaber *et al.*, 1999). Animals reared in subtropical and tropical regions are subject to high ambient temperature and/or high relative humidity for extended periods and have low productivity (Abi Saab and Sleiman, 1995). In these areas high ambient temperature and relative humidity exceeding the temperature-humidity index (THI) associated with heat stress persist for 3 to 4 months of each year.

The animals reared in arid and semi-arid region have to adapt by special physiological means to maintain their thermal equilibrium. Controlled vasoconstrictor tone of skin blood vessels, upper respiratory tract evaporative cooling and alterations in coat insulation (Copenland, 1983). Kuthe (1976) observed that high ambient temperature results in loss of water via skin and lung of animals leading to an increased water intake. The onset of sweating is the next rapid reaction of animals to heat exposure and increases linearly with the increase in ambient temperature (Kamal, 1975). Heat stress itself may increase in magnitude if continued global warming occurs (Hulme, 1997). Secondly, the world's human population is growing faster in tropical and subtropical regions of the world than in temperate regions (Roush, 1994) and it is reasonable to assume that a greater proportion of the world's food animals will live in hot, humid climates than is the case at present. Animals are being genetically manipulated to get higher productivity which reduces the animals thermoregulatory mechanisms against the warm environments. This is especially true for dairy cattle. Selection for milk yield reduces thermoregulatory ability in the face of heat stress (Berman *et al.*, 1985) and magnifies the seasonal depression in fertility caused by heat stress (Al-Katanani *et al.*, 1998).

In mammals, high environmental temperature may cause delayed puberty, delayed onset of sexual season, irregularity in cycle length, long duration of estrus, low ovulation rate, high frequency of anovulatory estrus, morphological abnormalities in ova, embryonic mortality, fetal abnormalities, abortion and undesirable semen characteristics. Animals have developed effective thermoregulatory mechanisms for the testis. The effects of extremes of temperatures on reproduction vary with the species, breed, age, stage of reproductive cycle, period of temperature exposure, nature of temperature fluctuation and altitude. When the body is challenged with elevated external thermal temperature, which begins to elevate body temperature beyond the physiologic homeostatic body temperature range, the animal is considered to be exposed to thermal stress. Thermal stress coupled with nutritional stress has a negative effect on most factors related to reproductive performance in sheep (Sawyer, 1979; Maurya *et al.*, 2004).

Heat stress is a major limitation to reproductive function of dairy cows because it can have major deleterious effects on both ovulation and conception. Different ameliorative measures may be adapted to reduce the magnitude of heat stress in animals. There is a need to improve the housing system of animals because existing housing systems are inadequate for eliminating the thermal stress which is resulting in depression in fertility (Hansen and Aréchiga, 1999). The hyperthermia coincident with heat stress occurs largely because animals cannot dissipate all the heat produced as a result of metabolism to the surrounding environment. Improper management such as exposure to water deprivation, nutritional imbalance and nutritional deficiency may exaggerate the impact of heat stress.

Therefore, some new strategies should be adapted for improving reproductive efficiency during heat stress by regulating the physiological and reproductive mechanisms of the animals or the cell biology of the tissues involved in reproduction. These approaches will depend upon greater knowledge of the cellular and physiological mechanisms by which heat stress leads to decreased estrous activity and embryonic survival as well as a better understanding of the physiological processes by which animal regulate body temperature.

2.0 PHYSIOLOGICAL REACTIONS TO THERMAL STRESS

The changes in the biological functions of animal due to exposure to heat stress include, the depression in feed intake and utilization, disturbance in the metabolism of water, protein, energy and mineral balances, enzymatic reactions, hormonal secretions and blood metabolites (Marai *et al.*, 2006; Maurya *et al.*, 2007).

2.1 Feed Intake

The most important reaction to heat exposure is the depression in feed consumption. High environmental temperature stimulates the peripheral thermal receptors to transmit suppressive nerve impulses to the appetite center in the hypothalamus causing decrease in the feed consumption. A higher heat increment is caused by the specific dynamic action that accompanies the metabolism of feed which is highest in the case of poor quality, fibrous feedstuffs (Marai *et al.*, 2001). Factors such as water deprivation, nutritional imbalance and nutritional deficiency may exaggerate the impact of heat stress. Studies show that dry matter intake decreases significantly following exposure to heat stress. Similarly, dry matter intake per kilogram live weight was lower and the maintenance requirements were higher at high ambient temperatures. The daily feed intake and feed conversion also significantly decreased in Suffolk lambs under hot conditions in a climatic chamber (30.5°C), compared to a group under shelter (19.3°C), during spring (Padua *et al.*, 1997). Digestion is the process of breaking down and dissolving feeds to be absorbed by the animal. Studies showed the digestibility of dry matter (DM), crude protein (CP), ether extract (EE) and energy (E) and metabolizable energy (ME) of diets to be depressed when the animals are exposed to high temperatures (Marai *et al.*, 2001). Ambient temperature affected feed digestibility by altering the volume of the gastro-intestinal tract and the rate of passage of the digesta through the system. The decrease in rate of passage or the increase in the retention time of the digesta apparently increased the digestibility (Conrad, 1985). In summary, exposure of sheep to heat stress is accompanied by changes in the biological functions. It include the depression in feed intake and utilization, disturbance in the metabolism of protein, energy and mineral balances, enzymatic reactions, hormonal secretions and blood metabolites.

2.2 Water Intake

Water is one of the most important nutrients required for the maintenance of life and is involved in many physiological functions essential for maximum performance of farm animals. Water requirements vary and are regulated by many factors such as intake of DM, environmental temperature and loss of water from the body tissues. Sheep consume less water per unit of dry matter intake (DMI) than cattle, but the intake increases with an increase in ambient temperature. Consumption of water is 9% and 11% of the total body weight during the winter and 19–25% during summer in 3-

year-old sheep (Khan and Ghosh, 1989). During the first few weeks of lactation, ewes drink approximately 4 kg water/kgDM at temperatures of between 15 and 30°C. Water intake increases markedly during pregnancy, due to the acute heat stress, with a corresponding water diuresis. Body temperature is, however, regulated during pregnancy under acute heat stress conditions by the loss of heat to ingested water and probably also via increased respiration (Conrad, 1985). Exposure of sheep to hot environmental conditions induces a marked increase in water turnover, as well as water intake (Padua *et al.*, 1997). During pregnancy, the water turnover is lower than during lactation, under heat stress conditions (Benlmlih, 1988). Water intake (frequency/h) was significantly higher in the summer than during the winter (Marai *et al.*, 2000). Water intake increase was determined to be 50%, while water loss decreased by approximately 25% in the faeces and 40% in urine, during heat stress. Insensible water loss (by diffusion of water from the skin) was two times higher at 35°C, than at below 18°C (Shafie *et al.*, 1994). In summary, exposure of sheep to hot environmental conditions induces a marked increase in water turnover, as well as, water intake.

2.3 Thermo-Regulatory Function

Evaporation becomes the most important avenue for heat dissipation, which can be performed mainly through sweating in animals and little amount by respiratory evaporation. Under long periods of exposure to heat, animals develop certain mechanisms to produce less body heat *i.e.* decreases in feed intake and in heat production—as well as stretching out the body to loose as much heat as possible (Marai *et al.*, 1997). When the environmental temperature rises to 42°C, the ears and legs of sheep dissipate a high proportion of the heat, as these areas contribute about 23% of the body surface area (Johnson, 1987). Respiration rate, pulse rate and rectal temperature of animal change as a response to thermo-regulatory function.

2.3.1 *Respiration Rate*

Increased respiration rate is the first reaction when animals are exposed to environmental temperatures above the thermoneutral zone (Kamal, 1975; Maurya *et al.*, 2007). This response ensures direct heat stimulation of the peripheral receptors, which transmit nervous impulses to the heat center in the hypothalamus. With the increasing environmental temperature, the respiration rate continues to rise linearly until a certain temperature, afterwards rate of respiration slows down and depth of breathing increases. At this level, the respiratory muscle reach a threshold of maximum activity and thus the respiration rate cannot increase further.

Respiration rate can be an indicator of heat stress (Habeeb *et al.*, 1992). In mammals, respiration is directed at the elimination of CO_2 from the tissues of the body and the provision of O_2 under thermo-neutral conditions, and evaporation of moisture from the respiratory tract and prevention of hypothermia under high ambient temperatures. During summer, the respiration rate is higher than in winter for sheep (Fahmy, 1994) and when walking (Abi-Saab and Saleim, 1995). With regard to the effect of humidity, when a load of high relative humidity was superimposed upon an already high ambient temperature, there was a further increase in respiratory frequency in animals. This was related to an increase in the perception of warmth (Marai *et al.*, 2002). Type of nutrition consumed seemed to affect the respiration rate under heat stress conditions. The increment in respiration rate also depends on the adaptive capability and age of animals.

The significance of the increase of the respiration rate under heat stress is that it enables the animal to dissipate the excess of body heat by vaporizing more moisture in the expired air, which accounts for about 30% of the total heat dissipation.

2.3.2 *Pulse Rate*

The pulse rate reflects primarily the homeostasis of circulation along with the general metabolic level. It increases on exposure to high environmental temperature (Aboul-Naga, 1987; Maurya *et al.,* 2007). This increases blood flow from core to surface area of body gives a chance to more heat to be lost by sensible (loss by conduction, convection and radiation) and insensible (loss by diffusion water from the skin) means. In some study it has been found that the pulse rate does not always changes appreciably under high environmental temperature (Marai *et al.*, 1991). It has been found that the heart rate of animals increase during the peak hour of the heat load. During hyperthermia (rectal temperature of up to 42°C), animal exhibited a greater increase in cardiac output due to increase in blood flow through the arteriovenous anastomose. It is suggested that the improvement of heat tolerance by physical fitness is caused by a greater cardiovascular capacity which not only permits greater perfusion of heat-loss from tissues, but also the maintenance of a better gastrointestinal tract blood supply—thereby better maintaining the normal barrier to movement of endotoxins from the gut lumen to the plasma. As the ambient temperature increases, the pulse rate, as well as, the circulation of blood increase to transfer heat from the core to the periphery.

2.3.3 *Rectal Temperature*

The significance of a body temperature increase has its importance in thermoregulation. It coincide with the impulses transmitted by peripheral thermal receptors which stimulate the central receptors of the hypothalamus to induce thermal polypnoea and sweating to increase heat loss (Kamal, 1975) The mammalian skin is an important pathway for heat exchange between the body surface and the environment. Body temperature is the result of the adjustment of the skin blood flow that ends with regulation of the heat between the body core and skin (Habeeb *et al.*, 1992). Body temperature was recorded to be highest during summer and lowest during winter in sheep (Fahmy, 1994). Generally, it can be stated that skin temperature of animals differ according to season of the year and time of the day and it becomes higher with an elevation in ambient temperature. In short, exposure of animals to elevated ambient temperatures coincides with an increase in the dissipation of excess body heat—in order to balance the excessive heat load. Dissipation of excess body heat is performed by evaporation of water from the respiratory tract and skin surface via panting and sweating, respectively. However, evaporation by panting in animals becomes the most important avenue for heat dissipation. When the environmental temperature increases to 45°C, the ears and legs of animals also dissipate a high proportion of the heat.

2.3.4 *Sweating Rate*

The onset of sweating is the next rapid reaction of animals to heat exposure and this increases linearly with the increase in the ambient temperature (Kamal, 1975). This increase in sweating is also controlled by the hypothalamus which is stimulated by the increase in core body temperature and by nervous impulses transmitted by the peripheral receptors. The significance of the increase in sweating is that the heat loss by sweating represents 60% of the total when the ambient temperature increase above normal value (Maurya *et al.*, 2007). Moreover, when the surrounding environmental temperature is equal to (or above) the animal's body temperature, the evaporative cooling (respiration and sweating) system represents the only mean of heat dissipation.

2.4 Biochemical Changes

The hemoglobin is a very important constitutes of the blood, which helps in the transport of oxygen to the tissues of the body. During thermal stress there is decrease in the hemoglobin concentration (Yousef, 1985; Maurya *et al.*, 2007) because of hemodilution and increase in the blood volume of animals. The packed cell volume (hematocrit percentage) also decreases in heat stressed animals due to red cell destruction or hemodilution (Marai *et al.*, 1991).

Several researchers have studied the effect of thermal stress on the blood glucose concentrations but the results reported by them are conflicting. Shaffer *et al.* (1981) reported 19% decrease of blood glucose in the Friesian cow exposed to thermal stress. The decrease in glucose level during heat exposure relates in decrease in concentration of insulin and thyroxin, which are closely associated to energy metabolism during heat exposure. Decrease in plasma glucose could also be due to the marked dilution of blood and body fluid as a whole or increase in the plasma glucose utilization to produce more energy for greater muscular expenditure required for high muscular activity (Habeeb, 1987). Other study show that glucose concentration is increased during thermal exposure of animals because there is decrease in the glucose utilization and depression of the catabolic and anabolic enzyme secretion and the subsequent reduction of metabolic rate and also due to breakdown to glycogen into free glucose.

Serum protein particularly albumin concentration, usually decrease in heat exposed cattle (Habeeb, 1987). A negative correlation between plasma protein and elevated environmental temperature has also been reported in some studies. The decrease in the plasma protein may be due to dilution of plasma proteins, decrease in protein synthesis as a result of depression of anabolic hormone secretion. In addition to this there is also increase in gucocorticoids concentration. The reduced feed intake in heat stressed animals also leads to decrease in nitrogen concentration in the body, which leads to low protein level. The serum total lipid concentration also decreased in the animals exposed to prolonged thermal stress which is due to increase in utilization of fatty acids for energy production as a consequence of decrease in glucose concentration (Daader *et al.*, 1989). Kamal *et al.* (1989) reported 16% decrease in the blood urea-N level in heat stressed cattle, which may be due to more resorption of the urea-N from the blood to the rumen to compensate for the decrease in ruminal ammonia-N as a result of decrease in feed intake due to thermal stress.

2.5 Blood Enzymes Secretions

The proper enzymes level in the body is very essential to run all the metabolic activities in the animal's body. External environment, herd management and seasonal changes drastically affect the blood enzymes of animals. The serum transaminase activity increases with the increase in the environmental temperature. In cattle, heat stress causes an increase in the activities of serum glutamic oxaloacetic transaminase (SGOT) and serum glutamic pyruvic transaminase (SGPT) due to increased stimulation of gluconeogenesis by corticoids (Kamal *et al.*, 1989; Hooda and Naqvi, 1990). The alkaline phosphatase enzyme has been mostly found to decrease significantly (El-Masry, 1987) by heat stress and may be attributed to the reduction in thyroid hormone which occurs under heat stress. However, acid phosphatase generally is not affected more in heat stressed animals.

2.6 Hormonal Secretions

The hormones plays very critical and important role in the thermoregulation of animals body. With prolonged heat exposure, release of hypothalamic hormones suppressed which inturn drastically affects the release of pituitary hormones. The main hormones related with thermal balance are insulin, thyroxin, cortisol and aldosterone.

Thermal stress decreases the plasma insulin concentration significantly in animals. This decrease may be due to decrease in heat production (Habeed, 1987). The thyroid hormone (either thyroxin T_4 or triiodothyronine T_3) declines during thermal exposure of animals and the decline ranges between 15% to 37% in different genotype of animals. Plasma cortisol level increases during acute heat stress and decrease during the chronic phase. The decline, which occurs under chronic phase of heat stress, is attributed to thermoregulatory protective mechanism preventing metabolic heat production in a hot environment (Alvarez and Johnson, 1973). Heat stress decreases aldosterone level significantly in cattle (Niles *et al.*, 1980) resulting large decrease in the potassium retention and blood electrolyte changes which leads to increase body fluid to cope up with stress.

2.7 Sexual Behaviour

Proper assessment of sexual behaviour and receptivity before breeding of animal may increase the possibility of fertile mating. High environmental temperature or the rapid and sudden fluctuations of temperature that often occur in arid and semi-arid regions have detrimental effect on sexual behaviour of animals. The implication of thermal stress is such that even if a ewe has a normal ovarian development leading to ovulation, there is a reduction in the full expression of estrus behaviour that might lead to the failure of ewe to mate and conceive. Estrus behaviour in animals is a part of a continuum of interaction between the sexes, which includes mutual sniffing and licking as well as motor pattern that are conducive to mating act. During the thermal stress the estrus incidence (onset of estrus and estrus duration) are significantly reduced thereby limiting the proper manifestation of estrus (Maurya *et al.*, 2005) and reproductive performance of ewe. Ehnert and Moberg (1991) reported that thermal stress affects the preovulatory release of LH, expression of estrus behaviour and causes delay in ovulation. The mechanism by which heat stress reduces expression of estrus may involve hormones. In several reports it has been reported that high ambient temperature reduces the circulating estradiol–17â concentration and some times increases the adrenocorticotropin secretion, which itself can block estradiol induced estrus behaviour (Hansen and Arechiga, 1999). The behavioral manifestation of estrus ranges from a phase of low intensity to a phase of high intensity occurring 8 to 12 h after the commencement and ends with another phase of low intensity with gradual disappearance as ewe enter in the diestrus stage. The incidence of estrus and receptivity of ewes are important factors, which can affect over all embryo production (Tompkins and Brayant, 1974). The preovulatory release of LH and the expression of estrus behaviour seem to be especially sensitive to stress (Dobson and Smith, 2000). The thyroid gland also plays a role in decreasing the reproductive activity during thermal stress (Farghaly, 1984) with a decrease in the level of triiodothyronine (T_3) and thyroxin (T_4) hormone.

2.8 Folliculogenesis and Ovulation

It has been reported that heat stress suppresses the functions of the largest follicle, *i.e.* the dominant follicle. Under heat stress, the size of the dominant follicle is reduced during the first and second follicular waves and the number of follicles of the next largest size is increased. In contrast, discrepancies still exist between previous studies concerning follicular recruitment under heat stress; it was found that heat stress caused either a decrease or an increase in the emergence of small follicles (2–5 mm in diameter). In addition, although it was reported that heat stress caused a decrease in the oestradiol concentration of the fluid of the dominant follicle, the steroidogenic activity in small and medium follicles under heat stress has been poorly investigated. Circulating concentrations of gonadotrophin, which plays an important role in regulating follicular dynamics, are also altered by heat stress. A previous study found that heat stress caused an increase in the plasma follicle-stimulating hormone (FSH) concentration. On the other hand, conflicting findings concerning the effect of heat stress on plasma luteinizing hormone (LH) have been reported; heat stress has been reported to cause the plasma LH concentration to decrease, remain constant, or increase. More recently, it has been reported that heat stress did not change the pituitary LH secretion in long-day-treated anoestrous goats that were injected with exogenous gonadotrophin-releasing hormone (GnRH) to induce artificial LH pulses, whereas it reduced the follicular responsiveness to LH. Follicular LH responsiveness is attributed to the expression of LH receptors on the follicles. Therefore, it appears probable that heat stress induces alterations in the follicular LH receptors. Putney *et al.* (1988) and Alfujairi *et al.* (1993) reported adverse effect of hot summer on the ovulation rate in cows. Thermal stress during folliculogenesis could lead to ovulation of low quality oocytes with lowered developmental competence.

3.0 IMPACT OF HEAT STRESS

3.1 Body Growth

Growth, the increase in live body mass or cell multiplication, is controlled genetically and environmentally. The available nutrients, hormones and enzymes, as well as, elevated ambient temperatures are considered as some of the environmental factors that can influence average daily gain (Hafez, 1987; Habeeb *et al.*, 1992). In addition to this, the thermal stress drastically affects the overall growth performance of animals *i.e.* growth rate, daily weight gain and live body weight (Marai *et al.*, 1991; Maurya *et al.*, 1994). The calculated loss in body solid due to heat stress conditions was found to be 14–29% in cattle (Kamal and Johnson, 1971). The effects of elevated temperature on growth performance are the products of the decrease in anabolic activity and increase in tissue catabolism. The decrease in anabolism is essentially caused by the decrease in voluntary feed intake of essential nutrients, particularly metabolizable energy for both maintenance and weight gain (Bianca, 1965). The increase in tissue catabolism occurs mainly in fat depots and lean body mass (Kamal and Johnson, 1971).

Exposure of the pregnant ewe during mid and late gestation to warm ambient temperatures greatly reduces the total embryo cell number and placenta size. In addition, lamb birth weight, live body weight gain or growth rate, as well as, total body weight and daily weight gain are impaired by exposure to elevated temperatures. In other words, it can be stated that lamb production is deleteriously affected by exposure to heat stress and this causes an economic loss. The percentage loss pertaining to heat stress has been estimated in other species (rabbits), to be more than 70% of live weight at weaning (Marai *et al.*, 2001). Such phenomena necessitate either to protect the ewes from exposure to high ambient temperatures during the breeding season or to carry out breeding of ewes during the milder periods of the year, in the warm climate zones.

3.2 Reproduction

3.2.1 *Female Fertility*

Most cells in the body produce heat shock proteins in response to heat stress that limit the damaging effects of elevated temperature on cell function. Unfortunately in cattle, around the time of ovulation, the oocyte and/or the resulting early embryo are unable to produce such proteins. Consequently, embryo viability is compromised resulting in lower conception rates. In large part, embryos survival depends upon the maternal environment. Embryonic loss can occur when there is disruption in the physiological regulation of oviductal and uterine function. Whether or not an embryo can survive perturbations in the maternal environment depends upon the degree to which the embryo can either 1) adjust its own internal physiology to allow for its survival and continued development or, 2) act on the mother to restore to some degree the microenvironment that the embryo resides in. The extent to which the embryo is capable of self-adjustment or adjustments of the maternal unit is not well understood for most maternal factors affecting fertility. The one exception is the case for heat stress, where abundant information on the ability of the embryo to maintain cellular function and development in the face of elevated temperature (*i.e.*, heat shock) is available. For this stress, at least, the embryo acquires mechanisms to become resistant to elevated temperature as it progresses through development. The biochemical and molecular nature of the thermo protective mechanisms that allow more advanced embryos to become more resistant to elevated temperature is still under investigation. In cattle, as for other species, exposure of pregnant females to heat stress during the embryonic period leads to embryonic loss (Ealy *et al.*, 1993). One cause of embryonic death is the disruption of embryonic development caused when the embryo is exposed to elevated temperature. Thus, exposure of embryos to heat shock by elevating culture temperature blocks or reduces subsequent development. Heat shock leads to embryonic death, at least in part, because protein synthesis is reduced (Edwards and Hanseen., 1997) and concentration of free radicals increases.

Fetal growth within the uterus is a complex biological event influenced by genetic, epigenetic, and environmental factors, as well as maternal maturity. These factors impact on the size and functional capacity of the placenta, uteroplacental blood flow, transfer of nutrients and oxygen from mother to fetus, conceptus nutrient availability, the endocrine milieu, and metabolic pathways (Reynolds *et al.*, 2005., Wu *et al.*, 2006). Heat stress drastically reduces the pregnancy rates (Hahn *et al.*, 2003). In addition to effects on embryonic mortality heat stress reduces the duration and intensity of sexual behaviour and estrus incidences (Naqvi *et al.*, 2004). The pregnant ewes exposed to thermal stress may result in developmental adaptations that permanently change the structure, physiology, metabolism and postnatal growth of the offspring. The reproductive efficiency of the animals in the environment they are kept should be fine-tuned to produce maximum number of offspring most efficiently (Lindsay, 1996). During heat stress redistribution of blood flow from the viscera to the periphery increases for dissipation of heat, which leads to reduce blood supply to placentas (Alexander and Williams, 1971) and retards fetal growth (Collier *et al.*, 1982). Fertility of animal is markedly affected by exposure to high ambient temperature immediately prior to estrus or immediately after mating. The lambs born from heat stressed ewes have lower birth weight than born from control ewes maintained under shed (Naqvi, 1987). Naqvi *et al.* (2004) reported that the ovulation in the ewes occurs towards the end of estrus and high ambient temperature may increase ovum wastage, which could lead to lower conception and lambing rate. Sawyer *et al.* (1979) reported low lambing rate in the Merino ewes that were exposed to thermal stress before estrus and insemination. In arid and semi-arid tropical environment most of the sheep are grazed in field and therefore exposed to high intensity of thermal stress because of long sunshine period and elevated temperature during summer.

3.2.1.1 *Measures to Increase Conception in Heat-Stressed Females*

Some of the effects of heat stress lead to reduced intensity of estrous expression. Therefore, it may be possible to improve reproduction in dairy herds by improving estrous detection methods. Some researchers demonstrated that using tail chalk could improve estrous detection rates in dairy cows during the summer. Other estrous detection aids including pressure activated patches or electronic devices placed on the tail head should theoretically improve reproductive performance of dairy cows during heat stress.

If estrous detection is a problem in heat-stressed dairy cows then it may be possible to improve reproduction by using timed AI. Timed AI is done by administering a series of gonadotropin releasing hormone (GnRH) and prostaglandin $F_{2á}$ (PGF) injections. Insemination is performed at a predetermined time following the last GnRH injection. Timed AI shortened the interval to first service and increased pregnancy rates in heat-stressed cows when compared to insemination at observed estrus. Therefore, timed AI is a good alternative when attempting to reproductively manage cows during heat stress.

Embryos are sensitive to the effects of heat stress. However, greatest sensitivity occurs early during embryonic development. Later during embryonic development (morula or blastocyst stage), embryos develop some tolerance for heat stress. It should be possible, therefore to improve pregnancy rates in heat-stressed cattle by using embryo transfer of frozen embryos collected from cows that are not heat stressed. Several investigators have used this approach. Embryo transfer nearly doubled conception rates when compared to Florida dairy cows inseminated artificially. Therefore, it is possible to by-pass early embryonic stages and improves conception rates during heat stress. The practicality of this approach is only limited by the availability of good quality embryos at an affordable price.

3.2.1 *Male Fertility*

The male animals are most important factor for over all reproductive efficiency and productivity of herd/flocks which some times overlooked. A highly fertile and adapted bull will settle a greater number of cows early in the breeding season, and will also be more likely to fertilize a higher proportion of eggs than less adapted and bull with poor fertility. Sexual performance of rams is highly variable (Maurya *et al.*, 1998) and strongly influences flock fertility (Matos and Thomas, 1992). Bull that exhibit relatively rapid ejaculation rates are capable of inseminating a greater number of cows per unit of time than males of having poorer libido or mating technique. Tests for ranking bull on sexual performance are repeatable and reliable predictors of sexual performance under field conditions, potentially allowing breeders to evaluate the mating competence of individual males before they are employed in a breeding programme.

In the fertile male harmonious interaction of gametogenic and endocrine testicular activity ensures the production of spermatozoa and seminal plasma in an integrated fashion. Under thermal stress one of these functions may be disrupted more seriously than the other, thus effecting spermatogenesis and steroidogenesis to different degrees. The stage of spermatogenesis that is most susceptible to elevated temperature is the primary spermatocytes although damage to B spermatogonia can occur in bulls and prolonged exposure to heat can damage dividing spermatocytes and spermatids (Thatcher and Collier, 1982). Heat stress occurs when the scrotum is not able to reduce the temperature of the testicles below normal body temperature. Marai *et al.* (2006) reported that during thermal stress, Merino rams with heavy skin wrinkle have a relatively poor ability to control testicular temperature resulted in loss of motility, increased proportion of abnormal sperm, decreased concentration of sperm and ultimately cessation of spermatogenesis. In addition to this there were a higher percentage of abnormal spermatozoa and acrosomal damage. Stress also caused the increase in temperatures of rectum, body skin and scrotal skin. Blanc *et al.* (1981) reported that heat stress on scrotum drastically affects the function of sertoli cells and release of androgen. Mating behavior is an important aspect of male reproductive function as it has a direct bearing on the number of females mated. Alterations in sperm production caused by heat stress commonly include decreased sperm numbers, decreased sperm motility and increased numbers of abnormal sperm (Meyerhoeffer *et al.*, 1985). If weather is extremely hot during the breeding season, rams may be removed from the ewes during the day and kept in a cool place until returning to the ewe flock at night for breeding. More emphasis should be placed on the adaptation of animals to their environment and on their behavioural response to different stresses. Genetic selection for certain traits could result in the development of breeds and strains of sheep better able to survive and thrive in such systems.

4.0 AMELIORATION OF THERMAL STRESS

Ruminants primarily adjust evaporative heat loss to maintain homeothermy during brief exposures to adverse weather, but will reduce feed intake to lower heat production during prolonged hot weather. In hot environment, energy exchanges by radiation are dominant. To alter the microclimate of an animal effectively through housing or environmental modification, one must consider altering one or more of the following factors: temperature of the surroundings; air temperature; air velocity; air vapor pressure; radiation or shade factors; and conductivity of surfaces that animals might contact.

Grazing animals or animals giving birth will seek shelter from strong winds. Structures or trees can markedly reduce wind-speed, and can be beneficial to the survival of exposed animals (especially newborn). However, windbreaks have an importance much beyond these benefits, especially in the tropical and subtropical regions. A windbreak acts as a barrier lowering the wind speed near the ground surface, deviating and splitting the air stream. The use of leguminous trees or shrubs can be a practical means to counteract the effects of the wind and the heat stress as well as to improve the animal diet.

Shades and other minimal measures should be thought of as a form of insurance for protecting farm animals in hot climates. As a matter of fact, when dairy cows are given access to adequate shades, milk production is increased (Davison *et al.*, 1988; Blackshaw and Blackshaw, 1994). The radiant environment in a shade has three constituent parts: the cold ground in the shade; the hot ground outside the shade; the lower (inner) surface of the roof. As for the materials used, hay or straw shades are the most effective and low cost artificial materials (Bond *et al.*, 1961). The ground cover around a shade is a factor of importance. The level of thermal radiation above the grass field is less than above a dirt ground (Bond *et al.*, 1969), thus shades are very important for animals in the corral in a hot, sunny environment. The most effective shades are trees, as they provide protection from sunlight combined with beneficial cooling as moisture evaporates from the leaves.

Many forms of environmental modification are available. Shades, sprinklers and fans are very effective methods of improving thermal environment for animals in hot, humid climates (Bucklin *et al.*, 1991). Cooling of roofs or other surrounding surfaces by evaporation of water can effectively reduce the radiant heat load on animals. The selection and use of a specific environmental modification practice or technique must be carefully evaluated, as not all will be cost-effective. In arid and semi-arid region frequent onset of droughts are mainly responsible for considerable losses of animals due to scarcity of fodder. It is vitally important to supplement pasture amelioration with fodder trees and shrubs in order to minimize such losses.

The temperature humidity index (THI) is used as a guide to measure heat stress. THI combines the effects of temperature and humidity into one value. There are three stress categories (temperatures given in Fahrenheit): Livestock alert is 75–78°F, Livestock danger is 79–83°F, and Livestock emergency is 84+°F. At 40% relative humidity (RH) the three levels are at 84°F, 90°F and 98°F, respectively. For 60% RH the three levels are 80°F, 84°F and 92°F, respectively. At 90% RH the three levels are 76°F, 80°F, and 86°F, respectively. The higher the humidity the lower the temperatures for the livestock alert danger and emergency levels to occur.

Acclimation and adaptation are different processes. Animals are considered acclimated to a given ambient environment when body temperature returns to pre-stress levels (Nienaber *et al.*, 1999). Systemic, tissue, and cellular responses associated with acclimation are coordinated, require several days or weeks to occur and are therefore not homeostatic in nature (Bligh, 1976). Furthermore, when stress is removed these changes decay. Adaptation, on the other hand, requires modifications of the genetic structure and is a process involving populations, not individuals. Genetic improvement is an evolutionary action; evolution should be defined as a continuous process of adaptation of the populations of organisms to the ever-changing geological, biological and climatic conditions (Dobzhansky, 1970). Because of the almost infinite number of combinations of environmental factors organisms must have a great variety of genetic types that can deal with a range of climatic, nutritional, or other conditions. In a word, any population must be genetically heterogeneous - *i.e.*, with a great genetic diversity - in order to be able to survive under the challenge of the changing environment. Therefore, any population in a specific ambient environment is composed by a majority of well-adapted individuals, while a minor number of individuals present genotypes that are not good enough for that environment, but are well suited for different conditions. This is the basis for the livestock genetic improvement. Rhoad (1940) was probably the first to propose the selection of livestock for traits related to heat tolerance. The genetic improvement of livestock in tropical countries should take into account not only production traits (milk yield, weight gain, egg, or wool production) but also those traits related to the interaction with environmental factors as the solar radiation, wind, air temperature and humidity.

5.0 CONCLUSIONS

Extended periods of high ambient temperature compromise the ability of the lactating dairy cow to dissipate excess body heat. Cows with elevated body temperature exhibit lower dry matter intake (DMI) and milk yield and produce milk with lower efficiency, reducing profitability for dairy farms in hot, humid climates. Although adequate cooling systems exist their efficiency in humid climates is less than in arid climates and these systems often lack the ability to maintain normal body temperature. Continued genetic selection for improved DMI and milk yield results in cows that are less heat tolerant, and coupled with the unknowns associated with global warming in the future, suggest that heat stress will become worse for dairies in the future. Improved cooling systems that are more efficient and that can cool cows at night when humidity is high are needed to meet challenges in the future.

There is genetic variation in cattle for cooling capability, which suggests that more heat tolerant cattle can be selected genetically, and cross-breeding may also offer opportunities. Continued advances in feeding are needed as cattle are selected for greater milk yield, but are subject to lower intake because of environmental stress. Developing nutritional strategies which support yield but which also address metabolic and physiologic disturbances induced by heat stress will help the cow to maintain a more normal metabolism which should enhance performance. Thermal stress alters the reproductive efficiency and productivity of animals. There is a need to further investigate the effect of thermal stress on early and late fetal development for augmenting the reproductive performance of animals. Development of suitable strategies is required to reduce the adverse affect of thermal stress and economic losses to the farmers for large and small ruminants reared in arid and semi-arid tropical environment.

6.0 REFERENCES

Abi-Saab, S. and Saleiman, F.T. 1995. Physiological responses of stress of filial crosses compared to local Awassi sheep. *Small Ruminant Research* **16(1)**: 55–59.

Aboul-Naga, A.I. 1987. The role of aldosterone in improving productivity of heat-stressed farm animals with different techniques. Ph.D. thesis. Faculty of Agriculture, Zagazig University, Zagazig, Egypt.

Alexander, G. and Williams, D. 1971. Heat stress and the development of the conceptus in domestic sheep. *Journal of Agricultural Sciences* **76**: 53–72.

Alfujairi, M.M., Albrahim, R.M. and Elnouty, E.D. 1993. Seasonal variations in superovulatory responses of Holstein cows treated with pregnant mare serum gonadotropin in Saudi Arabia. *Journal of Reproduction and Fertility* **11**: 75(Abstract).

Al-Katanani, Y.M., Webb, D.W. and Hansen, P.J. 1998. Factors affecting seasonal variation in nonreturn rate of lactating dairy cows. *Journal of Dairy Science* **81(Suppl. 1)**: 217(Abstract).

Alvarez, M.B. and Johnson, H.D. 1973. Environmental heat exposure on cattle plasma catecholamine and glucocorticoids. *Journal of Dairy Science* **56**: 189–194.

Benlmlih, S. 1988. Effect of heat stress and genotype onwater turnover in pregnant and lactating sheep. Isotope Aided studies on Livestock Productivity in Mediterranean and North Africa Countries. *In*: Proceedings of Final Research Co-ordination Meeting, Rabat, pp. 39–46.

Berman, A., Folman, Y., Kaim, M., Mamen, M., Herz, Z., Wolfenson, D., Arieli, A. and Graber, Y. 1985. Upper critical temperatures and forced ventilation effects for high-yielding dairy cows in a subtropical environment. *Journal of Dairy Science* **68**: 1488–1495.

Bianca, W. 1965. Reviews of the progress of dairy science. A physiology of cattle under hot environment. *Research in Veterinary Science* **3**: 38–49.

Blackshaw, J.K. and Blackshaw, A.W. 1994. Heat stress in cattle and the effect of shade on production and behaviour: a review. *Australian Journal of Experimental Agriculture* **34**: 285–295.

Blanc, M.R., Hochereau-de Reviers, M.T., Cahoreau, C., Courot, M. and Dacheux, J.L. 1981. Inhibin: effects on gonadotropin secretion and testis function in the ram and rat. *In*: Franchimont P, Channing CP (*Eds.*), Intragonadal Regulation of Reproduction. Toronto, ON: Academic Press, 1981. pp. 299–326.

Bligh, J. 1976. Introduction to acclimatory adaptation, including notes on terminology. In: *Environmental Physiology of animals* (Bligh, J., Cloudsley-Thompson, J.L. and Mcdonald, A.G. *Eds.*). J. Wiley & Sons, New York, pp. 219–229.

Bond, T.E., Kelly, C.F., Garrett, W.N. and Hahn, G.L. 1961. Evaluation of materials for livestock shades. *California Agriculture* **15**: 7.

Bond, T.E., Morrison, S.R. and Givens, R.L. 1969. Influence of surroundings on radiant heat load of animals. *Transactions of the A.S.A.E.* **12**: 246–248.

Bucklin, R.A., Turner, L.W., Beede, D.K., Bray, D.R. and Hemken, R.W. 1991. Methods to relieve heat stress for dairy cows in hot, humid climates. *Applied Engineering in Agriculture* **7**: 241–247.

Collier, R.J., Doelger, S.G., Head, H.H., Thatcher, W.W. and Wilcox, C.J. 1982. Effects of heat stress during pregnancy on maternal hormone concentrations, calf birth weight and postpartum milk yield of Holstein cows. *Journal of Animal Science* **54**: 309–319.

Conrad, J.H. 1985. Feeding of farm animals in hot and cold environments. *In*: Yousef, M.K. (*Ed.*), Stress Physiology in Livestock. CRC Press, Inc., Boca Raton, Florida, U.S.A., Place, Corvallis, OR 97330, USA.

Copenland, R.S. 1983. Responses to heat stress. *In*: Edey T.N. (*Ed.*) A Course Manual in Tropical Sheep and Goat Production. AUIDP, pp. 122–25.

Daader, A.H., Marai, I.F.M., Habeeb, A.A. and Yousef, H.M. 1989. Improvement of growth performance of Friesian calves under Egyptian sub-tropical conditions. 1. Internal cooling technique using diuretics and drinking cool water. *In*: Proceedings of 3rd Egyptian British Conference on Animal, Fish and Poultry Production, vol. 2, Alexandria, Egypt. pp. 595–605.

Davison, T.M., Silver, B.A., Lisle, A.T. and Orr, W.N. 1988. The influence of shade on milk production of Holstein-Friesian cows in a tropical upland environment. Australian *Journal of Experimental Agriculture* **28**: 149–154.

Dobson, H. and Smith, R.F. 2000. What is stress and how does it affects fertility? *Animal Reproduction Science* **60**: 743–752.

Dobzhansky, T. 1970. Genetics of the Evolutionary Process. Columbia Univ. Press, New York.

Ealy, A.D., Drost, M. and Hansen, P.J. 1993. Developmental changes in embryonic resistance to adverse effects of maternal heat stress in cows. *Journal of Dairy Science* **76**: 2899–2905.

Edwards, J.L. and Hansen, P.J. 1997. Differential responses of bovine oocytes and pre-implantation embryos to heat shock. *Molecular Reproduction and Development* **46**: 138–145.

Ehnert, K. and Moberg, G.P. 1991. Disruption of estrous behaviour in ewes by dexamethasone or management related stress. *Journal of Animal Science* **69**: 2988–2994.

El-Masry, K.A. 1987. The role of thyroxine in improving productivity of heat-stressed animals with different techniques. Ph.D. Thesis, Faculty of Agriculture, Zagazig University, Zagazig, Egypt.

Fahmy, S. 1994. Effect of crossing Romanov with Rahmani sheep on some physiological and productive performance. M.Sc. thesis. Faculty of Agriculture, Al-Azhar University, Cairo, Egypt.

Farghaly, H.A. 1984. Thyroid and some ovarian hormones and their relation to reproduction in Egyptian water buffalo. MSc. Thesis, Faculty of Agriculture Cairo, University, Cairo.

Habeeb, A.A. 1987. The role of insulin in improving productivity of heat-stressed farm animals with different techniques. Ph.D. thesis. Faculty of Agriculture, Zagazig University, Zagazig, Egypt.

Habeeb, A.A., Marai, I.F.M. and Kamal, T.H. 1992. Heat stress. *In*: Philips, C. and Piggens, D. (*Eds.*), Farm Animals and the Environment. C.A.B. International, pp. 27–47.

Habeeb, A.A., Marai, I.F.M. and Owen, J.B. 1997. Genetic improvement of livestock for heat adaptation in hot climates. International Conference on Animal Production & Health, Zagazig University, Zagazig, Egypt.

Hahn, G.L., Mader, T.I. and Eigenberg, R.A. 2003. Perspective on development of thermal indices for animal studies and management. *In*: Interactions between climate and animal production. EAAP Technical Series No.7, pp. 31–44. Wageningen Academic Publ., Wageningen, The Netherlands.

Hansen, P.J. and Aréchiga, C.F. 1999. Strategies for managing reproduction in the heat-stressed dairy cow. *Journal of Animal Science* **77(Suppl. 2)**: 36–50.

Hafez, E.S.E. 1987. Reproduction in Farm Animals, 5th edn. LEA & Febiger, Philadelphia.

Hooda, O.K. and Naqvi, S.M.K. 1990. Effect of thermal load and feed restriction on the relative adaptability of Malpura and Avikalin sheep in semi arid region. *Indian Journal of Animal Science* **60**: 608–611.

Hulme, M. 1997. Global warming. *Progress in Physiological Geography* **21**: 446–453.

Johnson, H.D. 1987. *In*: Johnson, H.D. (*Ed.*), Bioclimatology and the Adaptation of Livestock. Elsevier Science, Publishers BV5, Amesterdam/Oxford/New York/Tokyo.

Kamal, T.H. 1975. Heat stress concept and new tracer methods for heat tolerance in domestic animals. Proceedings of 1st Scientific Conference in Peaceful Uses of Atomic Energy for Scientific and Economic Development. Atomic Energy Commission,Baghdad, Iraq.

Kamal, T.H., Habeeb, A.A., Abdel-Samee, A.M. and Abdel-Razik, M.A. 1989. Supplementation of heat-stressed Friesian cows with urea and mineral mixture and its effect on milk production, in subtropics. Proceedings of International Symposium on the Constraints and Possibilities of Ruminant Production in the Dry Subtropics. Cairo, Egypt. pp. 183–185.

Kamal, T.H. and Johnson, H.D. 1971.Total body solid as a measure of short term heat stress in cattle. *Journal of Animal Science* **32**: 306–311.

Khan, M.S. and Ghosh, P.K. 1989. Physiological responses of desert sheep and goats to grazing during summer and winter. *Indian Journal of Animal Sciences* **59(5)**: 600–603.

Kuthe, G. 1976. Goat rearing in Africa-Its advantages and disadvantages. *Animal Research and Development* **15**: 97–05.

Lindsay, D.R. 1996. Environmental and reproductive behaviour. *Animal Reproduction Science* **42**: 1–12.

Marai, I.F.M., Abdel-Samee, A.M. and El-Gafaary, M.N. 1991. Criteria of response and adaptation to high temperature for reproductive and growth traits in rabbits. *Options Mediterraneennes Ser. A* **17**: 127–134.

Marai, I.F.M., Ayyat, M.S. and Abd El-Monem, U.M. 2001. Growth performance and reproductive traits at first parity of New Zealand White female rabbits as affected by heat stress and its alleviation, under Egyptian conditions. *Tropical Animal Health and Production* **33**: 457–462.

Marai, I.F.M., Bahgat, L.B., Shalaby, T.H. and Abdel-Hafez, M.A. 2000. Fattening performance, some behavioural traits and physiological reactions of male lambs fed concentrates mixture alone with or without natural clay, under hot summer of Egypt. *Ann. Arid Zone* **39**: 449–460.

Marai, I.F.M., Daader, A.M., Abdel-Samee, A.M. and Ibrahim, H. 1997. Winter and summer effects and their ameleoration on lactating Friesian and Holstein cows maintained under Egyptian conditions. *In*: Proceedings of International Conference on Animal, Poultry, Rabbits and Fish Production and Health, Cairo, Egypt.

Marai, I.F.M., El- Darawany, A.A., Abou-Fandoud, E.I. and Abdel-Hafez, M.A.M. 2006. Serun blood components during pre-oestrus, oestru and pregnancy phases in Egyptian Suffolk as affected by heat stress, under the conditions of Egypt. Egypt. *Journal of Sheep Goat Desert Animal Science* **1**: 47–62.

Marai, I.F.M., Habeeb, A.A.M. and Gad, A.E. 2002. Reproductive traits of female rabbits as affected by heat stress and light regime, under sub-tropical conditions of Egypt. *Journal of Animal Science* **75**: 451–458.

Matos, C.A.P. and Thomas, D.L. 1992. Physiology and genetics of testicular size in sheep: a review. *Livestock Production Science* **32**: 1–30

Mc Crabb, G.J. and Bortolussi, G. 1996. Placental growths and the ability of sheep to thermoregulate in hot environment. *Small Ruminant Research* **20**: 121–27.

Maurya, V.P., Naqvi, S.M.K. and Mittal, J.P. 1998. Reproductive behaviour of Malpura and Bharat Merino ewe lamb under semi-arid condition. *Indian Journal of Animal Science* **68**: 533–535.

Maurya, V.P., Naqvi, S.M.K. and Mittal, J.P. 2004. Effect of dietary energy level on physiological responses and reproductive performance of Malpura sheep in the hot semi-arid regions of India. *Small Ruminant Research* **55**: 117–122.

Maurya, V.P., Naqvi, S.M.K., Gulyani, R., Joshi, A. and Mittal, J.P. 2005. Effect of thermal stress on sexual behaviour of superovulated Bharat merino ewes. *Asian Australasian Journal of Animal Sciences* **18**: 1402–1405.

Maurya, V.P., Naqvi, S.M.K., Joshi, A. and Mittal, J.P. 2007. Effect of high temperature stress on physiological responses of Malpura sheep. *Indian Journal of Animal Sciences* **77**: 1244–1247.

Meyerhoeffer, D.C., Wetteman, R.P., Colleman and Wells, M.E. 1985. Reproductive criteria of beef bulls during and after exposure to increased ambient temperature. *Journal of Animal Science* **60**: 352–357.

Naqvi, S.M.K. 1987. Physiological adaptation of sheep during energy crisis and thermal stress. Ph.D. thesis submitted to Kurukshetra University, Kurukshetra, India.

Naqvi, S.M.K., Maurya, V.P., Gulyani, R., Joshi, A. and Mittal, J.P. 2004. Effect of thermal stress on superovulatory response and embryo production in Bharat Merino ewes. *Small Ruminant Research* **55**: 57–63.

Nienaber, J.A., Hahn, G.L. and Eigenberg, R.A. 1999. Quantifying livestock responses for stress management: a review. *International Journal of Biometeorology* **42**: 183–88.

Niles, M.A., Collier, R.J. and Croom, W.J. 1980. Effect of heat stress on rumen, plasma metabolite and plasma hormone concentration of Holstein cows. *Journal of Animal Science* **5(Suppl–1)**: 152(Abstract).

Padua, J.T., Dasilva, R.G., Bottcher, R.W. and Hoff, S.J. 1997. Effect of high environmental temperature on weight gain and food intake of Suffolk lambs reared in a tropical environment. *In*: Proceedings of 5th International Symposium, Bloomington, Minnesota, USA, pp. 809–815.

Putney, D.J., Drost, M. and Thatcher, W.W. 1988. Embryonic development in superovulated dairy cattle exposed to elevated ambient temperature between days 1–7 post- insemination. *Theriogenology* **30**: 195–209.

Reynolds, L.P., Borowicz, P.P., Vonnahme, K.A., Johnson, M.L., Grazul-Bilska, A.T., Redmer, D.A. and Caton, J.S. 2005. Placental angiogenesis in sheep models of compromised pregnancy. DOI: 10.1113/jphysiol.2004.081745

Rhoad, A.O. 1940. A method of assying genetic differences in the adaptability of cattle to tropical and subtropical climates. *Emperial Journal of Experimental Agriculture* **8**: 190–198.

Roush, W. 1994. Population - the view from Cairo. *Science* **265**: 1164–1167.

Sawyer, G.J., Lindsay, D.R. and Martin, G.B. 1979. The influence of radiant heat load on reproduction in the Merino ewe. III. Duration of oestrus, cyclical oestrus, activity, plasma progesterone, LH levels and fertility of ewes exposed to high temperature before mating. *Australian Journal of Agricultural Research* **30**: 1151–1162.

Shaffer, L., Roussel, J.D. and Koonce, K.L. 1981. Effect of age temperature sea son and breed on blood characteretics of dairy cattle. *Journal of Dairy Science* **64**: 62–70.

Shafie, M.M., Murad, H.M., El- Bedawy, T.M. and Salem, S.M. 1994. Effect of heat stress on feed intake, rumen fermentation and water turnover in relation to heat tolerance response by sheep. *Egypt Journal of Animal Production* **31(2)**: 317–327.

Stott, G.H. 1981. What is animal stress and how is it measured. *Journal of Animal Science* **52**: 150–153.

Thatcher, W.W. and Collier, R.J. 1982. *In*: Rechcigl, M. (*Ed.*) CRC Handbood of Animal Productivity. pp. 7– . CRC Press, Boca Raton, FL.

Tompkins, T. and Brayant, M.J. 1974. Oestrus behaviour of ewes and the influence of treatment with progesterone. *Journal of Reproduction and Fertility* **41**: 121–132.

Wu, G., Bazer, F.W., Wallace, J.M. and Spencer, T.E. 2006. Intrauterine growth retardation: Implications for the animal sciences. *Journal of Animal Science* **84**: 2316–2337.

Yousef, H.M. 1985. Studies on some environmental factors affecting production and reproduction in some farm animals. M.Sc. thesis. Faculty of Agriculture, Zagazig University, Zagazig, Egypt.

17

RESPONSE OF DROUGHT STRESS ON FODDER INSECT PESTS AND PATHOGENS AND THEIR MANAGEMENT

SHARMILA ROY AND PRADEEP SAXENA

ABSTRACT

The activity of insect pests and plant pathogens are governed by their environment. Severe and inconsistent weather conditions have influence on distribution and population fluctuations of pests. Likewise unpredictable weather conditions influence the physiology of host plants, leading to changes in its nutrient status, hence affecting the status of insect pests and pathogens that thrive on them. The scientific information on these lines is very scanty – more so in case of fodder crops. In this chapter the available information on the interactions of pests and pathogens with their environment and the possible impact of drought on their physiology and growth are reviewed. Diversity and dynamics of insects and pathogens are governed both by intrinsic and extrinsic factors. Crop damages by pests are a consequence of the complex ecological dynamics between two or more organisms and therefore are very difficult to predict. The drought has impact on the physiology of plants leading to changes in the nutrient status of plant which manipulate the pest attack condition. The effects of extreme weather events on crops may be either direct or indirect, or both. The elevated temperature and the changed nutrient status of drought stressed plants may change the pathogen scenario. The physiological changes in host plants may result in higher disease resistance under climate change scenarios but host resistance to disease may overcome more quickly by more rapid disease cycles, resulting in a greater chance of pathogens developing to overcome host plant resistance. Based on this knowledge generalized postulations are drawn to develop drought strategies to manage the pests of forage crops, grasses and trees. Also some management strategies are also suggested to counteract the drought impacts in forage crops.

Indian Grassland and Fodder Research Institute, Jhansi (UP), India

1.0 INTRODUCTION

In India a sizeable population depends on agriculture for its livelihood. Agricultural activity is adversely affected by any unforeseen weather conditions. Modernization and industrialization have disrupted the natural resource balance and responsible for climate change and global warming. One of the major repercussions that global warming has on agriculture is erratic rainfall and frequent occurrence of drought. A drought initiates when total rainfall is well below average for several months and temperature is high. The time of rainfall, its distribution are also contributing towards drought or draught like conditions.

The activity of insect pests and plant pathogens are governed by their environment. Severe and unpredictable weather conditions have impacts on distribution and population fluctuations of pests. Because of the extremely large variation of pest species' responses to meteorological conditions it becomes difficult to draw specific conclusions about the relationships between pests and weather.

It is well known that drought can change the physiology of host plants, leading to changes in the pests that feed on them and sometimes may influence insect pest outbreaks (Mattson and Haack, 1987). Abnormally cool, wet conditions can also bring on severe insect infestations, although excessive soil moisture may drown out soil-residing insects.

Climate factors that influence the growth, spread, and survival of crop diseases include temperature, precipitation, humidity, dew, radiation, wind speed, circulation patterns, and the occurrence of extreme events. Higher temperature and humidity and greater precipitation result in the spread of plant diseases, as wet vegetation promotes the germination of spores and the proliferation of fungi and bacteria, and influences the life cycle of soil nematodes (Garrett *et al.*, 2006). In regions that suffer aridity, however, disease infestation lessens, although some diseases (such as the powdery mildews) thrive in hot, dry conditions, as long as there is dew formation at night.

2.0 PEST INTERACTIONS WITH BIOTIC AND OTHER ENVIRONMENTAL FACTORS

Diversity and dynamics of insects and pathogens are governed both by intrinsic and extrinsic factors (Fig. 1). The external environment influence activity, abundance and distribution of an individual species and the community as a whole. The essential components of environment are weather (temperature, humidity, wind and light), food, habitat and interaction with other organisms. Severe and inconsistent weather

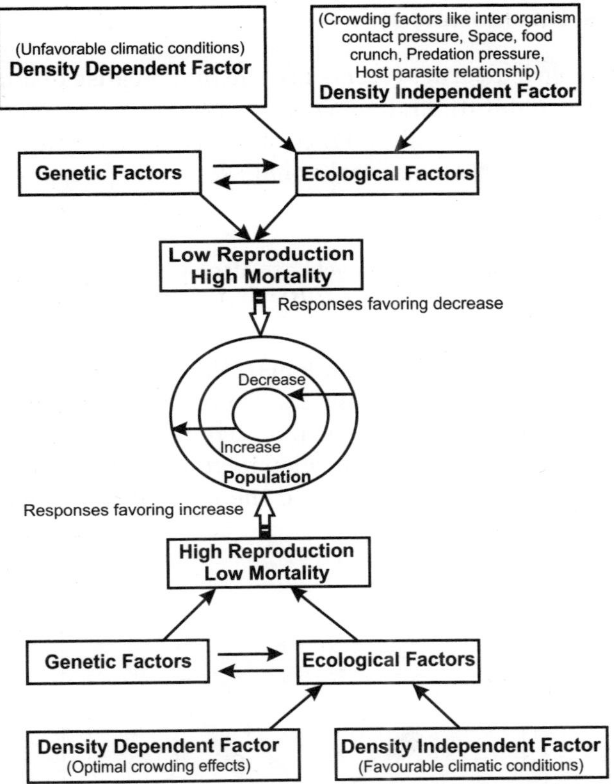

Fig. 1 Diagrammatic presentation of factors that affect pest and pathogen population (Adopted from Nayer *et al.*, 1976)

conditions can have outstanding impacts both on distribution and population fluctuations of insects and pathogens (Nayer *et al.*, 1976).

Several climatic factors as temperature, rain, etc. may affect the survival and reproductive behavior of insects and pathogens directly. In general, however, most pest species are favored with warm and humid conditions. These factors can affect the abundance of the pest populations indirectly by effecting on their food plant, other herbivores, predators, parasitoids and pathogens.

Crop damages by pests are a consequence of the complex ecological dynamics between two or more organisms and therefore are very difficult to predict. For example, dry conditions are unfavorable for sporulation of fungi, but are also unfavorable for the crop; a weak crop during a drought is sometimes more likely to become infected by fungi than when it is not stressed. Similarly, pest infestations often coincide with changes in climatic conditions, such as early or late rains, drought, or increases in humidity, which, in themselves, can reduce yields. In these circumstances, accurately attributing losses to pests may be difficult.

2.1 Insect's Environment

Insect pests in agricultural systems are the major cause of damage to yield quantity. Insect habitats and survival strategies are strongly dependent on patterns of climate. The abundance of insects is determined by weather, food (host plant) and their natural enemies.

2.1.1 *Weather*

The activities and population of insects are often guided by weather conditions specially temperature and humidity (Fig. 2). Light is also an important factor in the development, growth and survival of the insects. Insects are poikilothermes (cold-blooded organisms), the temperature of their body changes with that of the environment. Their fecundity, development and migration are directly dependent on climate temperature. Also temperature can impact insect physiology and development directly or indirectly through the physiology or existence of hosts and the natural enemies of the insects.

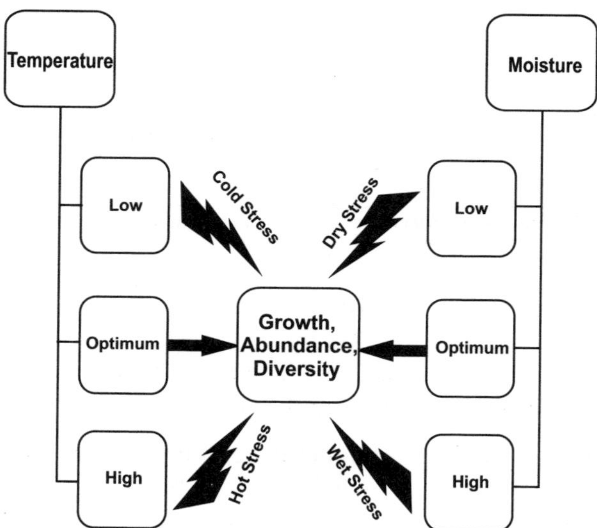

Fig. 2 Impact of temperature and moisture on development of insect population

The optimum range of temperature for oviposition (5°C-35°C), development and survival (15°C-35°C) is very narrow and varies with different species. The lives of majority of insects are influenced by this threshold level of temperature. The temperature above and below this level are treated as fatal high or fatal low temperatures. However, few insects have capacity to live in hot springs *e.g. Thermobia domestica* lives at 25-45°C. Similarly, some micro-arthropods *Folsomia quadrioculata, Camisia anomia* are found in arctic vegetations. Because of the direct dependence of insect development on temperature degree-day or phenology based approaches are used to predict the population dynamics of insects and their prospective to damage crops.

An excellent example of influence of temperature on the rate of migration and dispersal of insects is provided by locusts (*Schistocerca gregaria*). The formation of swarm directly depends on the prevailing temperature and the upward thermal current near the ground surface. Similarly, seasonal and daily trends in fluctuation of *Thrips inagins* was found influenced by temperature (Lewis, 1997).

Most of the terrestrial insects live in dry environment. Like other organisms insects too need to have to maintain water content of their body. The major source of water for insects is through their food, that is, water is maintained in insects is through the host plants upon which they feed. Therefore, insects develop a number of mechanisms to conserve water.

Like temperature humidity also influence rate of egg production and rate of development in insects. They die at lethal wetness or lethal dryness pointing towards the need of endurable range of moisture. The ability to withstand water reduction or unduly excess of water content depends on the individual and the species which varies from 50-90 per cent of the body weight. In general the lower water content leads to retardation in growth and metabolism rate. The requirement of water in different insects is fulfilled either by their food (sap sucking insects) or by direct absorption from their environment. The flies, cockroaches etc. drink water directly. Water is also absorbed by the cuticle of the insects or by some special structure like collophore in collembola and hydropyle cells in grasshopper and locusts. However, others retain tissue water which is lost as by product of respiration in the oxidation of fat and carbohydrate (Nayer *et al.*, 1976).

Moisture like temperature also influences the rate of egg production. Studies related to relative humidity and population of *Callosobruchus chinensis* increases at RH between 50-75 per cent (Utida, 1941). Similarly another study showed that the *Locusta migratoria* failed to produce eggs at 40 per cent RH and the rate of development is also influenced at this humidity level. Laboratory studies with 1st instar of southern green stink bug, *Nezara viridula* (L.) indicated that relative humidity (RH) greatly affected nymph emergence and survivorship up to the 2nd instar, reaching the maximum value (90%) with RH of $>$ 80%. At 0% RH only 15% of eggs hatched, and most nymphs died. In the presence of a gradient of humidity, nymphs tend to break up the colony with greater intensity and take longer time to regroup as they grew older and seek areas with greater RH, but do not tolerate wet areas that cause massive mortality at this period of development (Hirose *et al.*, 2006).

Light is a vital factor in response to insect behaviour. In many insects oviposition is stimulated by exposure to light or darkness. Similarly, day length influence the annual life cycle of others. The day length also influences the seasonal forms of insects. For example, the light colour butter flies come out from the caterpillars exposed to light for 12 h, while dark colour forms are frequent when the caterpillars are exposed to 16 h of light.

The variation in photoperiod also changes the reproduction forms from sexual to non sexual forms or vice versa.

2.1.2 *Host Plant*

This ability helps on their population growth. The phytophagus insects show various associations with the host plant (food plant). This association is mainly based on feeding stimuli which include gustatory, visual, olfactory, tactile etc. Some insects feed on diverse type of plants while some are monophagous. A number of insects choose to feed on related plants. Plant characters influence insect ability to select host plants under field conditions like wax on epidermis, leaf toughness, pubescence etc. The existence of qualitative and quantitative differences in sugars, amino acids and polyphenolic compounds at various stages of plant growth influence selection of host plant and feeding behaviour of insects.

The success of individual insects depends on their ability to locate good host plant. The plant density, plant quality, plant diversity affect the insect responses. Genetically uniform crops (monocultures) usually responsible for an out break of pest insects. While genetically diverse crops (polycultures) regulate the pest population. However, care must be taken to ensure that the crops should not serve as alternate hosts for the pests.

2.1.3 *Natural Enemies*

Predators and parasites are the major factors for regulating insect population. Environmental factors also affect the regulation and stabilization of the natural enemy's population as it affects the insect pest population. The parasites/predators disrupt the stability or homeostasis of the pest to a non economic level and help in maintaining this level.

The insects like other organisms also suffer by infectious diseases caused by bacteria, viruses, fungi etc. The diseases may be epizootic or enzootic in nature. However, the transmission of these diseases depends on climatic factors such as temperature, humidity, host density and host conditions.

2.2 Plant Pathogens and Their Environment

The development of a disease may occur when a susceptible plant, virulent pathogen and favorable environment are present together form "plant disease triangle" (Agrios, 1978). The susceptibility of plants and virulence of pathogens usually get affected by the climatic conditions. In general plant diseases develop during warm and wet season. A situation may occur where the host, parasite and environment factors occur; but if they don't occur at the right time then disease will not result. Diseases are often managed through the use of time, *i.e.* time of planting, time of harvest, timing of varieties, rotations, etc. (Fig. 3).

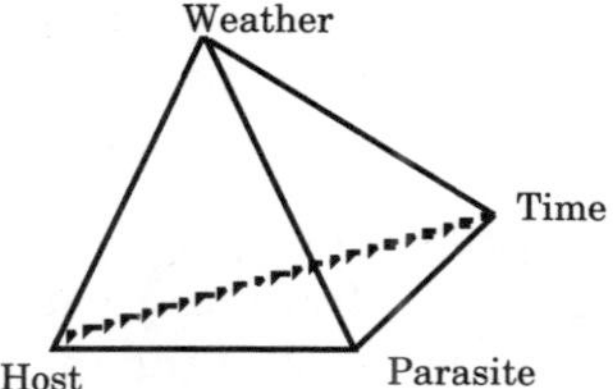

Fig. 3 of Plant disease tetrahedron (factors responsible for development of diseases

2.2.1 *Weather*

The environmental factors that affect the plant disease occurrence are temperature, humidity, wind velocity, soil pH, soil nutrients and host plant nutrients. The development of almost all plant pathogens is very dependent on the environmental conditions which occur during the time that a crop is grown. The most important environmental variables are temperature and humidity, followed by wind and rainfall.Temperature has potential impacts on plant disease through both the host crop plant and the pathogen (Fig. 4).

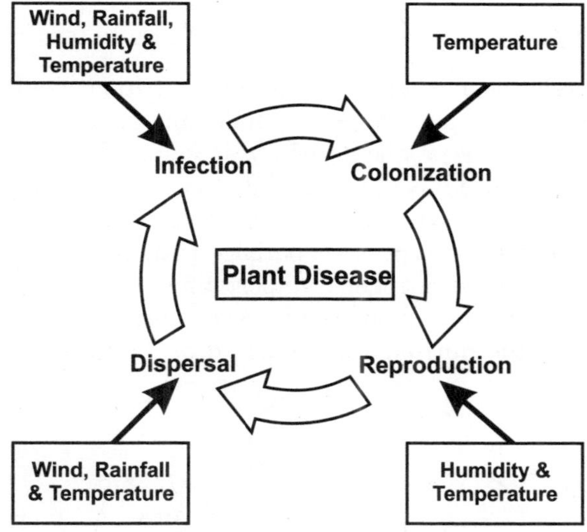

Fig. 4 Effect of climatic factors on disease progression

Temperature is important because it determines the rate at which all of the growth and dispersal processes occur in plant pathogens and the vectors such as aphids and thrips which commonly act as vectors for viral diseases. Rates of reproduction in plant pathogenic micro-organisms, for example, are highly dependent on temperature. Cell division in bacteria and hyphal growth and spore formation in fungi are strongly influenced by the temperature of the environment. Most plant pathogens have an optimum temperature at which they grow and multiply best; at temperatures above and below this optimum their growth and reproduction is slower, and in extreme cases may not occur at all. However, temperatures that inhibit host may not affect the parasite and *vice verse*. The daily temperature cycles play a modulatory role in plant disease progress.

Because the infective propagule spores of most plant pathogens are a single cell with little to protect it from desiccation, moisture in the form of humidity is important to the survival of the inoculum. The humidity of the atmosphere or soil moisture content affects the rate of development of plant disease epidemics because micro-organisms such as fungi and bacteria are often sensitive to dry conditions and will only grow when there is sufficient moisture. Abundant moisture on host surface is also needed for germination of infection propogules and successful establishment of infection. Sometimes, free water also play a role in dissemination of diseases. For example, zoospores of slime moulds need free water for motility, encystment and germination. The effect of moisture on the host should also be considered because of its effect on plant vigour. Low moisture may lead to wilting and collapse, while excessive moisture can lead to anoxia in roots thereby weakening the plant and increasing vulnerability to pathogenic activity.

2.2.2 *Host Plant*

There is a direct relationship between type, age, nutrition of plant and disease progress (Fig. 5). Plants that are grown under nutrient conditions, such as high nitrogen, results in the production of new succulent shoots and may prolong vegetative stage may also have increased disease susceptibility. Conversely, lack of nitrogen may result into slower growth, weaker plants and faster aging of plants making it more susceptible to pathogen attack. In this contest this is possible that the form of the nitrogen is important for the disease development. Even a balanced nutrition may affect the development of diseases when the concentration of all the nutrients is increased or decreased beyond specific

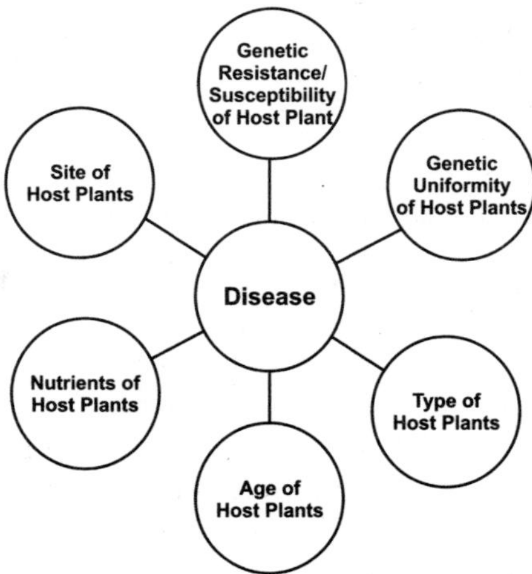

Fig. 5 Host factors that affect development of diseases

range. Plants genetic diversity too influences the resistance to infection and/ or pathogenesis of specific organisms. At the same time within a species, plants have variable reaction to the pathogens. The age of plants is also an important factor for disease occurrence. Some plants are more susceptible as juveniles while others are more susceptible as adults. The crop life cycle *i.e.* their perennial nature or annual nature too is important in disease progress. The location of the crop also has an important role. Often our crop species are grown far away from and under very different conditions than were their progenitor species; corn, potatoes, tomatoes, horticultural species, etc.

2.2.3 *Virulence of Pathogen*

The proximity and intensity of virulence of a parasite is important to turn out to be a pathogen. However, the ecology of pathogen and the type of reproduction of the parasite also have their role in spreading of diseases. Some parasites require free water, while others are inhibited by free water. Therefore, one must know the biology of the parasite in order to understand the potential for disease and be able to predict disease progress. All plant pathogens are dependent of external forces; *i.e.* wind, water, insects, humans, soil, etc, for their dissemination. Some are very specific in their needs for a vector while others are effectively disseminated as transients.

2.2.4 *Disease Progress in Time*

The effects of the environment on the temporal behaviour of disease lead to variations in the rate at which diseases increase. The disease progression curves (DPCs) are very useful because they summarise the combined effects of everything which affects the rate of disease progress. The parameters of DPCs illustrate a relationship between the way in which a disease spreads spatially (*i.e.* from plant to plant) and the rate at which the crop becomes diseased. However, an understanding of spatial component of plant diseases is also important. The pattern of diseased plants, the change in its pattern indicates the processes of dispersal of the pathogen or its vectors also the pattern of infection in the field which influence the rate of disease (Fig. 6).

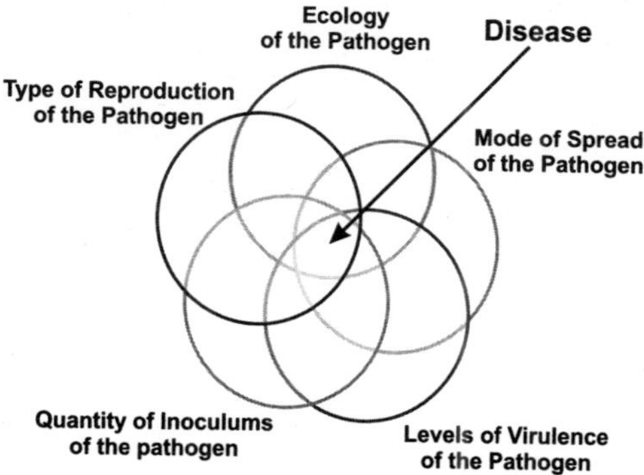

Fig. 6 Pathogen factors that affect development of disease

Table 1 Forage crops pest spectrum

Crop	Insect pests	Pathogens
Sorghum (*Sorghum bicolor*)	Shootfly (*Atherigona varia soccata*), Stem borer (*Chilo partellus*), Aphids (*Schizaphis graminum*), Sorghum midge (*Contarinia sorghicola*), Pyrilla (*Pyrilla perpusilla*), Pollen eating beetle (*Chiloba acuta*), Army worm (*Mythimna separata, Spodoptera exigua*), Cut worms (*Agrotis* sp.), Leaf roller (*Marsemia trapezalis*), Ear head bug (*Caocoris angustatus*), Ear head caterpillar (*Helicoverpa armigera)*	Anthracnose (*Colletotrichum graminicola*), Sooty stripe (*Ramulispora s orghi*), Zonate leaf spot (*Gloeocercospora sorghi*), Downy mildew (*Sclerospora sorghi*), Gray leaf spot (*Cercospora sorghi*), Rough leaf spot (*Arcochyta sorghi*), Turcicum leaf blight (*Helminthosporium turcicum*), Maydis leaf blight (*H. maydis*), Charcoal rot (*Macrophomina phaseo armigera*) Rust (*Puccinia purpurea*), Ergot (*Claviceps fusiformis*), Leaf spots (*Ramulispora sorghi*, *Pyricularia* sp.)
Oats (*Avena sativa*)	Bird cherry aphid (*Rhapalosiphum maidis*), Thrips (*Halothrips ganglbauen*)	Leaf blotch (*Helminthosporium avenae*), Crown rust (*Puccinia coronata avenae*), Stem rust (*Puccinia graminis avenae*), Sclerotial wilt (*Sclerotium rolfsii*)
Maize (*Zea mays*)	Shootfly (*Atherigona varia soccata*), Stem borer (*Chilo partellus*), Aphids (*Schizaphis graminum*), Sorghum midge (*Contarinia sorghicola*), Pyrilla (*Pyrilla perpusilla*), Pollen eating beetle (*Chiloba acuta*), Army worm (*Mythimna separata, Spodoptera exigua*), Cut worms (*Agrotis* sp.), Leaf roller (*Marsemia trapezalis*), Ear head bug (*Caocoris angustatus*), Ear head caterpillar (*Helicoverpa armigera*)	Brown stripe downy mildews (*Scleropthora rayssi* var. *zeae*), Bacterial stalk rot (*Erwinia carotovora* var. *zeae*), Turcicum leaf blight (*Helminthosporium turcicum*), Maydis leaf blight (*Dreschslera maydis*), Common Rust (*Puccinia sorghi*), Leaf spots (*Helminthosporium nodulosum, H. tetramera*), Seedling blight (*Fusarium moniliforme*), Stalk rot and seed rot (*Cephalosporium acrimonium, Diplodia maydis, Erwinia carotovora*)
Pearl Millet (*Pennisetum glaucum*)	Shootfly (*Atherigona varia soccata*), Stem borer (*Chilo partellus*), Aphids (*Schizaphis graminum),* Pollen eating beetle (*Chiloba acuta*), Blister beetle (*Zonabris pustulata*), Army worm (*Mythimna separata, Spodoptera exigua*), Cut worms (*Agrotis* spp.)	Downy mildew (*Scleropthora graminicola*), Ergot (*Claviceps fusiformis)*, Smut (*Tolyposporium penicillariae*), Rust (*Puccinia penniseti*), Leaf spots (*Cercospora fusimaculans, Curvularia penniseti, Helminthosporium australiense, H. rostratum, Pyricularia pennisetii*), Leaf blotch (*Xanthomonas annamalaiensis*), Twisted top (*Fusarium moniliforme*)

(Contd...)

(Contd...)

Crop	Insect pests	Pathogens
Berseem (*Trifolium alexandrinum*)	Gram caterpillar (*Helicoverpa armigera*), Bihar hairy caterpillar (*Spilosoma obliqua*), Hairy caterpillars (*Euproctis virguncola* and *E. lunata*), Green semilooper (*Plusia nigrisigna*, *Thysanoplusia orichalcea*), Aphid (*Aphis craccivora*), Dusky bug (*Oxycaraenus* sp.), Leaf miner (*Phytomyza* sp.)	Root rot complex (*Rhizoctonia solani, Fusarium semitactum* and *Tylenchorhynchus vulgaris*), Stem rot (*Sclerotinia trifoliorum*), Leaf spots (*Alternaria, Myrothecium* and *Epicoccum* spp.)
Cowpea (*Vigna unguiculata)*	Flea beetles (*Monolepta signata*), Hairy cater pillars (*Spilosoma obliqua*), semilooper (*Plusia nigrisigna*), Leaf hopper (*Empoasca kerri*), Blue butterfly (*Limpidness booties*), Pod borer (*Maruca testulalis*), Gram caterpillar (*Helicoverpa armigera*), Black aphid (*Aphis craccivora*), Blister beetle (*Mylabris phalerata*)	Root rot (*Macrophomina phaseolina anamorph, Rhizoctonia bataticola*), cowpea mosaic virus (CPMV), Leaf spots (*Colletotrichum lendmuthianum*), Bacterial blight (*Xanthomonas compestris* pv. *vignicola*), Seedling rot (*Rhizoctonia, Pythium, Sclerotium, Phytopthora* sp.), Top necrosis (*Fusarium equiseti*), Wilt (*Fusarium* spp.), Powdery mildew (*Oidium* spp.)
Lucerne (*Medicago sativa*)	Lucerne weevil (*Hypera postica*), Hairy caterpillar (*Spilosoma obliqua*), Aphid (*Acyrthosiphon pisum* and *Therioaphis trifolii f. maculata*), Leaf hoppers (*Exitianus indicus, Empoascanara* sp., *Austroagalia* sp., *Macrosteles* sp. and *Empoasca* sp.), Alfalfa caterpillar (*Spodoptera exigua*), Tobacco caterpillar (*S. litura*), Lucerne butterfly (*Terias hecabe*), Semilooper (*Thysanoplusia nigrisigna*), Thrips (*Halothrips tenuipennis*), Seed chalcid (*Bruchophagus roddi*)	Downy mildew (*Peronospora trifoliorum*), Rust (*Uromyces striatus*), Common leaf spot (*Pseudopeziza medicagenis*), Anthracnose (*Colletotrichum trifolii*), Crown warts (*Physoderma alfalfae*), Leaf spots (*Cercospera mediagenis, Stemphylium botryosum, Alternaria medicagenis, Xanthomonas alfafae*), Powdery mildew (*Erysiphe polygoni*), Mosaic (Alfalfa mosaic virus), Crown and root rot (*Sclerotinia trifoliorum*), Damping off or Root rot (*Phytopthora megasperma.*), Root rot (*Rhizoctonia* and *Fusarium* sp., *Corynobacterium insidiosum*)
Guar (*Cymopsis tetragonaloba*)	Leaf hoppers (*Empoasca* sp., *Exitianus indicus*), Yellow mite (*Polyphagotarsonemus latur* and *P. datus*)	Bacterial blight (*Xanthomonas compestris* pv. *Cyamopsidis*), Leaf spots (*Alternaria cyamopsidis*), powdery mildew (*Erysiphe polygoni*), Root rot (*Rhizoctonia solani*) and Anthracnose (*Colletotrichum lindmuthianum*)

(Contd...)

(*Contd...*)

Crop	Insect pests	Pathogens
Stylo (*Stylosanthes* spp.)	Pod borer (*Helicoverpa armigera*), Hairy caterpillars (*Spilosoma obliqua*) Leaf hopper (*Empoasca kerri*), Black aphid (*Aphis craccivora*)	Anthracnose (*Colletotrichum gloeosporiodes* and *C. dematium* f. *truncata*), Leaf spot (*Cercospora* spp.), Leaf and stem spot (*C. truncatum*), Seedling rot and blight (*Corticium rolfsii*), Foliar blight (*Rhizoctonia solani*), Inflorescence blight (*Botrytis cinerea*, *Rhizopus stplonifer*)
Leucaena (*Leucaena leucocephala*)	Psyllids (*Heteropsylla cubana*)	Root rot (*Ganoderma lircidium*), Gumosis (*Fusarium solani*), Pink disease (*Corticum salamanicolor*), Necrotic lesions (*Botryodiplodia theobromus*), Seedling rot (*Colletotrichum* spp.), Leaf spot (*Camptomeris leucaenae*), Black spot(*Colletotrichum gloeosporiodes*), Rust (*Uredo leucaenae- glaucae*), Damping off (*Pythium solani*, *Rhizoctonia solani* and *Fusarium oxysporum*), Seedling rot (*Colletotrichum* spp.)
Grasses	Grasshopper (*Acrida crenulata, Thisoicetrus pulcher, Cyrtocanthacris tatarica, Hieroglyphus nigrorepletus, Catantops pinguis, Acrida exaltata, Oedaleus abruptus, Chrotogonus trachypterus, Aelopus tamulus, Colemania* sp., *Oxya* sp., *Locusta migratoria, Melanoplus spretus, Attractomorpha crenulata, Acrida exaltata),* Yellow hairy caterpillars *(Psalms pentacle*), Hairy caterpillar (*Euproctis* sp., *Creatonotus gangis*), Red hairy caterpillar (*Amsacta* spp.), Army worms *(Cirphis unipuncta, C. albistigma, Spodoptera mauritia, Euxoa spinifera*), Termites (*Microtermes* sp., *Odontotermis* sp.), Aphid (*Rhapalosiphum maidis*), Leaf hoppers (*Nephotattix* sp., *Empoasca* sp., *Empoascanara* sp., *Exitianus indicus*)	Rust (*Puccinia* and *Uromyces* spp.), Leaf spots (*Pyricularia*, *Phylochora*, *Dreschslera*, *Rhynchosporium*, *Sorosporium*, *Colletotrichum* and *Helminthosporium* spp.), Ergot (*Claviceps purpurea*), Smut (*Ustilago* sp.), Blight (*Rhizoctonia solani*)

3.0 KEY INSECT PESTS AND PATHOGENS OF FODDER CROPS

Forage crops are subjected to the depredation of pests and diseases in the same manner as in other agricultural produce (Table 1). These pests and diseases hamper crop establishment, impair forage quality and reduce green fodder and seed yield (Saxena *et al.*, 2002). Besides this, they also cause indirect losses, like knot formation in legumes ultimately resulting to reduction of nitrogen fixation capacity. Nematodes, aphids, jassids and other sucking pests inflict direct losses to the crop, yet are also indirectly responsible for the transmission of a number of viral and other pathogens. Further disease and pest interactions in the plant produce toxins, adversely affecting the animal health, which sometimes can be fatal.

4.0 HOST - PEST RESPONSE TO DROUGHT

4.1 Plant Response to Drought

The effects of extreme weather events on crops may be either direct or indirect, or both. Higher temperatures increase moisture stress on crops directly by increasing evaporation and transpiration as well as the atmospheric holding capacity for water vapor. An indirect effect is produced when higher temperatures accelerate the breakdown of organic matter in soils, which in turn leads to lower soil organic matter levels, culminating in less soil-moisture retention and additional crop moisture stress. Symptoms of drought will vary widely depending on the plant species and the severity of the water deficit. There are also both short term and long term effects of drought. Initial symptoms of drought, is expressed as wilt, curl, roll, yellow, marginal scorch, interveinal necrosis and chlorosis, and premature leaf drop, dieback, poor growth, and stunting. Both direct and indirect effects threaten yields. Plants in the early stages of development are especially vulnerable to extreme weather events. This chaotic nature of the climate has strongly detrimental effects on crop yields (Table 2).

4.1.1 *Effect on Morphology and Physiology*

The scanty water, high temperature and low relative humidity affect physical and physiological conditions of plants. The bulk of roots and root system of crops are in the top soil layer *i.e.* 12-15 inches of soil, the lack of water initially damage root system of plants. In the absence of optimum moisture and temperature the feeder roots dried up and the water absorbing capability of plants declines drastically, which in turn effects transpiration and nutrient uptake by plants and gaseous exchange in plants. Most plants lower their osmotic potential during drought by accumulating inorganic ions, amino acids, sugars, sugar alcohols, and organic acids (Kramer, 1983;

Table 2 High temperature and soil moisture effects on major field crops in the U.S.

Crop	Effects
Corn	• Temperature higher than 36ºC causes pollen to lose viability. Extremely sensitive to soil-moisture deficits. Four days of visible wilting in (a) the period before tasseling reduces yield by 10–25%; (b) between the week before tasseling and the milk stage reduces yield by more than 50%; in (c) the soft dough stage, decreases yield by 40%. • Aflatoxin concentration rises when the crop has a water deficit. • Very intolerant to flooding except after the silking stage; the effect of flooding depends on temperature. Before the 6th leaf stage the crop does not survive more than 4 days of flooding if the temperature is less than 25ºC; and less than 24 h if the temperature is more than 25ºC. When the crop is less than 6 inches high, 24 h of flooding reduces yield by 18% at any temperature. • Continuous soil saturation causes long-term problems related to rot development and increased damage by diseases (*e.g.*, crazy top and common smut).
Soybean	• Soil temperature higher than 35ºC at planting causes seedling death. Very sensitive to temperatures above 35ºC during the first three weeks after bloom. Great ability to recover from temperature stress at other times. • Sensitive to soil moisture deficits and drought at planting and from bloom to pod-fill. Very sensitive to soil moisture deficits during pod-filling and seed enlargement. • Relatively tolerant to excess soil humidity, but saturated soils increase the risk of seedling diseases especially at temperatures above 32ºC.
Wheat	• Temperature above 30ºC for more than 8 h can reverse vernalization. • Flowering, pollination, and grain-filling sensitive to water stress. • Excess soil moisture causes water logging, and increases risk of fungal infestations.
Cotton	• Temperature above 40ºC for more that 6 h cause bolls to abort. • Relatively tolerant to temperatures under 40ºC. • Sensitive to soil moisture deficits and drought at planting and flowering. • Excess rainfall at maturity damages quality of crop.

Source: Rosenzweig *et al*., 2000

Mattson and Haack, 1987). The roots become cork like under water stress and high temperature conditions thus the uptake of mineral nutrient is altered during drought (Kramer, 1983). Several minerals "(calcium, chlorine, potassium, magnesium, nitrogen, sodium) can accumulate to higher than normal levels in the above ground tissues of drought-stressed plants (Mattson and Haack, 1987).

The damage becomes severe when drought occurs during periods of active plant growth. During the growth period cell elongation, expansion of cell membrane, metabolic activities in cell protoplasm and thickening of cell wall demands high water availability. In water deficit condition all the growth processes get arrested. Similarly, in the water stress conditions stomata openings are low in frequency thus CO_2 intake gets reduced hence the rate of photosynthesis gets decreases.

The low photosynthesis is responsible for reduced carbohydrate and secondary metabolites production. All these affect growth and maintenance of crops. Due to lesser or no production of secondary metabolites *viz.*, alkaloids, tannins etc. by plants which are mainly responsible for plant self defense mechanism against pests and pathogens, the crops becomes vulnerable for pest attacks.

4.1.2 *Effect on Gene Expression of Plant*

Under drought like conditions changes occur at the molecular level also. Jacobson *et al.* (1986) and Heikkila *et al.* (1984) have recorded quantitative and qualitative changes in protein synthesis in barley and maize crops. The high temperature prevalent in drought condition was found to induce the production of heat shock proteins in plants (Walbot and Cullis, 1985). Way *et al.* (2005) found that genes up-regulated in coding for aldehyde dehydrogenase (associated with osmoregulation), delta pyrroline-5-carbixylate synthetase (responsible for biosynthesis of proline (protect plant cell from dehydration) and fatty acid alpha oxidase (involved in repair of damage cell membrane, permeability and fluidity of membrane) get affected by short term and long term drought stress. The up-regulated genes (involved in cellular metabolism, cellular transport, signal transduction and transcriptional regulation as well hydrophilic and heat soluble proteins) and down-regulated genes (involved in cell wall, cellulose and germin like protiein synthesis, responsible for leaf growth, changes in leaf structures, stomata closures and change in root: shoot ratio) are affected by water stress (Bray, 2004).

4.2 Effect of Drought on Insects

The drought induces unusual climatic changes and its impact on plants morphology, and nutrient status affect the pest populations in several complex ways. The response under the stress conditions on the insect species and individuals within the same species (Lombardero *et al.*, 2000; Glynn *et al.*, 2003) vary depending on plant type (Mutikainen *et al.*, 2000).

Depending on the development pattern of pest species, temperature can have different effects. For example, cicadas take several years to complete one life cycle and this way they moderate the temperature variability required for their growth. In case of low relative humidity (40%) and high temperature (35°C) the *Locusta migratoria* were seen to produce lower number of eggs and length of life was shortened. However, the fossil record studies by some researchers conclude that the diversity of insect species and the intensity of their feeding have increased historically with increasing temperature (Bale *et al.*, 2002). The warmer temperatures in temperate climates may result in increase in population as well diversity of insects in particular habitat or crop. It has been estimated that with a 2°C temperature increase, insects might experience one to five additional life cycles per season (Yamamura and Kiritani, 1998). Lower winter mortality of insects due to warmer winter temperatures could be important in increasing insect populations (Harrington *et al.*, 2001). Insects that spend important parts of their life histories in the soil may be gradually affected by temperature changes than those that are above ground simply because soil provides an insulating medium that will tend to buffer temperature changes more than the air (Bale *et al.*, 2002). The insects living in dry atmosphere can survive the drought like conditions by fulfilling the water requirement through metabolic water. These insects are capable of altering their metabolism by oxidizing fats and carbohydrates in excess. The grasshoppers in dry conditions regulate water evaporation from body by controlling spiracles opening and closing cycles. However, other insects such as pea aphids are not tolerant of drought (Macvean and Dixon, 2001).

The changes in temperature effects might tend to depress insect populations by altering the insect predator – prey interactions. The environmental factors that impact pest insects can impact their predators and parasites as well as the disease organisms that infect the pests, resulting in increased attack on insect populations. At warmer temperature aphid population in alfalfa is reduced, as their predator lady bird beetles are more active than their prey and require more food (Cammell and Knight, 1992). At the same time at higher temperatures, aphids were found less responsive to the aphid alarm pheromone they release when under attack by insect predators and parasitoids – resulting in the potential for greater predation (Awmack *et al.*, 1997). Some times it was found that the pest species tolerate hotter climate while the predators may not. The brown plant hoppers were found 17 times tolerant to 40°C temperature than its hymenoptera predator *Cyrtorrhynus lividipennis* but wolf spider *Paradosa pseudoannulata* is tolerant to this temperature (Ramanjaneyulu and Raghunath, 2009).

The *drought or draught* like conditions may reduce the water contents of plants and there by make the plants vulnerable to sucking insects or the reduced water content may discourage the chewing insects like caterpillars (Bethan *et al.*, 2005). *Decreased foliar water content may also have contributed to lower insect performance* (Scriber, 1977). In contrast, drought stress had no effect on the growth of tussock moth, perhaps because this species is less sensitive than gypsy moth to changes in phenolic glycosides, as tussock moth has been shown to be relatively insensitive to other phenolic compounds (Karowe, 1989; Kopper *et al.*, 2002).

The incidence of Rice Hispa in Telengana region of Andhra Pradesh, India, has increased in the last two years due to dry weather. A swing from the leaf eating caterpillars to sucking pest has been noticed in recent years in various crops. For example, incidence of mealy bugs and jassids in cotton, aphids incidence in groundnut, thrips and yellow mites attack in chillies, eriophids in coconut. This swing of pest scenario has increased the incidence of various plant diseases as these sucking insects are vectors of viral diseases *e.g.* bud necrosis in groundnut, tobacco streak virus in cotton etc. (Ramanjaneyulu and Raghunath, 2009).

The changing weather conditions may indirectly alter insect abundance by altering the relationship and condition of host plants. The drought stressed plants are more acceptable morphologically and physiologically by pests. The yellowing leaf colour attracts buprestid beetles and many other insects (Altner and Loftus, 1985). Drought stress alters the qualitative and quantitative nutrient compositions of plants hence are acceptable to the herbivores (Miller and Stickler, 1984). The increase in sugars and salts stimulates feeding of spruce budworm *Choristoneura Fumiferana* (Albert *et al.*, 1982) high sucrose level stimulates feeding in *Locusta migratoria* (Mattson and Haack, 1987). Increased levels of proline, inositol, ethylene, ethane, acetaldehyde and ethanol stimulate the host finding ability of the insects (Dunn *et al.*, 1986; Visser, 1986). The drought stress lowers the terpenes emission due to stomatal closure in plants, results in increased insect pests (Chararas, 1979; Cates and Alexander, 1982). The bark beetle and needle feeding lepidoptera larva colonize in the drought stress conifers as the production of oleoresin decreases in the stressed plants. While in many others cases all the terpenes are not reduced like alpha pinene may increase and resulting in bark beetle attack in the drought affected conifers (Mattson and Haack, 1987).

Some studies have shown drought stress increase secondary metabolism and decrease host quality of trees for insect herbivores (Ross and Berisford, 1990; Craig *et al.*, 1991; Mopper and Whitham, 1992), whereas others have found the opposite (Wagner, 1986; Cobb *et al.*, 1997; Roth *et al.*, 1997). The drought may also affect insect detoxification system. Hinks (1985) reported that increasing temperature reduces the susceptibility of *Melanopus sanquinipes* and other insects to pyrethroids, DDT and some carbamates. Mattson and Haack (1987) reported that due to elevated nutrient status and temperature of drought stressed plants, insect's detoxification mechanism may be enhanced in respect of some toxicants.

Environmental variation does not affect all secondary metabolites within a plant in the same way (Muzika, 1993; Reichardt *et al.*, 1991; Kainulainen *et al.*, 1996), which could also generate complex patterns of herbivore response. In two willow species, condensed tannin levels were decreased by mild drought, whereas simple phenolic compounds were not affected (Glynn *et al.*, 2004). Additionally, positive effects of drought on nutritional quality, such as increases in free amino acids, soluble proteins, and soluble carbohydrates (Brodbeck and Strong, 1987; Mattson and Haack, 1987), may counteract negative effects of increased secondary metabolite levels (Glynn *et al.*, 2004).

Outbreaks of phytophagous insects have been linked to droughts (Table 3), although effects may be specific to particular feeding guilds (Larsson, 1989; Koricheva *et al.*, 1998; Huberty and Denno, 2004). The study of White (1984) did not provide any evidence to support the hypothesis that drought stress enhances folivore performance. In contrast, drought stress decreased the growth of gypsy moth, perhaps because drought also decreased foliar nitrogen and increased total phenolic glycoside concentrations, both of which are key determinants of gypsy moth performance on quaking aspen (Lindroth and Hemming, 1990; Hemming and Lindroth, 2000; Osier *et al.*, 2000). Nutrient availability had a positive linear effect on tussock moth growth and foliar nitrogen, but had no effect on gypsy moth. This overall pattern of treatment effects on these two species is consistent with relatively high sensitivities of tussock moth to nutritional quality and gypsy moth to phenolic glycosides. Differential effects of resource availability on constitutive *vs.* induced resistance may be an important source of variation in herbivore performance (Lewinsohn *et al.*, 1993; Lerdau *et al.*, 1994; Lombardero *et al.*, 2000).

Table 3 Selected genera of forest and range insects that historically have reached outbreak proportions following drought

Insect			Genera of woody hosts
Order	**Family**	**Genus**	
Coleoptera	Buprestidae	*Agrilus Tetropium*	*Betula, Populus, Quercus*
	Cerambycidae	*Corthylus*	*Abies*
	Scolytidae	*Dendroctonus*	*Acer*
		Ips	*Picea, Pinus*
		Scolytus	*Abies, Carya*
Homoptera	Aphididae	*Aphis*	*Crataegus*
	Psyllidae	*Cardiaspina*	*Eucalyptus*
Hymenoptera	Diprionidae	*Neodiprion*	*Pinus*
Lepidoptera	Geometridae	*Bupalus*	*Pinus*
		Lambdina	*Abies, Picea*
		Selidosema	*Pinus*
	Lymantriidae	*Lymantria*	*Picea*, several hardwoods
	Tortricidae	*Choristoneura*	*Abies, Picea, Pinus*
Orthoptera	Acrididae	Several genera	Many grasses and range plants

Source: Mattson and Haack, 1987

4.3 Effect of Draught on Pathogens

The elevated temperature and the changed nutrient status of drought stressed plants may change the pathogen scenario. The physiological changes in host plants may result in higher disease resistance under climate change scenarios but host resistance to disease may be overcome more quickly by more rapid disease cycles, resulting in a greater chance of pathogens developing to overcome host plant resistance. The type of disease-related variables (incidence *vs.* severity) and the intensity and timing of water stress were significant factors affecting the drought- infection interaction. Direct effects of drought on pathogens are generally negative, although most fungal pathogens exhibit an important plasticity and can grow at water potentials well below the minimum for growth of their host plants. Pennypacker *et al.* (1991) while studying plant pathogen interaction in alfalfa plants found fewer symptoms of *Verticillium albo-atrum* under drought stress. The spread of *Phytophthora cinnamomi* from the tropical and subtropical regions to the Mediterranean region in the oak plantation is explained by Braiser (1992) as an effect of climate change which has onset the drought stress.

Positive drought-infection effects can mostly be explained by indirect effects of drought on host physiology. The enhanced nutrition may permit to colonize root fungi on the roots of stressed trees. Pathogens that take advantage of weakened trees include canker diseases, vascular wilts, root rots, and wood rots. *Armillaria* root rot, *Botryosphaeria*, *Cytospora* and *Nectria* cankers, *Diplodia* tip blight, *Rhisosphaera* needlecast, and *Verticillium* wilt are commonly associated with stressed trees. It is the inability of the drought stressed plant to produce sufficient amounts of secondary metabolites to be used for defense, and reduced water and carbohydrates needed to compartmentalize wounds, that allows for development of these secondary diseases. According to Wargo (1981), root colonizing fungi, *Armillaria mellea* succeeds in stressed trees because the increased concentrations of sugars and amino acids increase its ability to oxidize the normally inhibitory phenols and even to utilize the phenolics as a carbon source. Warmer winters under drought may allow larger populations of flea beetles to survive would be expected to increase the frequency of growing seasons with severe Stewart's wilt (Whitney *et al.*, 2000; Baniecki and Dabaam, 2000a, b).

Research has shown that host plants such as wheat and oats become more susceptible to rust diseases with increased temperature; but some forage species become more resistant to fungi with increased temperature (Coakley *et al.*, 1999). Many mathematical models that have been useful for forecasting plant disease epidemics are based on increases in pathogen growth and infection within specified temperature ranges. Generally, fungi that cause plant disease grow best in moderate temperature ranges. Temperate climate zones that include seasons with cold average temperatures are likely to experience longer periods of temperatures suitable for pathogen growth and reproduction if climates warm. For example, predictive models for potato and tomato late blight (caused by *Phytophthora infestans*) show that the fungus infects and reproduces most successfully during periods of high moisture that occur when temperatures are between 7.2°C and 26.8°C (Wallin and Waggoner, 1950).

The impact of drought varied with disease types also. In some cases, drought conditions will impact disease by making the environment more or less favorable for infection, disease development, and/or disease spread. In other cases, drought may not impact the pathogen at all, but may exacerbate the damage caused by disease in drought stressed plants (Table 4).

Table 4 General effects of drought on different diseases

Pests/Diseases	**Drought intensity**		
	Less severe	**Severe**	**Very severe**
Root diseases	Downy mildew of sunflower, Phytophthora root rot of soybeans, Aphanomyces root rot of sugarbeet, and damping off of many plants caused by *Pythium* specie	*Phytophthora*, *Pythium*, *Aphanomyces*, and *Plasmopara*	Root rots of legume forages (*Fusarium* and *Verticillium*)
Leaf diseases		Fungal or bacterial pathogens causing leaf spots	Leaf spot pathogens, rusts, powdery mildew
Stem diseases	Infections on leaves or florets (*Phomopsis* and *Phoma*, white mold)	White mold (*Sclerotinia sclerotiorum*) infects many broadleaf,	Charcoal rot (*Macrophomina*) is a stem disease of corn, soybean, sunflower, edible beans and other crops

5.0 MANAGEMENT OF INSECT PESTS AND PATHOGENS UNDER DROUGHT CONDITION

The effect of drought on pest and pathogens tend to be different for different pathosystem in different locations. Although the specific impacts of drought on plant insect pests and diseases are difficult to predict given our present knowledge, it is possible to make some generalizations for management of pests and pathogens of fodder crops and trees in drought conditions: a) increased temperatures may likely mean higher populations of pests and pathogens survive to initially attack or infect plants; b) increased temperatures may likely result in spread of pests and pathogens to newer regions; c) Due to change in climate conditions and introduction of new developed varieties to newer regions, the pests may spread to the non infected zones; and d) erratic rainfall events may tend to favor some types of pests and pathogens over others.

To reduce the impact of drought in the crops and fodder trees following measures may be helpful to maintain the plants:

- Maintenance of moisture level: by slow and regular irrigation to maintain the optimum moisture in upper root zone so as the feeder roots may not be damaged
- Organic manure helps in maintenance of soil moisture. But there are evidences of enhanced root feeding insects or rots
- Optimization in fertilizer application is important as the fertilizer contain salts that can aggravate drought problems
- Removal of diseased plants may help to minimize problems with secondary pests and pathogens.

Since the onset of drought is a gradual process it gives some opportunity to adapt and to understand how these changes will affect crops, insects, diseases, and the relationships among them. The problem of pest and diseases may be tackled by understanding crop ecosystem and by modifying the cropping system. It is possible that new pests will become established be able attack plants in new regions. It is expected that plants in some regions will be attacked more frequently by certain pests. A few pests may be less likely to attack crops as change occurs.

It is anticipated that the actual impacts of climate change on pests may not be known until they occur. Therefore it is important to be conscious of crop pest trends in particular region/s and flexibility in choosing both their management methods and in the crops selection. Close monitoring of the occurrence of pests in the fields and maintenance of records of the severity, frequency is the essence of ecological pest management. Use of farmer's knowledge, skills and use of local resources and natural processes help to keep the natural balance of pests and cost of managing pests.

Pesticides and their applications have significant economic costs for growers and environmental costs for society. As a result of increased population and generation of pests more pesticide applications are required to maintain populations below economic damage thresholds. Simultaneously some pesticides may be affected negatively by physiological changes of the host plants (smaller stomatal opening or thicker epicuticular waxes in crop plants grown under higher temperatures) that slow down the uptake rates of the pesticides. At the same time increased plant metabolic rates due to dry weather conditions could increase fungicide uptake. In this condition insecticide resistant may develop (Shelton *et al.*, 2001; Georghiou and Taylor, 1986). To avoid the development of pesticide resistance it is necessary to apply a range of insecticides of varied mode of action. Additionally, some classes of pesticides (pyrethroids and spinosad) have been shown to be less effective in controlling pests at higher temperatures (Musser and Shelton, 2005).

To prevent disaster caused by drought following measures are suggested for pest and pathogen management for fodder crops and trees:

- Crop diversification
- Use of drought tolerant crops
- Use of resistant cultivars
- To prevent dispersion of disease and pest; an early warning/ forecasting of disease and insect pests
- Epidemiological surveillance for improved awareness and knowledge of the distribution of disease or infection
- Effective management of pest or disease outbreak by promoting regional/ localized planning and initiative
- Procurement of early and bulk of pesticides for the abundant and diverse crop pests scenario

The best use of the basics of integrated pest management (IPM) such as field monitoring, pest forecasting, record keeping, and choosing economically and environmentally sound control measures will be most likely to be successful in dealing with the effects of climate change (Petzoldt and Seaman, 2006).

6.0 REFERENCES

Agrios, G.N. 1978. Effect of environment on development of infectious plant diseases. *In*: Plant Pathology, Academic Press Inc. pp. 104–114.

Albert, P.J., Cearley, C., Hanson, E. and Parisella, S. 1982. Feeding responses of eastern spruce budworm larvae to sucrose and other carbohydrates. *J. Chem. Ecol*. **8**: 233–239.

Altner, H. and Loftus, R.1985. Ultrastructure and function of insect thermo-and hygroreceptors. *Ann. Rev. Entomol.* **30**: 273–296.

Awmack, C.S., Woodcock, C.M. and Harrington, R.1997. Climate change may increase

Bale, J.S., Masters, G.J., Hodkinson, I.D., Awmack, C. and Bezemer, T.M. 2002. Herbivory in global climate change research: direct effects of rising temperature on insect herbivores. *Global Change Biol.* **8**: 1–16

Baniecki, J. and Dabaan, M. 2000a. Crop Profile for Corn (sweet) in West Virginia.

http://www.ipmcenters.org/CropProfiles/docs/wvcorn-sweet.html

Baniecki, J. and Dabaan, M. 2000b. Crop Profile for Potatoes in West Virginia.

http://www.ipmcenters.org/CropProfiles/docs/WVpotatoes.html

Bethan, K.H., Daniel, A.H., Hansen, R.C., Thomas, P.C. and Arnold, Danielle 2005. Effects of drought stress and nutrient availability on dry matter allocation, phenolic glycosides, and rapid induced resistance of poplar to two *Lymantriid defoliators*. *J. Chem. Ecol.*, **31(11)**: 2601–2620.

Brasier, C.M. 1992. Oak tree mortality in Iberia. *Nature* pp. 360–539.

Bray, E.A. 2004. Genes commonly regulated by water deficit stress in *Arabidopsis thaliana*. *J. Experimental Biol.* **55**: 2331–2341.

Brodbeck, B. and Strong, D. 1987. Amino acid nutrition of herbivorous insects and stress to host plants. *In*: Insect Outbreaks. Barbosa, P. and Shultz, J.C. (*Eds.*) Academic Press, San Diego, CA. pp. 347–364.

Cammell, M. and Knight, J. 1992. Effects of climatic change on the population dynamics of crop pests. *Adv. Ecol. Res.* **22**: 117–162.

Cates, R.G. and Alexander, H. 1982. Host resistance and susceptibility. *In*: Bark Beetles of North American Conifers. Mitton, J.B. and Sturgeon, K.B. (*Eds.*) University of Texas Press, Austin. pp. 212–263.

Chararas, C. 1979. Ecophysiologie des Insectes Parasites des Forets. 38 Bis, Avenue Rene Coty, 75014, Paris, France. (Cross Reference)

Coakley, S.M., Scherm, H. and Chakraborty, S. 1999. Climate Change and Disease Management. *Ann. Rev. Phytopatho.* **37**: 399–426.

Cobb, N.S., Mopper, S., Gehring, C.A., Caouette, M., Christensen, K.M. and Whitham, T.G. 1997. Increased moth herbivory associated with environmental stress of pinyon pine at local and regional levels. *Oecologia* **109**: 389–397.

Craig, T.P., Wagner, M.R., McCullough, D.G. and Frantz, D.P. 1991. Effects of experimentally altered plant moisture stress on the performance of Neodiprion sawflies. *Forest Ecol. Mgmt.* **39**: 247–261.

Dunn, J.P., Kimmerer, T.W. and Nordin, G.L. 1986. Attraction of the twolined chestnut borer, *Agrilus bilineatus* (Weber) (Coleoptera: Buprestidae), and associated borers to volatiles of stressed white oak. *Can. Entomol.* **118**: 503–509.

Garrett, K.A., Dendy, S.P., Frank, E.E., Rouse, M.N. and Travers, S.E. 2006. Climate change effects on plant disease: Genomes to ecosystems. *Ann. Rev. Phytopathol.* **44**: 489–509.

Georghiou, G.P. and Taylor, C.E. 1986. Factors influencing the evolution of resistance. *In*: Pesticide Resistance: Strategies and tactics for management. National Research Council, National Academy Press. pp. 143–157.

Glynn, C., Herms, D.A., Egawa, M., Hansen, R. and Mattson, W.J. 2003. Effects of nutrient availability on biomass allocation as well as constitutive and rapid induced herbivore resistance in poplar. Oikos 101:385–397.

Glynn, C., Ronnberg-Wastljung, A.C., Julkunen-Tiito, R. and Weih, M. 2004. Willow genotype, but not drought treatment, affects foliar phenolic concentrations and leaf-beetle resistance. *Entomol. Exp. Appl.* **113**: 1–14.

Harrington, R., Fleming, R. and Woiwod, I., 2001. Climate change impacts on insect management and conservation in temperate regions: can they be predicted? *Agric. For. Entomol.* **3**: 233–240.

Heikkila, J.J., Papp, J.E.T., Schultz, G.A. and Bewley, J.D. 1984. Induction of heat shock protein messenger RNA in maize mesocotyls by water stress, abscisic acid, and wounding. *Plant Physiol.* **76**: 270–274.

Hemming, J.D.C. and Lindroth, R.L. 2000. Effects of phenolic glycosides and protein on gypsy moth (Lepidoptera: Lymantriidae) and forest tent caterpillar

(Lepidoptera: Lasiocampidae) performance and detoxication activities. *Environ. Entomol.* **29**: 1108–1115.

Hinks, C.E. 1985. The influence of temperature on the efficacy of three pyrethroid insecticides against the grasshopper *Melanoplus sanguinipes* (Fab.) (Orthoptera: Acrididae) under laboratory conditions. *Can. Entomol.* **117**: 1007–1012.

Huberty, A.F. and Denno, R.F. 2004. Plant water stress and its consequences for herbivorous insects: a new synthesis. *Ecology* **85**: 1383–1398.

Hirose, E., Panizzi, A.R. and Cattelan, A.J. 2006. Effect of relative humidity on emergence and on dispersal and regrouping of first instar *Nezara viridula* (L.) (Hemiptera: Pentatomidae). *Neotropical Entomol.* **35(6)**: 757–761.

Jacobsen, J.V., Hanson, A.O. and Chanyores, P.C. 1986. Water stress enhances expression of amylase gene in barley leaves. *Plant Physiol.* **80**: 350–359.

Kainulainen, P., Holopainen, J., Palomaki, V. and Holopainen, T. 1996. Effects of nitrogen fertilization on secondary chemistry and ectomycorrhizal state of Scots pine seedlings and on growth of grey pine aphid. *J. Chem. Ecol.* **22**: 617–636.

Karowe, D. 1989. Differential effects of tannin acid on two tree-feeding Lepidoptera: implications for theories of plant anti-herbivore chemistry. *Oecologia* **80**: 507–512.

Kopper, B.J., Jakobi, V.N., Osier, T.L. and Lindroth, R.L. 2002. Effects of paper birch condensed tannin on whitemarked tussock moth (Lepidoptera: Lymantriidae) performance. *Environ. Entomol.* **31**: 10–14.

Koricheva, J., Larsson, S. and Haukioja, E. 1998. Insect performance on experimentally stressed woody plants: a meta-analysis. *Annu. Rev. Entomol.* **43**: 195–216.

Kramer, P.J. 1983. Water Relations of Plants. Academic Press, Orlando, FL.

Larsson, S. 1989. Stressful times for the plant stress-insect performance hypothesis. *Oikos* **56**: 277–283.

Lerdau, M., Litvak, M. and Monson, R. 1994. Plant-chemical defense-monoterpenes and the growth - differentiation balance hypothesis. *Trends Ecol. Evol.* **9**: 58–61.

Lewinsohn, E., Gijzen, M., Muzika, R.M., Barton, K. and Croteau, R. 1993. Oleoresinosis in grand fir (*Abies grandis*) saplings and mature trees-modulation of this wound response by light and water stresses. *Plant Physiol.* **101**: 1021Y1028.

Lewis, T. 1997. Thrips as crop pests. CAB International, Cambridge: University Press. pp. 740.

Lindroth, R.L. and Hemming, J.C.D. 1990. Responses of the gypsy moth (Lepidoptera: Lymantriidae) to tremulacin, an aspen phenolic glycoside. *Environ. Entomol.* **19**: 842–847.

Lombardero, M.J., Ayers, M.P., Lorio, P.L. Jr. and Ruel, J.J. 2000. Environmental effects on constitutive and inducible resin defences of *Pinus taeda*. *Ecol. Lett.* **3**: 329–339.

Macvean, R. and Dixon, A.F.G. 2001. The effect of plant drought-stress on populations of the pea aphid *Acyrthosiphon pisum*. *Ecol. Entomol.* **26**: 440–443.

Mattson, W.J. and Haack, R.A. 1987. The role of drought in outbreaks of plant-eating insects. *Bioscience* **37**: 110–118.

Miller, J.R. and Strickler, K.L. 1984. Finding and accepting host plants. *In*: Chemical Ecology of Insects. Bell, W.J. and Carde, R.T. (*Eds.*) SinauerAssociates, Sunderland, MA. pp. 127–157

Mopper, S. and Whitham, T.G. 1992. The plant stress paradox: effects on pinyon sawfly sex ratios and fecundity. *Ecology* **73**: 515–525.

Musser, F.P. and Shelton, A.M. 2005. The influence of post-exposure temperature on the toxicity of insecticides to *Ostrinia nubilalis* (Lepidoptera: Crambidae). *Pest Mgmt. Sci.* **61**: 508–510.

Mutikainen, P., Wallis, M., Ovaska, J., Keinanen, M., Julkunen-Tiitto, R. and Vappaavuori, E. 2000. Herbivore resistance in *Betula pendula*: effect of fertilization, defoliation and plant genotype. *Ecology* **81**: 49Y65.

Muzika, R.M. 1993. Terpenes and phenolics in response to nitrogen fertilization: a test of the carbon/nutrient balance hypothesis. *Chemoecology* **4**: 3–7.

Nayer, K.K., Ananthakrishnan, T.N. and David, B.V. 1976. General and Applied entomology. Tata McGraw Hill Publ. Ltd. New Delhi. pp. 589.

Osier, T.L., Hwang, S.Y. and Lindroth, R.L. 2000. Effects of phytochemical variation in quaking aspen *Populus tremuloides* clones on gypsy moth, *Lymantria dispar*, performance in the field and laboratory. *Ecol. Entomol.* **25**: 197–207.

Pennypacker, B.W., Leath, K.T. and Hill, R.R. Jr. 1991. Impact of drought stress on the expression of resistance to *Verticillium albo-artum* in alfalfa. *Phytopathology* **81**: 1014–1024.

Petzoldt, C. and Seaman, A. 2006. Climate Change Effects on Insects and Pathogens. *In*: A report on Climate Change and Agriculture: Promoting Practical and Profitable Responses. New York State IPM Program, 630 W. North St., New York State Agricultural Extension Station, Geneva, NY 14456.

Ramanjaneyulu, G.V. and Raghunath, T.A.V.S. 2009. Pest Shifts are observed with changes in the ecological balance. The Hindu: Survey of Indian Agriculture 2009. pp. 115–118.

Reichardt, P.B., Chapin, F.S., Bryant, J.P., Mattes, B.R. and Clausen, T.P. 1991. Carbon nutrient balance as a predictor of plant defense in *Alaskan balsam* poplar V potential importance of metabolite turnover. *Oecologia* **88**: 401–406.

Rosenzweig, C., Iglesias, A., Yang, X.B., Epstein, P.R. and Chivian, E. 2000. Climate Change and U.S. Agriculture: The Impacts of Warming and Extreme Weather Events on Productivity, Plant Diseases and Pests. A report Published by Centre for Health and the Global Environment, Harvard Medical School, 260 Longwood Avenue, Boston, MA 02115.

Ross, D.W. and Berisford, C.W. 1990. Nantucket pine tip moth (Lepidoptera: Tortricidae): response to water and nutrient status of loblolly pine. *For. Sci.* **36**: 719–733.

Roth, S., McDonald, E.P. and Lindroth, R.L. 1997. Atmospheric CO_2 and soil water availability: consequences for tree Y insect interactions. *Can. J. For. Res.* **27**: 1281–1290.

Saxena, P., Pandey, K.C., Faruqui, S.A., Hasan, N., Bhaskar, R.B., Roy, S., Azmi, M.I., Shah, N.K. and Padmavati, C. 2002. *Forage Plant Protection*. Technical Bulletin. IGFRI, Jhansi, India. pp. 38.

Scriber, J.M. 1977. Limiting effects of low leaf-water content on the nitrogen utilization, energy budget, and larval growth of *Hyalophora cecropia* (Lepidoptera: Saturniidae). *Oecologia* **28**: 269–287.

Shelton, A.M., Wilsey, W.R. and Soderlund, D.M. 2001. Classification of insecticides and acaricides for resistance management. Dept. of Entomology, NYSAES,

Geneva, NY 14456. pp. 315–787–2352. http://www.nysaes.cornell.edu/ent/faculty/shelton/pdf/res_mgmt.pdf

Utida, S. 1941. Studies on experimental population of the azuki bean weevil *Callosobruchus chinensis* (L.) I. The effect of population density on the progeny population. *Mem. Coll. Science,* Kyoto Univ. **48**: 1–30.

Visser, J.H. 1986. Host odor perception in phytophagous insects. *Annu. Rev. Entomol.* **31**: 121–144.

Wagner, M.R. 1986. Influence of moisture stress and induced resistance in ponderosa pine, *Pinus ponderosa* Dougl. Ex. Laws, on the pine sawfly, *Neodiprion autumnalis* Smith. *For. Ecol. Manag.* **15**: 43–53.

Walbot, V. and Cullis, C.A. 1985. Rapid genomic changes in higher plants. *Annu. Rev. Plant Physiol.* **36**: 367–396.

Wallin, J.R. and Waggoner, P.E. 1950. The influence of climate on the development and spread of *Phytophthora infestans* in artificially inoculated potato plots. *Plant Dis. Reptr. Suppl.* **190**: 19–33.

Wargo, P.M. 1981. Defoliation and secondary action organism attack: With emphasis on *Armillaria mellea*. *J. Arboric.* **7**: 64–69.

Way, H., Chapman, S., McIntyre, L., Casu, R. and Xue, G.P. 2005. Identification of differentially expressed genes in wheat undergoing gradual water deficit stress using a subtractive hybridization approach. *Plant Sci.* **68**: 661–670.

White, T.C.R. 1984. The abundance of invertebrate herbivores in relation to the availability of nitrogen in stressed food plants. *Oecologia* **63**: 90–105.

Whitney, S., Whalen, J., VanGessel, M. and Mulrooney, B. 2000. Crop Profile for Corn (sweet) in Delaware. http://www.ipmcenters.org/CropProfiles/docs/DEcorn-sweet.html

Yamamura, K. and Kiritani, K. 1998. A simple method to estimate the potential increase in the number of generations under global warming in temperate zones. *Appl. Entomol. Zoo.* **33**: 289–298.

18

HANDLING VETERINARY EMERGENCY DURING NATURAL DISASTER

P. Das

ABSTRACT

The primary role of veterinary service in any disaster is to i) ensure high standard of animal health, ii) reduce mortality iii) keep watch on animal welfare and iv) restore the animal productivity at the earliest. The risk of epidemics increases after natural disasters because of exposure to many kinds of disease causing bacteria, viruses, fungi, parasites spread by flood waters debris and stress that lowers the immunity. The traditional methods of controlling and eventually managing epidemics are movement control, vaccination and slaughter of infected herds (wherever possible). In the present circumstances in India the effective deployment of all three methods is difficult and in some cases impossible. Livestock owners may be reluctant to accept movement controls and the necessary forces may not be available to enforce them. Vaccination is acceptable provided that there be no untoward losses post vaccination; but to achieve good vaccination cover of the population requires careful planning and execution. The quarantine of infected herds is needed till the infection has died out. The slaughter of infected herds is unacceptable to most livestock owners unless there is immediate and full compensation, but the compensation payments are costly to the authorities. Early and accurate warning of new outbreaks of epidemic diseases and particularly the spread of such diseases to new areas is an essential pre-requisite to the effective containment and control of these diseases. Some of the common problems in early warning systems include lack of farmer awareness, absence of active disease surveillance, inadequate training of staff, failure to establish confirmatory diagnosis, inadequate liaison with international reference laboratories. Veterinary hospitals in disaster prone area should make a list of essential list of medicines and equipments for use during disaster.

Indian Veterinary Research Institute, Izatnagar (UP), India

1.0 INTRODUCTION

Veterinarians have a role to play in all stages of disaster mitigation and management, but it is during relief effort that they can play a major role in increasing survivability of animals that are victims of disaster and of those that are deployed in rescue teams (*e.g.* dogs engaged in rescue team). The primary role of veterinary service in any disaster is to i) ensure high standard of animal health, ii) reduce mortality iii) keep watch on animal welfare, and iv) restore the animal productivity at the earliest. Veterinarians can play a major role in promoting local disaster planning at community level which places a high priority on facilitating livestock and pet evacuation (Heath, 2000). Veterinarians can also instruct their clients on first aid for livestock and pet animals and advice on the content and appropriate use of first aid kids. Veterinarians also contribute significantly in 'Search and Rescue' team which generally employs dogs to detect the injured and dead persons/animals. Veterinarians provide the necessary care (like first aid and treatment) to the dogs injured in the rescue operation. In rural areas the number of animals could be greater than the number of humans and in urban areas pets could form a sizeable population. In the USA, special Veterinary Medical Assistance Team (VMAT), comprising veterinarians, veterinary pathologists, epidemiologists, toxicologist, technicians etc. actively is involved in emergency situations to assess veterinary needs of the animals. The VMAT teams also include private citizens who are activated in times of disaster and they are paid a salary and protected under the Federal Toirt Claims Act against personal liability.

Singh and Sivakumar (2006) outlined the responsibilities of the veterinarians. The primary duties are:

- Attend the injured animals and save their life
- Maintenance of animal health
- Providing shelter and food to the rescued animals
- Disposal of animal waste and dead animals (carcasses)
- Disinfection and disinfestations
- Intervention in case of disease outbreak
- Preventive vaccination
- Capture and care of stray animals including their housing and feeding
- Storage and preservation of food of animal origin
- Training and updating personnel
- Appropriate intervention to sustain the animal productivity

2.0 DISEASES DURING DISASTER

After a disastrous event, many problems concerning animals need to be addressed like care and treatment of injured animals, stray animal control, obtaining suitable housing for livestock, return of the strays to their owners, providing food to abandoned animals, determination if animals that die during the disaster, died due to the event or some other cause as an infectious disease and disposal of animal carcasses. The risk of epidemics increases after natural disasters because of exposure to many kinds of disease causing bacteria, viruses, fungi, parasites spread by flood waters debris and stress that lowers the immunity. Animals housed in emergency animal shelters where disease agents build up in the environment and pass to others. After natural disasters, vectors (which transmit disease from one living thing to another, often by bite) like rodents (rats, mice) which are associated with many diseases like plague, salmonellosis, leptospirosis and mosquito population increase after floods and cyclones which are associated with Malaria, encephalitis. Common diseases after disasters are Rabies, Leptospira, *E. coli*, plague, ring worm, internal parasites (*Toxocara*, *Ancylostoma*, *Giardia*), external parasites (fleas, scabies, mites).

2.1 During Flood

- **Poultry**: Avian influenza and Newcastle disease.
- **Pig**: Classical swine fever, Foot and Mouth Disease (FMD), Japanese encephalitis
- **Sheep and Goat**: PPR, Pasteurellosis, fleas, scabies and mites
- **Cattle**: PPR, FMD, Hemorrhagic septicemia, Anthrax, Black leg, Malignant Edema, Tetanus (Lockjaw), Botulism, Brooder pneumonia, Foot rot and Mastitis.

These diseases should be monitored and where appropriate, animal disease control procedures should be implemented to curb the spread of diseases. Vaccination against epidemics and anthelmintics and treatment campaigns should be made. It needs adequate supply of vital veterinary medicines, food and water that can make difference between life and death for animals in the aftermath of the disaster.

2.2. During Drought

There are certain diseases, which are more common during drought period. These diseases need more attention so as to prevent' its outbreak. Infectious diseases (bacterial, fungal, viral and parasitic) are more common. The most common diseases are Foot and Mouth Disease, Hemorrhagic septicaemia, Black Quarter, Anthrax Enterotoxaemia, Colibacillus, Surra, Babesiosis,

Theileriosis, Anaplasmosis, Pox diseases, Mastitis, Brucellosis, Ring worm, Ascariasis, Fascioliasis, Microfilariasis, Tick infestation and mange etc.

3.0 AVAILABILITY OF MEDICINE DURING DISASTER

Veterinary hospitals in disaster prone are should make a list of essential list of medicines for use during disaster. A list of medicines is given in the Table 1 as ready reference. However the list is only an indicative one, not an exhaustive list. Moreover the choice of medicines depends on the availability in the local market.

Table 1: List of common veterinary medicines for emergency use during natural disasters

S. No.	Name of the drug	Dose	Mode of administration
Antibiotics			
1	Procaine penicillin G	25000 IU	I/M bid
2	Trimethoprim–sulphamethoxazole	20 mg/kg	PO bid
3	Gentamycin	6.6 mg/kg	I/V or I/M bid
4	Metronidazole	20 mg/kg 7.5 mg/kg	PO as initial dose followed by PO 6 hourly
5	Oxytetracycline	10 mg /kg	I/V sid
6.	Ointments and creams	–	Tropical application
Sedative and anesthetics			
1	Xylazine	0.25–1.0 mg/kg	I/V or I/N
2	Butorphanol	0.05–1.0 mg/kg	I/V or I/M
3	Acepromazine	0.02–0.05 mg/kg	I/V or I/M
4	Ketamine	2.2 mg/kg	I/V, 5 min after Xylazine
5	Lidocaine		Local anesthetic
6	Mepivicaine		Local anesthetic
Anti-inflammatory			
1	Phenylbutazone	2.2–4.4 mg/kg	I/V bid or sid
2	Flunixin meglumine	0.25–1.0 mg/kg	I/V or I/M bid or sid
3	Dexamethasone	0.02–0.2 mg/kg	I/M or I/V
4	Prednisolone	0.02–1.0 mg/kg	PO bid
Fluids			
1	0.9% Sodium Chloride Soln.		I/V
2	5% Dextrose		I/V
3	Lactated Ringers soln.		I/V
4	Balance electrolyte soln.		I/V
Miscellaneous			
1	Furosemide	1–3 mg /kg	I/V or I/M bid
2	Tetanus antitoxin	1500 Units	I/M or I/V
3	Tetanus toxoid	1 ml	I/M
4	Pentobarbital	100 mg/kg	I/V for euthanasia
5	Ivermectin	0.2 mg/kg	I/M

Source: Dwivedi, 2006

4.0 VETERINARY VACCINES

The objective of vaccination is to provide effective immunity by establishing adequate levels of antibody and a primed population of cells, which can rapidly expand on renewed contact with antigen. The first contact with antigen during vaccination obviously should not be injurious and the maneuver is to avoid the pathogenic effect while maintaining protective antigens.

4.1 Bacterial Vaccines

4.1.1 *Anthrax Spore Vaccine (Live)*

- Anthrax is an acute febrile diseases of mammals caused by *Bacillus anthracis.* Mortality rate is very high. There will be edema, enlargement of spleen and oozing of dark color blood from natural body openings, which fails to clot. Anthrax spore vaccine is a suspension of living spores of an attenuated, non-capsulated strain of *B. anthracis* in 50% glycerin saline solution. Vaccination is done in areas with recent outbreak or in areas with past report of outbreak.
- *Issue*: Each dose contains 1.5 million spores per ml. Vaccine is given by subcutaneous route (sheep and goats 0.5 ml and cattle 1.0 ml).
- *Immunity*: About one year.
- *Storage*: Store the vaccine in a cool (2 to 4°C) and dark place. Do not freeze it. Self-life of the vaccine is 9 months.

4.1.2 *Avian Spirochaetosis Vaccine (Inactivated)*

- Avian spirochaetosis is an acute, febrile, septicemic, bacterial disease of poultry birds. The causative organism, *Borrelia anserina,* is a spirochete. Mortality rate is very high and birds will die within 3 to 4 days of infection. Fowl tick, *Argas persicus,* transmits the organism. Body temperature rise up to 43 to 44°C. Clinically there will be marked cyanosis, paralysis, enteritis, and emaciation. No satisfactory method is available for its (*Borrelia anserina*) laboratory multiplication.
- *Reconstitution*: The freeze-dried vaccine is reconstituted in sterile normal saline solution (NSS). Route 1/M, birds below 1 kg - 0.5 ml, above - 1.0 ml.
- *Immunity*: About one year.
- *Storage*: At –20°C storage, the product can be used for 1 year and at 4°C storage, the product can be used for 1 month from the date of manufacture.

4.1.3 *Black Quarter Vaccine (Inactivated)*

- Black quarter (blackleg, quarter ill), caused by *Clostridium chauvoei* or in association with *Clostridium septicum,* is an infectious, fatal disease of cattle and is characterized by swelling (commonly in shoulder & hindquarters), edema and emphysema of the muscle and subcutaneous tissues of the infected part.

- *Issue*: The product is bottled in 25 ml, 50 ml and 100 ml volumes. It is recommended for the prevention of black quarter (*Clostridium chauvoei*) and malignant oedema (*Clostridium septicum*) infection in cattle and other ruminants. Vaccine is given by subcutaneous route. Dose - sheep and goats - 2.0 to 3.0 ml, cattle - 5.0 ml.
- *Immunity*: About one year.
- *Storage*: At 2 to 4°C. Self-life - 2 years from the date *of* manufacture.

4.1.4 *Brucella abortus Strain-19 Vaccine (Live)*

- Brucellosis in cattle is caused by *Brucella abortus.* DNA hybridization analysis shows that the genus *Brucella* contains only one species *i.e. Brucella melitensis. Brucella abortus* is not a separate species but the biovariety *of Brucella melitensis.* Conventional name is still in use. Clinically there will be abortion, retained placenta, orchitis, epididymitis and rarely arthritis. Organisms excrete through uterine discharge and milk. It is economically one *of* the most important diseases in cattle. In addition to abortion there will be sterility and decrease in milk yield. *Brucella abortus* strain 19 vaccine, is a suspension *of* live organisms in buffer saline solution
- *Issue*: Finished product is bottled in single dose vials, each containing 5 ml *of* vaccine. Vaccine is given by subcutaneous route. Dose - 5.0 ml, age 6 to 9 months.
- *Immunity*: For 2 pregnancies.
- *Storage*: Store the vaccine in a cool (2 to 4°C) and dark place. Do not freeze it. Self-life *of* the vaccine is 3 months.

4.1.5 *Enterotoxaemia Vaccine (Inactivated)*

- Enterotoxaemia (pulpy kidney disease, overeating disease) is an acute, highly fatal disease of sheep, particularly lambs, caused by absorption of epsilon toxin produced by *Clostridium perfringens* type D in the intestinal tract of the animal. The disease is characterized by cerebral symptoms like depression, excitement, and convulsions. The animals die in convulsions.
- *Issue*: The product is bottled in 250 ml quantities. Used for prophylactic vaccination of sheep. Lambs above 3 months of age can be vaccinated safely. Vaccine is given by subcutaneous route and repeated after 14 days. Dose - –2.5 ml.
- *Storage*: At 2 to 4°C. Self-life - 12 months from the date of manufacture.

4.1.6 *Haemorrhagic Septicaemia Vaccine (H. S. Oil Adjuvant vaccine)*

- Haemorrhagic septicaemia (HS), caused by *Pasteurella multocida,* is an acute infectious disease of cattle and buffaloes. The disease is characterized by septicaemia, haemorrhages, and pneumonia.
- *Issue*: The vaccine is bottled in 100 doses packing. Route - intermuscular. Dose - 2 ml per animal weighing up to 150 kg and 3.0 ml per animal weighing over 150 kg.
- *Immunity*: About 1 year. Pre-'monsoon vaccination is performed.
- *Storage*: At 2 to 4°C. Self-life - about 1 year.

4.1.7 *Salmonella abortus Equi Vaccine (Inactivated)*

- *Salmonella abortus equi* causes genital infection in equines. The organism is specially associated to horses and donkeys (primary salmonella). In mare, it causes late abortion (7–8 months), retention of placenta. In foals, born from infected mother, it produces acute septicaemia in the first week of life.
- *Issue*: Vaccine is distributed in 20 ml and 200 ml quantities. Route I/M, 1st dose – 10 ml, 2nd dose (after 10 days) - 10 ml, 3rd dose (10 days after 2nd dose) - 20 ml, subsequently 20 ml each at 4th, 6th and 8th pregnancies. Vaccination should be done in studs showing 10% or above infection as judged by agglutination test.
- *Immunity*: Low-grade immunity, lasting for 2 months.
- *Storage*: At 4°C, Self-life: 6 months from the date of manufacture.
- *Precaution*: A slight swelling may develop at the site of inoculation, which may persist for 3 to 4 days.

4.2. Viral Vaccines

4.2.1 *Canine distemper Vaccine, Live*

- Canine distemper is an acute, highly infectious viral disease of canines; most animals show involvement of respiratory system. Catarrhal enteritis is usually present.
- Vaccine is prepared from modified (chick embryo adopted) 'Ladevlae' strain of *canine distemper* virus (ssRNA), a Morbillivirus under Paramyxoviridae family. Other Morbillivirus members are measles virus and rinderpest virus. Freeze-dried ampoule is reconstituted to 2 ml and single dose is given to young pup (4–16 weeks) by SIC or 11M route.

4.2.2. *Foot and Mouth Diseases Vaccine, Inactivated*

- Foot and Mouth disease (FMD) is a highly contagious viral disease of cloven-footed animals, characterized by vesicular condition of feet, buccal mucosa and in females, the mammary glands.

- Vaccine is prepared from 7 types of FMD virus (ssRNA), of Aphthovirus under Picornaviridae family. The types are 0, A, C, SAT-I, SAT- 2, SAT-3, ASIA- 1. Virus is grown on bovine kidney cell or BHK-21 cell line and inactivated with formalin. Aluminium hydroxide gel is used as adjuvant. Route S/C, dose 10 ml. First dose during 4 months of age, 2nd dose 4 months after the 1st dose, revaccinate yearly.

4.2.3 *Fowl Pox Vaccine, Live*

- Fowl pox is a contagious disease of chickens and turkeys. It is slow spreading and characterized by the formation of proliferate lesions and scabs on the skin, and diphtheritic lesions in the upper parts of the digestive and respiratory tracts.
- Fowl poxvirus (dsDNA) of the genus *Avipoxvirus* under Poxviridae family is the cause of fowl pox. But the vaccine is produced from chick embryo adopted modified live pigeon poxvirus. Vaccine is used through feather follicle method below 6 weeks of age. Immunity last for 6 months.

4.2.4 *Infectious Bursal Disease Vaccine, Inactivated*

- Infectious bursal disease (IBD, Gumboro disease), caused by ssRNA virus of Birnaviridae family, is mainly a disease of chicken. Severe acute disease of 3 to 6 weeks old birds is associated with high mortality, but a less acute or sub clinical disease is common in 0 to 3 weeks old birds. This can cause secondary problems due to the effect of virus on bursa of febricius as it causes lymphoid depletion of bursa. Infectious bursal disease vaccine is an attenuated live vaccine introduced by BioMed. Vaccination of chicks is through oral route.

4.2.5 *Marek's Disease Vaccine, Live*

- Marek's disease, caused by a dsDNA virus of Gammaherpesviridae subfamily, is a lymphomatous and neuropathic disease of young domestic chicken.
- Marek's disease vaccine is prepared from turkey herpes virus strain FC-126. Route I/M, doses 0.2 ml to day old chicks, immunity life long.

4.2.6 *Peste Des Petits Ruminants (PPR) Vaccine, Live*

- Peste des petits ruminants (PPR), caused by ssRNA Morbillivirus of Paromyxodiridar family, is an acute febrile viral disease of sheep and goats, characterized by fever, nasal discharges, necrotizing and erosive stomatitis, diarrhoea and pneumonia.
- Peste des petits ruminants (PPR) vaccine is prepared from an attenuated (vero cell line) live virus. The vaccine is reconstituted in chilled and sterile 1 M $MgSO_4$ 10.85% NSS. Route SIC, dose - 1 ml (in neck region), immunity - 4 years.

4.2.7 *Rabies Vaccine (Cell culture), Inactivated/Live*

- Rabies is caused by a neurotropic ssRNA Lyssavirus of Rabdoviridae family and is trans- missible to all mammals, including humans.
- Production of BPL inactivated antirabies vaccine: Sheep (9–12 months) is inoculated intra-cranially with rabies fixed virus. After 7 days brain is harvested and triturated. Titration of virus (minimum antigen value 0.3) is done and finally inactivated with 5% BPL.
- Production of LEP or HEP living vaccine with Flury strain.
- Production of human diploid cell culture vaccine (Serum Institute, Pune). NDDB and Indian Immunological are producing killed vaccine from CV strain after growing it on BHK-21 cell line.

Pre-expose immunization:	0 – 7 – 21 – 28 day(s) followed by booster yearly.
Post-expose immunization:	0 – 3 – 7 – 14 – 28 day(s).
Post-expose immunization:	0 – 3 days(s).
	(With previous immunization)

4.2.8 *Ranikhet (New Castle) Disease Vaccine, Live*

- Ranikhet disease (RD), caused by a ssRNA paramyxovirus of Paramyxoviridae family, is an acute, highly contagious fatal disease of chicken and occasionally of other fowl, associated with respiratory symptoms (coughing, sneezing and gasping), paralysis of legs or wings or often torticollis (twisting of neck).
- Two types of RD vaccines are prepared, one from F (Lento genic) strain and other from R2B (Mesogenic) strain in 10–day–old embryonated egg. Freeze-dried ampule contains 100 doses. After reconstitution with NSS, it should be used within 1 to 2 h. Day-old chicks are vaccinated with F strain through I/M route and revaccination is done with vaccine containing R2B strain through I/ M or S/C route.

4.2.9 *Rinderpest Vaccine (Tissue culture), Live*

- Rinderpest (RP), caused by a Morbillivirus (ssRNA), under Paramyxoviridae family, is an acute, highlyy contagious fatal disease of cattle chiefly, and occasionally of other ruminants, characterized by "shooting diarrhoea". Other symptoms include marked depression, stupor, anorexia, emaciation, conjunctivitis and mucopurulent nasal discharge.
- Rinderpest vaccine is a freeze-dried preparation of a live strain (Kabete- "0") of RP virus, attenuated after passaging 99 times on BK cells. Route S/C, dose 1 ml. Self-life 3 years at -20ºC, 6 months at 4ºC. Calf hood vaccination. Life long (6 years) immunity.

4.2.10 *Sheep Pox Vaccine (Tissue culture), Live*

- Sheep pox, caused by a capripoxvirus (dsDNA) of Poxviridae family, is disease of sheep, characterized by fever, generalized papules or nodules, vesicles (rarely), internal lesions (lungs) and death.
- Sheep pox vaccine is a live vaccine, obtained by growing an attenuated virus strain (Romanian RM-65 strain) on lamb testicular cells. Route S/C, dose 1 ml, immunity 1 year. Self-life, 1 year at -20°C.

4.2.11 *Swine Fever Vaccine (Lapinised), Live*

- Swine fever (Hog cholera), caused by a pestivirus (ssRNA) of the family Flaviviridae, is contagious viral disease of pigs, begins with a rise in temperature up to 107°F and remains high until death. Depression, anorexia, stilted gait, in coordination, purulent conjunctivitis and diarrhoea are usually characteristic of the disease. The disease is accompanied by generalized hemorrhages.
- Swine fever vaccine is a live vaccine. It is produced after growing an attenuated (rabbit adopted) strain of the virus. It is a 10% suspension of spleen, taken from infected rabbit. The suspension is made in a yolk buffer base. Route S/C, dose 1 ml, immunity 1 year.

4.3 Veterinary Diagnostics

They are mainly used for diagnosis of diseases of animal.

1. *Brucella abortus* antigens and sera
    - *Brucella abortus* Rose Bengal plate test (RBPT) antigen
    - *Brucella abortus* milk ring test (MRT) antigen
    - *Brucella abortus* serum agglutination test (SAT) antigen
    - *Brucella* abiortus standard serum
2. *Salmonella abortus equi* - H antigen
3. *Salmonella pullorum* colored antigen
4. *Salmonella pullorum* plain antigen
5. *Salmonella pullorum* standard serum
6. *Salmonella* poly '0' sera
7. Johnin purified protein derivative (Johnin PPD).
8. Mallien purified protein derivative (Mallein PPD).
9. Tuberculin purified protein derivative (Tuberculin PPD).

4.3.1 *Brucella abortus Milk Ring Test (MRT) Antigen*

- An efficient means of screening dairy herds is by testing milk from the bulk tank. When a positive test result is obtained, all cows contributing milk should be blood tested. False positive reactions may occur in cattle vaccinated less than 4 months.

- MRT antigen is prepared from concentrated, killed *B. abortus* S99 cell suspension. The test is performed by adding 30–50 ml of antigen to 1–2 ml of whole milk in a narrow tube (so that the column should be at least 25 mm) followed by violent shaking. The tube is incubated at 37°C for 1 h. A strongly positive reaction is indicated by formation of a dark red ring above a white milk column. A red layer at the interface of milk and cream is also considered as positive. The test is considered negative if the color of the underlying milk exceeds that of the cream layer.

4.3.2 *Brucella abortus Serum Agglutination Test (SAT) Antigen*

- The antigen represents a bacterial cell suspension in phenol saline (NaCl 0.85% and phenol 0.5%) as has been described previously.
- The test may be performed in tubes. The mixture of antigen and serum dilutions should be incubated for 16–24 h at 37°C. The degree of *Brucella* agglutination in a serum is expressed in IU per ml. A serum containing 30 or more IU per ml is considered to be positive.

4.3.3 *Brucella abortus Standard Serum*

- It is a hyperimmune serum raised against *Brucella abortus* strain 99 in rabbit. This show a titer of 50% agglutination at between 1/640 to 1/1280 when tested by standardized serum agglutination test (SAT) antigen. The serum is preserved with merthiolate (1: 1000).

4.3.4 *Salmonella abortus equi "H" Antigen*

- *Salmonella abortus equi* causes genital infection in equines.
- *Salmonella abortus equi* strains *S–155* is used for production of antigen. The antigen is bottled in 250 ml quantity and stored at 4°C.

4.3.5 *Salmonella pullorum Coloured Antigen*

- Pullorum disease or bacillary white diarrhoea is almost worldwide in its distribution and young chicks are highly susceptible to this disease during the first 'fortnight after hatching. Deaths may occur suddenly and amount to some 95 percent of the total hatch in severe outbreaks. The causative is pullorum variety of *Salmonella gallinarum* (S. *pullorum*). *Salmonella pullorum* differ from *S. gallinarum* on biochemical reactions. Pullorum disease produced by *S. pullorum* is an enteric disease while fowl typhoid produced by *S. gallinarum* is a septicemic disease. These two biovars differ from other salmonella in being non-motile.
- The product is bottled in 20 ml quantities in amber colored bottle and stored at 4°C. Self-life: 6 months from the date of manufacture.

4.3.6 *Salmonella pullorum Plain Antigen*

- The product is bottled in 20 ml quantities in amber colored bottle and stored at 4ºC. Self-life: 6 months from the date of manufacture.

4.3.7 *Salmonella pullorum Standard Serum*

- It is a hyperimmune serum raised against *Salmonella pullorum* strain S.P.17 in rabbit. The serum should give a titer of 1: 1000 and is preserved with merthiolate (1: 1000).
- The serum is tested with *Salmonella pullorum* plain antigen. It should have titer of at least 1: 1000. It can be used as *Salmonella pullorum* poly '0' sera.

4.3.8 *Salmonella Poly '0' Sera*

- Salmonella poly '0' sera is the sera of rabbits, hyperimmunized with *Salmonella pullorum,* S. *essen,* S. Oslo, S. Virginia, S. London, and S. Illinois antigens.

4.3.9 *Johnin Purified Protein Derivative (PPD)*

- Paratuberculosis (Johne's disease) is a chronic infectious disease of cattle, sheep, goats and occasionally other animals caused by *Mycobacterium avium* subsp. *paratuberculosis* and characterized by recurrent diarrhoea, progressive emaciation and thickening and corrugation of the intestinal mucosa. In cattle and other ruminants, the gastrointestinal tract is the primary site of infection (predominantly the small intestine). Young animals (up to a few week of age) acquire the infection by ingestion of *M paratuberculosis* in feed or milk. Adult animals not manifesting clinical signs of paratuberculosis shed the causative agent in the milk. The disease can also be passed on to fetus by intra-uterine infection. *Mycobacterium paratuberculosis* is excreted at high level in the faeces and reflect the number of organisms in the intestinal wall.
- Immunodiagnostic procedures are run in parallel with culture of organism and examination of direct smear. There are three types of delayed hypersensitivity tests (i) the skin test using intradermal johnin PPD, (ii) systemic tests using intravenous johnin PPD and (ill) leukocyte migration test. The intradermal test requires the intradermal injection of johnin, followed by examination of the injection site for swelling 48 to 72 hours later. The intravenous johnin test is more specific. Reacting animals develop fever following intravenous injection of johnin. In leukocyte migration inhibition test peripheral blood lymphocytes are collected and exposed to johnin. Sensitized lymphocytes will release substances that inhibit macrophage migration.

- Johnin PPD is the preparation of precipitated protein of *M avium* subsp. *paratuberculosis* strain. The antigen contains 1 mg of PPD per ml and preserved in 0.5% phenol.
- *Bottling, storage and expiry*: One ml of the product is filled in a 2 ml capacity ampoule. The ampoule is sealed labeled and stored at 4°C. The antigen is expected to retain its potency for 2 years if stored at 4°C.
- How johnin test is performed:
  1. Identify the animals
  2. Select a site on the side of the neck.
  3. Hair is clipped around the site to a radius of about 2 cm.
  4. Measure the fold of he skin and records the measurement.
  5. Johnin (0.1 ml) is injected intradermally through a fine, sharp hypodermic needle (tuberculin syringe). After 72 h, measure the same fold of skin and record the thickness of the skin.
- Interpretation of single intradermal johnin test:
  1. An increase in skin-fold thickness no more than 2 mm and without clinical signs of inflammation and edema - Negative.
  2. An increase in skin-fold thickness more than 2 mm and less than 4 mm and without clinical signs of inflammation and edema - Inconclusive.
  3. An increase in skin-fold thickness more than 4 mm or more and/or with clinical signs of inflammation and edema - Positive

4.3.10 *Mallein Purified Protein Derivative (PPD)*

- Glanders is a contagious and fatal disease of horses, donkeys and mules. It is caused by *Burkholderia mallei,* a category B bio-terrorism agent. Humans are highly susceptible to glanders infection. The disease causes nodules and ulcerations in the upper respiratory tract and lungs. A skin form also occurs and is known as 'farcy'. There is chronic sticky yellow nasal discharge from one or both nostril, inflammatory nodule and ulcer formation on mucus membrane of upper air passages, enlargement of submaxillary lymph nodes, nodule and ulcer along the lymph channel of the skin, loss of weight, respiratory difficulty and lameness. The disease is more acute and more frequently recorded in donkeys and mules. In case of horses it is more chronic but spontaneous recovery is rare.
- *Bottling, storage and expiry*: One ml of the product is filled in a 2 ml capacity ampoule. The ampoule is sealed labeled and stored at 4°C. The antigen is expected to retain its potency for 2 years if stored at 4°C.
- How mallein test is performed:
  1. Identify the animals.

2. Mallein is used by intradermo-palpebral route. The dose of mallein PPD for this test is 0.1 ml and is injected through a fine, sharp hypodermic needle (tuberculin syringe) into the skin of the one lower eyelid. A minute circumscribed bleb will appear at the site of injection. The area is examined at 24 and 48 h.

- Interpretation of single intradermo-palpebrally mallein test:
  1. Marked swelling and edema of the injected eyelid, purulent conjunctivitis, light sensitivity, pain, and depression characterize a positive mallein test. Initially, there will be watery discharge: from the eye. It will turn into mucopurulent. Palpebral tissue becomes swollen around the orbit. The eyelashes become matted with discharge and usually well observed within 24 h after injection. The changes may persist for three to four days. In normal animals no changes of this kind are seen.
  2. Animal that gives undoubted reaction to this simple test may be dealt with as being infected, while those that give suspicious reaction may be segregated and re-tested after six weeks.
  3. It is an immunological test and false positive may occur due to cross-reaction with *B. pseudomallei.* Also, in advance clinical cases in horses and acute cases in donkeys and mules may give inconclusive result, requiring additional method (CFT) of diagnosis to be employed.

4.3.11 *Tuberculin Purified Protein Derivative (PPD)*

- Tuberculosis, caused by *Mycobacterium* species, is of the oldest recorded disease in the history of mankind. Tuberculosis is a chronic and infectious disease of mammals and is characterized by the formation of granulomatous lesions with central caseation and calcification. The most common sites are the lungs but also affect other organs. Delayed-type hypersensitivity (DTH) skin testing has been the most common test for the diagnosis of bovine tuberculosis. Various skin testing reagents have been developed but the most widely used is purified protein derivative (PPD), originally produced by Seibert and Glenn (1941). The standard method for detection for bovine tuberculosis is the tuberculin test, which involves intradermal injection of bovine tuberculin (PPDB) and subsequent detection of swelling (delayed hypersensitivity) at the site of injection 3 days later.
- Bovine tuberculin PPD is the preparation of precipitated tuberculoprotein of *M bovis* production strain AN5. The antigen contains 1 mg of PPD per ml equivalent to 32,500 International Units per ml and preserved in 0.5% phenol.
- *Bottling, storage and expiry*: One ml of the product is filled in a 2 ml capacity ampoule. The ampoule is sealed labeled and stored at 4ºC.

The antigen is expected to retain its potency for 2 years if stored at 4°C.

- How tuberculin test is performed:
  1. Identify the animals
  2. Select a site on the side of the neck.
  3. Hair is clipped around the site to a radius of about 2 cm.
  4. Measure the fold of the skin and records the measurement.
  5. Tuberculin (0.1 ml) is injected intradermally through a fine, sharp hypodermic needle (tuberculin syringe). After 72 h, measure the same fold of skin and record the thickness of the skin.
- Interpretation of single intradermal tuberculin test:
  1. An increase in skin-fold thickness no more than 2 mm and without clinical signs of inflammation and edema - Negative.
  2. An increase in skin-fold thickness more than 2 mm and less than 4 mm and without clinical signs of inflammation and edema - Inconclusive.
  3. An increase in skin-fold thickness more than 4 mm or more and/or with clinical signs of inflammation and edema – Positive

4.4 Parasitic Infestation

To check the parasitic infestation regular deworming should be followed. This practice must be started at the age of two weeks, followed after 21 days and repeated 3–4 times a year at regular interval. There are many chemical compounds available; hence, a different compound should be used each time to avoid resistance development. All calves at 0–3 months should be given Coccidiostat for 3–5 successive days to control infection of new born. For this purpose sulfonamides are frequently used the first dose is 220 mg/kg body weight following by 110 mg/kg body weight for the next four days. For control of ticks, flies, mosquitoes, lice etc. various insecticides like kilex, carbars, methrin, melathion, aldrin, etc. may be used for this purpose. These can be used at least 3–4 times a year and dose rate to be used as mentioned by the manufacturer. Livestock owners should take proper care during spray or application of these compound on animals because these are toxic substances so all precautions should be taken which have been suggested by the manufacturer.

5.0 TEMPORARY VETERINARY HOSPITALS

The equipment and other infrastructure are needed along with stockpile of emergency equipment and drugs. Early phases would need more work force and in all departments of activity *viz.* specialists like surgeons, anaethesiologists (pain managers), diagnosticians as well as task force like para-veterinary personnel and physical workers. All veterinary personnel may not be used at a time. It would be important to plan shifts to avoid mental and physical fatigue. As search and rescue would be over, relief and rehabilitation would acquire importance. The impact of disaster being calamitous one may come across a number of cases, which are beyond mere physical injuries or contagious disease. Many of the cases of extensive trauma that cannot be handled may have to be put to sleep (mercy killing). Doing so under full public view where the public is already surcharged with losses of life and property may set in depression or even public outcry. There is a role for Veterinary specialists to handle such situations. For the information exchange and co-ordination of veterinary support, control rooms are used. These handle, co-ordination information among directorate of Animal Husbandry, district Vet. The control room would be working on feed back from affected area on the extent and nature of emergency, the need for special equipment, emergency equipment (pain killer, sedatives, antibiotics, fracture equipments etc.). The role of a veterinary public relation man (Vet PR) assumes significance.

5.1 Actions within the Infected Areas

The traditional methods of controlling and eventually managing epidemics are:

- Movement control
- Vaccination
- Slaughter of infected herds (where possible)

In the present circumstances in India the effective deployment of all three methods is difficult and in some cases impossible. Epidemics are spread by direct close contact between animals and also due to the movement of infected livestock. Livestock owners may be reluctant to accept movement controls and the necessary forces may not be available to enforce them. Control long distance movements of livestock by ensuring that cattle do not move without health certificates (indicating that they have been examined and are free of clinical disease) and vaccination certificates (indicating vaccination within the last 3 months). Vaccination is acceptable provided that there be no untoward losses post vaccination but to achieve good vaccination cover of the population requires careful planning and execution. The quarantining of infected herds is needed till the infection has died out. This may take several months. The slaughter of infected herds is unacceptable to most livestock owners unless there is immediate and full compensation and compensation payments are costly to the authorities. This is the quickest way of eradicating the disease provided it is accompanied by effective movement control. Herds containing clinically affected cattle are identified by inspection and sub clinically infected livestock detected by serological tests. The cost of these measures means that slaughter should be a strategy of last resort to be used in critical epidemiological situations *e.g.* in the case of outbreaks in the free area or the surveillance zone (of a cordon sanitaire) or on major trade routes.

5.2 Warning and Notification of Epidemics

During any disaster or when there is an outbreak of epidemics, information must be conveyed through out the country via mass media like T.V., Radio and Newspapers. The movement of the animal should be restricted in that area. Detection of sub clinical cases of important infectious diseases like brucellosis, tuberculosis, johne's disease and mastitis and slaughter of positive cases will prevent the outbreak of these diseases. Field test like CMT, MBR, tuberculin and johnin test etc. are helpful. Early, accurate warning of new outbreaks of epidemic diseases, and particularly the spread of such diseases to new areas is an essential pre-requisite to the effective containment and control of these diseases. Some of the common problems in early warning systems for serious epidemic livestock diseases after disaster include:

- Lack of farmer awareness programs on high threat epidemic livestock diseases.
- Disease reporting systems which are based primarily on passive reporting of outbreaks rather than active disease surveillance.
- Inadequate training of veterinary and para-veterinary staff. Poor coordination of field and laboratory veterinary services.
- Failure to establish confirmatory diagnostic capabilities for the target diseases within national laboratories. Lengthy and over-complicated routine disease reporting chains and failure to institute an emergency reporting system for serious disease outbreaks.
- Lack of an epidemiology unit and expertise to analyse new disease outbreaks, including traceback and traceforward activities.
- Inadequate liaison with international reference laboratories. Failure to report new disease occurrences to appropriate international organizations, *e.g.* OlE, within an acceptable time.

6.0 VETERINARY INSTITUTIONAL PREPAREDNESS

Veterinary colleges in Indian are affiliated to both agricultural and veterinary universities. They are located on large campuses housing experimental fields and laboratories, live animals and plants, genetic material, museum specimens, expensive equipments and infrastructure, computer and other devices of storing research information, libraries, housing books, possibly rare manuscripts and innumerable other resources of national importance. To protect these resources, steps should be taken both before and after the disasters happen. There is a need to make an institutional disaster plan. The first step would be to undertake a risk assessment based on impact and probability of various events that may occur. Detailed inventories of all the above mentioned asses should be prepared and updated regularly. More attention should be given to proper storage of archival materials and methods of emergency rescue of materials in laboratories, science museums, laboratories or the need for maintenance of important data off-site.

The clinics of veterinary colleges generally include facilities for treatment and diagnosis of diseases and large number of trained manpower which can be used during the disaster. Resources like national network of veterinary diagnostic laboratories, referral laboratories, vaccine production centers and similar establishments can also be brought into the network of disaster management. Further these resources need to be modernized and strengthened to meet the future challenges like bio-terrorisms. Vaccine production facilities should have high standard of security arrangement which ensure that biological materials cannot be stolen or released into the environment inadvertently or intentionally.

Veterinary intuitions and research institutions also have the responsibility to help the communities to overcome the impact of disaster either through direct intervention or through long term interventions like research and development of technological innovations and extension activities. Information about preparedness and mitigation can be disseminated through their extension network (Dipak, 2002).

7.0 REFERENCES

Dipak, N. 2002. Coping with new disasters: Need for veterinary institutional preparedness and special disaster management teams in India. National Seminar on Biological Threats and Disaster Management. 9–11, August, 2002, Punjab Agriculture University, Ludhiana.

Dwivedi, S.K. 2006. Experience of post-disaster problems in equine and need for the resettlements. National symposium on Technolocial Interventions for Livestock Improvement and Production. 17–19, February, 2006. Delhi.

Heath, S.E. 2000. The role of veterinarians in disasters. World Veterinary Association Press Release, Washington, DC.

(http://www.worldvet.org/press73.htm)

Singh, N. and Shivakumar, P. 2006. Health care and management of livestock during natural disaster and calamities. National Symposium on Technological Intervention for Livestock Improvement and Production. 17–19, February, 2006. New Delhi.

VMAT Website: http://www.vmat.org/

19

LIVESTOCK BREEDS PREVALENT IN DISASTER PRONE AREAS

B.P. Kushwaha

ABSTRACT

Livestock genetic resources take millions of years to evolve through the process of natural selection. Various species of livestock have been domesticated throughout the world to meet local or regional needs. Animals have been selected by farmers, for their particular characteristics and at the same time being genetically adapted to local conditions, diseases, available feeds, climate, predators and many other variables imposed by the surrounding environment. Consequently it results in the development of the breeds suitable for a particular environment. Therefore, each breed/ type should be regarded as a clan specially designed genetic pool to serve specific purpose in a particular agro-climatic zone. These local breeds exhibit a distinct superiority in utilizing poor quality feed and are adapted to withstand heat and show better resistance to tropical diseases. According to World Watch list, there are 140 well defined breeds of livestock and poultry in our country. As per the 2003 livestock census data, there were 185.18 million cattle, 97.92 million buffaloes, 61.47 million sheep, 124.36 million goat, 13.52 million pig, 0.75 million horse, 0.63 million camel, 0.28 million mithun, 0.07 million yak and 440.69 million poultry in the country. Only 20–25% of these animals have been classified as recognized breeds and the rest are generally called non-descript breeds. The topics covered in this chapter include: importance of livestock in flood and drought prone areas, unique characteristics and physiological adaptation of the breeds to these areas, livestock breeds (cattle, buffalo, sheep and goat) of flood and drought prone areas and their production performance and their suitability to the specific areas, breeding of livestock in disaster prone areas.

Senior Scientist, Animal Genetics Breeding, Central Institute for Research on Buffaloes, Hisar (Haryana), India

1.0 INTRODUCTION

Due to its geographical location and resultant climatic conditions, India is often affected by severe natural calamities. The frequency and severity of extreme weather events like droughts, floods, cold and heat waves, cyclone and tsunamis have markedly increased in the Asian subcontinent during the past 15 years. On an average 57% geographical area of India is vulnerable to earthquakes, 28% to droughts, 12% floods and 8% to cyclones. These natural calamities produce cascading adverse effect on agriculture, livestock, surface ground water resources, employment and livelihood security of the people. The crops are badly damaged if the intensity of the flood or drought is more and that losses can not be recovered. However, in case of livestock the loss is not that severe, some animals may die, their production may go down due to direct or indirect effect (non availability of feed and fodder and water) of natural calamities. Losses in livestock productivity due to natural calamities can further be reduced through keeping livestock breeds/species which can sustain during natural calamities, their production can be restored and economic losses can be reduced to a great extent. The livestock rearing activity provides subsidiary occupation to a large section of society particularly to the people living in the drought/flood prone areas where crop production on its own may not be very productive and engage them fully or partially. The main assets of desert landowners are large herds of livestock. In absence of favorable conditions for agriculture, livestock rearing is the only alternative source of livelihood for majority of the rural population in flood or drought affected areas. The livestock to some extent compensate the losses in crop production due to weather- induced uncertainties. The contribution of the livestock sector in national economy can be assessed from the fact that the gross value of output from livestock sector alone, at current price of 2004-05 was about Rs. 1239 billion which was equal to about 24.72% of the value of total output from agriculture sector. This excludes the contribution of animal draught power.

Livestock genetic resources take millions of years to evolve through the process of natural selection. Various species of livestock have been domesticated throughout the world to meet local or regional needs. Animals have been selected by farmers, for their particular characteristics and at the same time being genetically adapted to local conditions, diseases, available feeds, climate, predators and many other variables imposed by the surrounding environment. Consequently it results in the development of the breeds suitable for a particular environment. Therefore, each breed/type should be regarded as a clan specially designed genetic pool to serve specific purpose in a particular agro-climatic zone. These local breeds exhibit a distinct superiority in utilizing poor quality feed and are adapted to withstand heat and show better resistance to tropical diseases. According to World Watch list, there are 140 well defined breeds of livestock and poultry in our country. As per the 2003 livestock census data, there were 185.18 million cattle, 97.92 million buffaloes, 61.47 million sheep, 124.36 million goat, 13.52 million pig, 0.75 million horse, 0.63 million camel, 0.28 million mithun, 0.07 million yak and 440.69 million poultry in the country. Only 20–25% of these animals have been classified as recognized breeds and the rest are generally called non-descript.

Cattle and buffaloes are the most preponderant species among all the livestock in India and they account for over 75% of the total output value from livestock sector and the rest 25% comes from other livestock including sheep and goats. The livestock breeds (cattle, buffalo, sheep and goat) prevalent in the flood and drought prone areas or the breeds which are suitable for rearing in these areas have been presented in Tables 1 and 2, respectively.

2.0 LIVESTOCK PREVALENT IN DROUGHT PRONE AREAS

The arid and semi-arid region of northern and southern India is characterized by low to moderate rainfall with recurring drought. Most of the area in these regions is drought prone and often put sustainability of livestock production at stake due to long stretches of dry periods and non availability of feed and fodder. Rainfall distribution is highly varied and nearly 70-80% area of the region gets low rainfall. In arid region rainfall is scanty and erratic and the crop production is extremely low. This region encountered drought almost every (alternate) year but many times for a longer duration *i.e.* 2-4 years together. States like Rajasthan, Gujarat, Andhra Pradesh, Tamil Nadu, Karnataka and Maharashtra experience drought almost in alternate year or at least once in three years. Over the period, the inhabitants settled in these regions have learnt to live with the drought. They have developed their own ways to negate/moderate drought impacts and maintain livelihood security. The livestock rearing is considered more suited than crop farming in the drought prone areas. The ruminants usually do not compete with man for food. Being herbivores they can utilize large quantities of cellulose and convert it into milk, wool and meat. The natural grazing lands, which are not suited for crop cultivation may be successfully utilized for livestock rearing.

The frequent spell of drought in the region has transformed the animals to adapt to vagaries of the adverse conditions under extensive system of animal production. The animals of these regions gradually loose body weight during the lean summer season and regain the weight once the greens are in plenty following a favorable rainy season. The animals of different species not only withstand the condition, but also multiply, grow

Table 1 Livestock breeds of Drought prone districts

State	District	Cattle	Buffalo	Sheep	Goat
Andhra Pradesh	Anantapur	Ongole		Deccani	
Andhra Pradesh	Chittoor	Ongole, Punganur			
Andhra Pradesh	Hyderabad				
Andhra Pradesh	Kadapa				
Andhra Pradesh	Kurnool	Ongole		Deccani	
Andhra Pradesh	Mahbubnag	Nellore			
Andhra Pradesh	Nalgonda	Nellore		Deccani	
Andhra Pradesh	Prakasam				
Andhra Pradesh	Rangaredd				
Bihar	Aurangaba				
Bihar	Begusarai				
Bihar	Bhojpur				
Bihar	Buxar				
Bihar	Gaya			Sahabadi	
Bihar	Jamui				
Bihar	Jehanabad				
Bihar	Kaimur				
Bihar	Lakhisara				
Bihar	Munger				
Bihar	Nawada				
Bihar	Rohtas				
Bihar	Sheikhpur				
Gujarat	Ahmedabad	Kankrej	Mehsana		Mehsana
Gujarat	Amreli	Gir	Jaffarabadi		Gohilwadi
Gujarat	Anand				
Gujarat	Banaskantha		Mehsana		Mehsana
Gujarat	Bharuch		Surti		
Gujarat	Bhavnagar	Gir	Jaffarabadi		Gohilwadi

(Contd...)

(Contd...)

State	District	Cattle	Buffalo	Sheep	Goat
Gujarat	Dahod				
Gujarat	Jamnagar				
Gujarat	Kutch	Kankrej	Banni	Patanwadi	Kutchi
Gujarat	Kheda		Surti		
Gujarat	Mehsana	Kankrej	Mehsana	Patanwadi	Mehsana
Gujarat	Panchmahal				
Gujarat	Rajkot	Gir			Zhalawadi
Gujarat	Sabarkant	Kankrej	Mehsana		
Gujarat	Surendran				
Haryana	Bhiwani		Murrah		Beetal
Haryana	Faridabad		Murrah		Beetal
Haryana	Gurgaon		Murrah		Beetal
Haryana	Jhajjar		Murrah		Beetal
Haryana	Mahendrag		Murrah		Beetal
Haryana	Rewari		Murrah		Beetal
Haryana	Rohtak		Murrah	Nali	Beetal
Jammu	Doda				
Jammu	Udhampur				
Jharkhand	Garhwa				
Jharkhand	Palamu			Chhotanagpuri	
Karnataka	Bagalkot				
Karnataka	Bangalore	Hallikar, Amritmahal			
Karnataka	Bangalore				
Karnataka	Belgaum	Hallikar, Krishna valley	Pandharpuri		
Karnataka	Bellary			Bellari	
Karnataka	Bijapur	Hallikar, Krishna valley		Deccani	
Karnataka	Chamarajn				
Karnataka	Chikmagal	Amritmahal			

(Contd...)

(Contd...)

State	District	Cattle	Buffalo	Sheep	Goat
Karnataka	Chitradur	Hallikar, Amritmahal			
Karnataka	Dharwad	Khillari			
Karnataka	Gadag Dis				
Karnataka	Gulbarga			Deccani	
Karnataka	Hassan	Hallikar, Amritmahal		Hassan	
Karnataka	Haveri				
Karnataka	Kolar				
Karnataka	Mandya	Hallikar		Mandya	
Karnataka	Mysore	Hallikar	South Kanara	Mandya	
Karnataka	Raichur	Krishna valley		Deccani, Kenguri	
Karnataka	Tumkur	Hallikar			
Madhya Pradesh	Anuppur				
Madhya Pradesh	Betul				
Madhya Pradesh	Datia				
Madhya Pradesh	Dewas	Malvi			
Madhya Pradesh	Dhar				
Madhya Pradesh	Jhabua				
Madhya Pradesh	Khandwa	Nimari			
Madhya Pradesh	Khargone	Nimari			
Madhya Pradesh	Shahdol				
Madhya Pradesh	Shajapur	Malvi			
Madhya Pradesh	Sidhi				
Madhya Pradesh	Ujjain	Malvi			
Maharashtra	Ahmed Nagar	Dangi		Deccani, Sangamneri	
Maharashtra	Aurangabad				
Maharashtra	Beed		Marathwada		
Maharashtra	Jalna				
Maharashtra	Latur	Deoni, Red Kandhari	Marathwada		Osmanabadi

(Contd...)

(Contd...)

State	District	Cattle	Buffalo	Sheep	Goat
Maharashtra	Nashik	Dangi			
Maharashtra	Osmanabad	Deoni			Osmanabadi
Maharashtra	Pune			Deccani, Sangamneri	
Maharashtra	Sangli	Khillari, Krishna valley	Pandharpuri		
Maharashtra	Satara	Khillari, Krishna valley	Pandharpuri		
Maharashtra	Solapur	Khillari, Krishna valley	Pandharpuri		
Orissa	Boudh (Ba				
Orissa	Kalahandi		Kalahandi		
Orissa	Kandhamal				
Orissa	Nabarangp				
Rajasthan	Ajmer			Marwari, Malpura	Marwari
Rajasthan	Banswara			Malpura	
Rajasthan	Barmer	Tharparkar		Marwari, Jaisalmeri	
Rajasthan	Bikaner	Rathi, Nagauri		Chokla, Magra, Pugal	Marwari
Rajasthan	Churu			Chokla, Nali, Magra	Marwari
Rajasthan	Dungapur			Sonadi	
Rajasthan	Jaisalmer	Rathi, Tharparkar		Magra, Jaisalmeri, Pugal	Marwari
Rajasthan	Jalore				Marwari
Rajasthan	Juhnjhunu			Chokla, Nali	
Rajasthan	Jodhpur	Nagauri		Marwari, Jaisalmeri	Marwari
Rajasthan	Nagaur	Nagauri		Chokla, Marwari	Marwari
Rajasthan	Pali			Marwari	
Rajasthan	Rajsamand				
Rajasthan	Udaipur		Surti	Marwari, Sonadi	
Tamil Nadu	Coimbatore	Alambadu		Mecheri, Coimbatore	
Tamil Nadu	Dharmapur	Kangayam		Trichi black	
Tamil Nadu	Dindigul				
Tamil Nadu	Erode	Bargur, Kangayam			

(Contd...)

(*Contd...*)

State	District	Cattle	Buffalo	Sheep	Goat
Tamil Nadu	Kanyakuma				
Tamil Nadu	Karur	Kangayam			
Tamil Nadu	Madurai			Kilakarsal, Coimbatore	
Tamil Nadu	Namakkal				
Tamil Nadu	Perambalu				
Tamil Nadu	Ramanatha			Kilakarsal	Kannaiadu
Tamil Nadu	Salem			Mecheri, Trich balck	
Tamil Nadu	Sivaganga				
Tamil Nadu	Theni				
Tamil Nadu	Thoothuku				
Tamil Nadu	Tirunelve			Vembur, Ramnadwhite	Kannaiadu
Tamil Nadu	Tiruchira				
Tamil Nadu	Virudhuna				
Uttar Pradesh	Allahabad				
Uttar Pradesh	Banda	Kenkatha		Jalauni	
Uttar Pradesh	Chandauli				
Uttar Pradesh	Chitrakoo				
Uttar Pradesh	Hamirpur	Kenkatha		Jalauni	
Uttar Pradesh	Jalaun		Bhadawari	Jalauni	Jamnapari
Uttar Pradesh	Mahoba	Kenkatha			
Uttar Pradesh	Mirzapur				
Uttar Pradesh	Sonbhadra				
Uttar Pradesh	Sant Ravi				
Uttar Pradesh	Varanasi				
West Bengal	Bankura	Bengali			
West Bengal	Midnapore				
West Bengal	Purulia				

Table 2 Livestock breeds of flood prone districts

State	District	Cattle	Buffalo	Sheep	Goat
Andhra Pradesh	Adilabad				
Andhra Pradesh	East Godavari		Godavari		
Andhra Pradesh	West Godavari		Godavari		
Andhra Pradesh	Krishna				
Andhra Pradesh	Guntur				
Andhra Pradesh	Prakasam			Nellore	
Andhra Pradesh	Cuddapah				
Arunachal Pradesh	West Siang				
Assam	Dhemaji		Assamese or Mongoor		
Assam	Dibrugarh				
Assam	Lakhimpur				Black Bengal
Assam	Jorhat		Swamp		Black Bengal
Assam	Sonitpur				Black Bengal
Assam	Golaghat				Black Bengal
Assam	Nagaon				Black Bengal
Assam	Nalbari				Black Bengal
Assam	Barpeta				Black Bengal
Assam	Bongaigaon				Black Bengal
Assam	Dhubri				
Bihar	Pashchim Champaran				
Bihar	Purbi Champaran				
Bihar	Sitamarhi	Bachaur			Black Bengal
Bihar	Madhubani	Bachaur			Black Bengal
Bihar	Kishanganj				Black Bengal
Bihar	Supaul				Black Bengal

(*Contd...*)

(*Contd...*)

State	District	Cattle	Buffalo	Sheep	Goat
Bihar	Gopalganj				Black Bengal
Bihar	Araria				Black Bengal
Bihar	Muzaffarpur				Black Bengal
Bihar	Darbhanga	Bachaur			Black Bengal
Bihar	Saran	Sahabadi			Black Bengal
Bihar	Siwan				Black Bengal
Bihar	Saharsa				Black Bengal
Bihar	Madhepura				Black Bengal
Bihar	Samastipur	Bachaur			Black Bengal
Bihar	Vaishali				Black Bengal
Bihar	Katihar				Black Bengal
Bihar	Begusarai				Black Bengal
Bihar	Bhojpur				Black Bengal
Bihar	Buxar				Black Bengal
Bihar	Khagaria				Black Bengal
Bihar	Patna	Taylor		Sahabadi	Black Bengal
Bihar	Bhagalpur				Black Bengal
Bihar	Nalanda				Black Bengal
Bihar	Munger				Black Bengal
Bihar	Lakhisarai				Black Bengal
Bihar	Rohtas				
Bihar	Jehanabad				
Bihar	Banka				
Bihar	Sheikhpura				
Bihar	Nawada				
Bihar	Aurangabad				
Bihar	Purnia	Purnea			

(*Contd...*)

(*Contd...*)

State	District	Cattle	Buffalo	Sheep	Goat
Delhi	Delhi		Murrah	Muzzafarnagari	
Gujarat	Banas Kantha	Kankrej			Mehsana
Gujarat	Sabar Kantha	Kankrej			
Gujarat	Patan			Patanwadi	
Gujarat	Ahmedabad	Kankrej			Mehsana
Gujarat	Panch Mahals				
Gujarat	Kheda		Surti		
Gujarat	Vadodara		Surti		Surti
Gujarat	Anand				
Gujarat	Bharuch		Surti		
Gujarat	Narmada				
Gujarat	Surat		Surti		Surti
Gujarat	Navsari				
Haryana	Panchkula				Beetal
Haryana	Ambala				Beetal
Haryana	Yamuna Nagar				Beetal
Haryana	Kurukshetra				Beetal
Haryana	Kaithal	Haryana	Murrah		Beetal
Haryana	Karnal	Haryana	Murrah		Beetal
Haryana	Jind	Haryana	Murrah		Beetal
Haryana	Panipat	Haryana	Murrah		Beetal
Haryana	Hissar	Haryana	Murrah		Beetal
Haryana	Sonipat	Haryana	Murrah		Beetal
Haryana	Rohtak	Haryana	Murrah		Beetal
Haryana	Jhajjar	Haryana	Murrah		Beetal
Haryana	Gurgaon	Haryana	Murrah		Beetal
Haryana	Faridabad	Haryana	Murrah		Beetal

(*Contd...*)

(*Contd...*)

State	District	Cattle	Buffalo	Sheep	Goat
Kerala	Kasaragod	Vechur			
Kerala	Kannur				
Kerala	Kozhikode				
Kerala	Malappuram				
Kerala	Palakkad				
Kerala	Thrissur				
Kerala	Ernakulam				
Kerala	Alappuzha				
Kerala	Kottayam	Vechur			
Kerala	Pathanamthitta				
Kerala	Kollam				
Kerala	Thiruvananthapuram				
Manipur	Senapati				
Manipur	Imphal				
Manipur	Thoubal				
Manipur	Bishnupur				
Orissa	Baleshwar				
Orissa	Bhadrak				
Orissa	Jajpur				
Orissa	Kendrapara				
Orissa	Cuttack	Binjharpuri, Ghumsuri	Kujang		
Orissa	Jagatsinghpur	Binjharpuri, Ghumsuri			
Orissa	Puri			Ganjam	Ganjam
Punjab	Gurdaspur	Sahiwal	Nili-Ravi, Murrah		Beetal
Punjab	Amritsar	Sahiwal	Nili-Ravi, Murrah		Beetal
Punjab	Hoshiarpur				
Punjab	Kapurthala				

(*Contd...*)

(Contd...)

State	District	Cattle	Buffalo	Sheep	Goat
Punjab	Jalandhar		Murrah		
Punjab	Nawan Sahar		Murrah		
Punjab	Rupnagar				
Punjab	Firozpur	Sahiwal	Nili-Ravi, Murrah	Munjal	
Punjab	Ludhiana				
Punjab	Faridkot				
Punjab	Monga		Murrah	Munjal	
Punjab	Muktsar		Murrah	Munjal	
Punjab	Patiala		Murrah	Munjal	
Punjab	Sangrur		Murrah	Munjal	
Punjab	Bathinda		Murrah	Munjal	
Punjab	Mansa		Murrah	Munjal	
Punjab	Fatehgarh				
Uttar Pradesh	Saharanpur			Muzzafarnagri	
Uttar Pradesh	Muzaffarnagar		Murrah	Muzzafarnagri	
Uttar Pradesh	Bijnor		Murrah	Muzzafarnagri	
Uttar Pradesh	Baghpat		Murrah	Muzzafarnagri	
Uttar Pradesh	Meerut		Murrah	Muzzafarnagri	
Uttar Pradesh	Moradabad				
Uttar Pradesh	Rampur				
Uttar Pradesh	Jyotiba Phule Nagar				
Uttar Pradesh	Ghaziabad		Murrah		
Uttar Pradesh	Bareilly				Bareilly
Uttar Pradesh	Pilibhit	Ponwar			
Uttar Pradesh	Bulandshahr				
Uttar Pradesh	Gautam Budh				

(Contd...)

(Contd...)

State	District	Cattle	Buffalo	Sheep	Goat
	Nagar				
Uttar Pradesh	Kheri	Kherigarh, Ponwar			
Uttar Pradesh	Budaun				
Uttar Pradesh	Shahjahanpur				
Uttar Pradesh	Bahrraich				
Uttar Pradesh	Aligarh				
Uttar Pradesh	Etah		Bhadawari		Barbari
Uttar Pradesh	Mathura	Mewati or Kosi	Bhadawari		Barbari
Uttar Pradesh	Mahamayanagar				
Uttar Pradesh	Shravasti				
Uttar Pradesh	Sitapur				
Uttar Pradesh	Balarampur				
Uttar Pradesh	Hardoi				
Uttar Pradesh	Farrukhabad				
Uttar Pradesh	Siddharthnagar				
Uttar Pradesh	Maharajganj				
Uttar Pradesh	Mainpuri				
Uttar Pradesh	Agra	Mewati	Bhadawari		Barbari, Jamnapari
Uttar Pradesh	Gonda				
Uttar Pradesh	Kanauj				
Uttar Pradesh	Barabanki				
Uttar Pradesh	Kushi Nagar				
Uttar Pradesh	Gorakhpur				
Uttar Pradesh	Basti				
Uttar Pradesh	Lucknow				
Uttar Pradesh	Unnao				

(Contd...)

(Contd...)

State	District	Cattle	Buffalo	Sheep	Goat
Uttar Pradesh	Sant Kabir Nagar				
Uttar Pradesh	Etawah		Bhadawari		Jamnapari
Uttar Pradesh	Auraiya		Bhadawari		Jamnapari
Uttar Pradesh	Kanpur Dehat				
Uttar Pradesh	Faizabad				
Uttar Pradesh	Deoria				
Uttar Pradesh	Kanpur Nagar				
Uttar Pradesh	Sultanpur				
Uttar Pradesh	Rae Bareli				
Uttar Pradesh	Ambedkar Nagar				
Uttar Pradesh	Jalaun		Bhadawari	Jalauni	Jamnapari
Uttar Pradesh	Azamgarh				
Uttar Pradesh	Mau				
Uttar Pradesh	Fatehpur				
Uttar Pradesh	Jaunpur				
Uttar Pradesh	Pratapgarh				
Uttar Pradesh	Jhansi		Bhadawari	Jalauni	
Uttar Pradesh	Banda	Kenkatha			
Uttar Pradesh	Ghazipur				
Uttar Pradesh	Kausambi				
Uttar Pradesh	Allahabad				
Uttar Pradesh	Varanasi				
Uttar Pradesh	Chitrakut				
Uttar Pradesh	Sant Ravidas Nagar				
Uttar Pradesh	Chandauli				
Uttar Pradesh	Mirzapur				
Uttar Pradesh	Ballia				

(Contd...)

(*Contd...*)

State	District	Cattle	Buffalo	Sheep	Goat
Uttar Pradesh	Hamirpur	Knkatha		Jalauni	
Uttaranchal	Nainital	Kumauni	Tarai	Rampur bushair	Gaddi
Uttaranchal	Udhamsinghnagar	Kumauni	Tarai	Rampur bushair	
Uttaranchal	Hardwar				Gaddi
West Bengal	Jalpaiguri				
West Bengal	Koch Bihar	Bengali			Black Bengal
West Bengal	Uttar Dinajpur	Bengali			Black Bengal
West Bengal	Dakshin Dinajpur	Bengali			Black Bengal
West Bengal	Maldah	Bengali			Black Bengal
West Bengal	Murshidabad	Bengali			Black Bengal
West Bengal	Birbhum	Bengali			Black Bengal
West Bengal	Nadia	Bengali			Black Bengal
West Bengal	Barddhaman	Bengali			Black Bengal
West Bengal	Puruliya	Bengali			Black Bengal
West Bengal	Bankura	Bengali			Black Bengal
West Bengal	North 24 Parganas	Bengali			Black Bengal
West Bengal	Hugli	Bengali			Black Bengal
West Bengal	Medinipur	Bengali			Black Bengal
West Bengal	Haora	Bengali			
West Bengal	Calcutta	Bengali			
West Bengal	South 24 Parganas	Bengali			

and produce economically with low input and low output thus making them suitable over the crossbreds.

Under drought situation where water is a limiting factor, the erratic monsoon further adds to the harshness of climate. Under such a climate, the animal should possess the characteristics having high efficiency of feed utilization, the ability to promote heat loss efficiency, the ability to conserve heat production and allow productive process to proceed at a normal level even when the air temperature is high, insulation against solar radiation, the ability to withstand a high degree of dehydration and rise of body temperature, and possess a high degree of resistance against the common diseases.

The indigenous cattle have prominent hump, drooping ears, upright horns, and dewlap and relatively lower basal metabolic rate and high capacity of heat dissipation, which in turn facilitate adaptation under heat stress. Hair coat of zebu cattle reduces the absorption of solar radiation and enhances convective and conductive heat losses (Sharma and Kumar, 2008). Indigenous cattle have higher true digestibility, more extensive ruminal digestion, more efficient protein synthesis and poor lower fecal excretion than the most temperate breeds (Hunter and Siebert, 1985). Frisek and Vercoe (1984) observed that *Bos indicus* cattle are more resistant to gastro-intestinal nematodes than European *Bos tourus* cattle.

Small ruminant has been considered to possess an important mechanism for coping up drought because of better adaptability and mobility as compared to crops and large ruminants. Sheep and goats have developed some unique characteristics to adjust themselves in harsh environment. Low body mass and low metabolic requirement of goat can be regarded as important asset to them. The ability to reduce metabolism allows goats to survive even after prolonged periods of severe limited food availability. A skillful grazing behavior and efficient digestive system enable goats to attain maximal food intake and maximal food utilization in a given condition. The rumen plays an important role in the adaptation by serving as a huge fermentation vat and water reservoir. The water stored in the rumen is utilized during dehydration, rumen serve as a container, which accommodates the indigested water upon rehydration. The rumen, the salivary glands and the kidney coordinately function in the regulation of water intake and water distribution following acute dehydration and rehydration (Silanikove, 2000).

2.1 Livestock Breeds in Drought Prone Areas

2.1.1 *Cattle Breeds*

The milch breeds like Sahiwal, Red Sindhi, Rathi, Gir and Tharparkar have evolved in the arid region/the drought prone area of the country where agriculture is not developed and the animal milk is the main source of livelihood for the farmers. Gir, the main cattle breed of Kathiawar region of Gujarat is known for stress tolerance capacities. Rathi is an other important cattle breeds of desert widely found among nomads who maintain their animals on ranges. Similarly, other breeds or non-descript cattle of the region are hardy and resistant to many diseases. In spite of facing the harsh climatic conditions, they are capable enough to maintain their production. Sahiwal, Red Sindhi, Gir, Rathi and Tharparkar are admittedly the best milch breeds of our country, which can produce an average of 5-7 litres of milk daily and the animals of these breeds are efficient converter of roughages. Under good management conditions they can produce up to 11-12 kg of milk in a day. A farmer who is maintaining a small herd of Tharparkar at his farm in Tamil Nadu has recorded 11 litres of milk per day from a Tharparkar cow. In general, cows yielded between 8 to 10 litres each on an average every day (The Hindu, 2004).

In semi-arid region, where some irrigation is available and agriculture is the main occupation for which draught animals are required, cattle breeds like Kangayam, Ongole, Kankrej, Haryana, Nagauri have evolved to provide good bullocks for draught. The milk yield of these animals is reasonably good. The bullock of Nagauri breed are well known for their draught power in the deep sandy soil. The breeds like Mewati, Malvi, Kenkatha, Nimari etc. have evolved in the dry lands of Uttar Pradesh, Madhya Pradesh. These breeds are mainly reared with zero or least purchased input. Though the milk yield of these breeds is comparatively poor but the male are very active and are good for draught purpose. Heavy and powerful cattle breeds like Amritmahal, Khillari, Krishna valley, Hallikar etc. were evolved in the southern part of the country, where the land is hard and rocky. Some breeds which have adapted to harsh climatic conditions of the drought prone areas of the country have been described below:

2.1.1.1 *Amritmahal*

Amritmahal is a special breed of Karnataka and has a history of over 500 years. Animals of this breed are found in Hassan, Chikmanglur and Chitradurga district of Karnataka. It is a famous draught breed known for its power and endurance. Cows are poor milker. Animals are able to withstand extremely hot, humid climatic conditions and are resistant to many diseases prevailing in the region (Prabhu *et al.*, 2008). However, the population of this breed is declining at a faster rate and there is an urgent need to make efforts for conservation of this breed before it becomes extinct.

2.1.1.2 *Bargur*

Found in Coimbatore district of Tamil Nadu. Animals of this breed are of light built and are developed mainly for carrying out agricultural operations in uneven and hilly terrain.

2.1.1.3 *Deoni*

Is a dual purpose breed. Animals of this breed are resistant to harsh and dry climate of Maharashtra (Mitkari and Thombre, 2006).

2.1.1.4 *Gaini Cattle*

Gaini is a dwarf breed of Rajasthan found in Jaipur, Malpura, Tonk and Kishangarh (Ajmer). Animals are resistant to almost all diseases prevalent amongst most of the crossbred animals in the vicinity. Average adult body weight in males 220 to 270 kg and in females is 200 to 240 kg (Banerjee Sandeep, 2006).

2.1.1.5 *Gangatiri*

Gangatiri is a dual purpose breed, found mainly in the "Doaba" (Gangages) belt of eastern Uttar Pradesh and adjacent area of Bihar. Its breeding tract mainly includes Chandauli, Ghazipur and Ballia district of U.P., and Rohtas, Sahabad and Bhojpur district of Bihar. Gangatiri cows are average milkers and bullocks are good for agricultural operations. The animals of this breed have good tolerance to drought and heat (Prakash *et al.*, 2008).

2.1.1.6 *Gaolao*

Found in Vidarbha region of Maharashtra. Cows are average milker. Bullocks are useful for ploughing. There is a close similarity between Ongole and Gaolao except that the latter is much lighter with greater agility.

2.1.1.7 *Gir*

Animals of this breed are found in Saurashtra region of Gujarat. Gir cows are high milk yielder and yield ranges from 2000 kg to 6000 kg per lactation. Bullocks are heavy and powerful draught animals. Gir animals are known for heat and drought tolerance.

2.1.1.8 *Haryana*

Home tract of this breed is Haryana state but it is also found in U.P., Bihar and some parts of Rajasthan. It is an excellent dual purpose breed known for draught power and heat tolerance.

2.1.1.9 *Kangeyam*

Kangeyam is a famous breed of Tamil Nadu, found in Coimbatore, Dindigul, Erode and Namakkal districts. These animals are known for its superior draught power, adaptation to poor nutrition, drought conditions and longevity (Gunasekaran and Reddy, 2006). They have docile temperament and can thrive on scanty ration (Ghotge, 2004).

2.1.1.10 *Kankrej/Banni*

It is the heaviest cattle breed of India. Kankrej animals are commonly found in Mehsana, Kutch, Ahmedabad, Kaira, Sabarkantha and

Banaskantha districts of Gujarat and Barmer and Jalore districts of Rajasthan. Cows of this breed have excellent milk production capacity (1500-2500 litres per lactation) and bullocks are excellent draught animals. Kankrej cattle are known for their hardiness, heat and tick infestation resistance and high fat content in their milk. These cattle are ideal for growing in drought conditions.

2.1.1.11 *Kenkatha*

The breed Kenkatha derives its name from the River Ken which flows through the Bundelkhand region of Madhya Pradesh (MP) into Uttar Pradesh (UP) and terminates in the Yamuna river in UP. Animals of this breed are also known as Kenwaria or Ajaigarhi. The animals of this breed are small in size, cows are poor milker but the males are strong and powerful and are good for cultivation in rocky areas. These animals are generally maintained on low input and known for heat resistance. The Bundelkhand region, home tract of Kenkatha, is marked by extremes of temperature, reaching up to 48°C during the summer months and dropping to as low as 1°C in winter. Hot breezes locally known as *loo* are common during summer.

2.1.1.12 *Khillari*

It is the most popular draft breed of north-western India. Khillari bullocks from Karnataka, are famous for very fast travel without any fatigue. This breed is found in Kolhapur, Solapur, Sangli and Satara districts of Maharashtra and Belgaum, Bijapur and Dharwad districts of Karnataka.

2.1.1.13 *Krishna Valley*

The breeding tract of this breed lies in the southern border of Maharashtra and Andhra Pradesh. Bullocks are powerful and good for heavy ploughing. Cows are fairly good milkers.

2.1.1.14 *Malvi*

It is a draught purpose breed having breeding tract in Indore, Dewas, Ujjain, Shajapur, Ratlam and Rajgarh district of Madhya Pradesh and Jhalawar district of Rajasthan.

2.1.1.15 *Mewati*

Mewati cattle are in general sturdy, powerful and docile and useful for heavy ploughing, carting and drawing water from deep wells. Cows are fairly good milkers. Animals of this breed are found in Alwar and Bharatpur districts of Rajasthan.

2.1.1.16 *Nagori*

This breed is a famous draught breed and capable of working efficiently in deep sand and tolerant to heat stress and poor nutrition. Breeding tract of this breed is Nagaur district in Rajasthan.

2.1.1.17 *Nimari*

A draught breed found in Khandwa, Khargaon and Barwani districts of Madhya Pradesh. Cows are poor milker.

2.1.1.18 *Punganur*

Ponganur is one of the shortest cattle breeds of our country. This breed is found in Chittoor district of Andhra Pradesh. The breed is reputed to have an excellent feed conversion ratio. Females are fairly good milkers needing extremely small quantities of fodder to sustain them. It is good for rearing by the resource poor farmers in drought prone areas. However, the number of these animals is very less which put this breed under endangered category.

2.1.1.19 *Rathi*

Rathi is one of the milch breed of our country. Milk yield ranges from 1800 to 3500 kg per lactation. This breed is found in Bikaner, Ganganagar and Jaisalmer districts of Rajasthan. This breed is resistant to adverse climatic conditions of the desert area. The heat tolerance co-efficient value of free grazing animals of this breed was found (89.87 ± 2.51) to be quit high (Kaushish, 2006).

2.1.1.20 *Red Sindhi*

The milk yield varies from 2000 to 4000 kg per lactation. It is known for milk production, heat and drought tolerance and resistant to many diseases.

2.1.1.21 *Red Kandhari*

The cows of this breed are good milkers. They are used for draught purpose and are hardy in nature. Bullocks are good draught animal. The breeding tract is in Marathwada region of Maharashtra.

2.1.1.22 *Sahiwal*

It is the best Indian dairy breed and the most economic milk producer. Sahiwal cows are well known for their milking ability. Milk yield varies from 2000 to 4000 kg per lactation. The animals of this breed are heat and drought tolerant. It is found in Punjab, Haryana and Uttar Pradesh. As compared to Holestien calves, the Sahiwal calves were found more resistant to tick borne tropical theileriosis infection (Glass *et al*., 2005).

2.1.1.23 *Tharparkar*

Tharparkar is a dual purpose breed raised primarily for its milking potential. In India these animals are found along the Indo-Pak border covering western Rajasthan and up to Rann of Kutch in Gujarat. Animals with typical characteristics of breed are found in Jodhpur, Barmer, Jaisalmer districts of Rajasthan and Kutch region of Gujarat. Tharparkar breed is well adapted to stresses of arid environment and known for its heat tolerance. It is a special breed of cattle that can cross the Thar Desert by walking continuously for more than 24 h. The heat tolerance co-efficient (88.88 ± 1.92) of free grazing animals of this breed was found to be equal to animals of Rathi breed (89.87 ± 2.51). Pulse and respiration rate of the animals was reported to increase during the summer season. No adverse effect of summer stress on the fertility was observed (80% conception between April to August) in Tharparkar cows (Kaushish, 2006).

2.1.2 *Buffalo Breeds*

It is interesting to note that in spite of buffaloes being a water animal, some of the important breeds of buffaloes like Jaffrabadi, Surti, Bhadawari have been evolved under semi arid region where water is a major limitation. Bhadawari buffaloes are known for high fat content and can sustain on grazing alone. Banni is another breed of buffaloes, which has evolved in the arid region of the Kutch of Gujarat. Pandharpuri and Nagpuri buffaloes were evolved in drought prone areas of Maharashtra. Kalahandi, Paralakhemundi and Manda breeds of buffaloes are found in Orissa and are known for their heat tolerance and less water requirement. Buffalo breeds found in drought prone areas are described below.

2.1.2.1 *Banni*

Banni buffalo is a unique breed, which is locally adapted under extensive system and they are only source of livelihood for Maldharies in Kachchh region of Gujarat. Banni buffaloes are maintained under conditions that are close to those prevailing in the wild environment and harbors the physical fitness of high performance stock, resistance to drought and diseases and resilience in harsh climate (Technical bulletin/folder No. 07/04).

2.1.2.2 *Bhadawari*

Bhadawari is one of the recognized buffalo breeds of India and is famous for high fat content in their milk. Bhadawari buffaloes are found in the ravines of Yamuna and Chambal rivers spread over in Uttar Pradesh and Madhya Pradesh. The main breeding tract is Agra and Etawah in Uttar Pradesh and Bhind and Morena district in Madhya Pradesh. The tract was a part of the erstwhile Bhadawar estate from where the name of this animal originated. Bhadawari, the brown beauty of ravines and predominantly the only breed to have adapted to the harsh conditions of the ravines with undulating topography, thorny and scanty bushes, climatic stress and draught conditions. They are said to be resistant to many tropical bovine diseases. They are also known for eating less and producing more. Calf mortality is significantly lower than the other breeds of buffaloes. Male animals are good for agriculture work especially for ploughing in rice fields. The buffaloes are of medium size with medium milk yield but the fat content in their milk may go as high as 13 percent.

2.1.2.3 *Kalahandi / Peddakimedi*

This breed is known as Peddakimedi in the eastern hill region of Andhra Pradesh and Kalahandi in the adjoining areas of Orissa. The usual color of these buffalo is grey or ash grey. These buffaloes can tolerate solar heat better than the black colored buffaloes because of their light colour. Feed cost of these animals is low and water requirement is less. Milk yield is low but the cost of production is low. These characteristics make this breed suitable for drought prone areas (Nevsarkar *et al.*, 2000).

2.1.2.4 *Paralakhemundi*

Paralakhemundi buffaloes are found in Paralakhemundi and Gunpur areas of Orissa and around Mandasa, Bobli and Srikakulum of Andhra Pradesh, mostly in the down hill plains of Eastern Ghats (Patro and Kornel, 1987). These buffaloes are hardy, resistant to heat and live on road side grazing with straw supplementation. These animals can thrive in diversified conditions.

2.1.2.5 *Manda*

Animals of this breed are found around Kakriguma, Laxmipur, the hils of Damanjodi, the Patraput area and Petta hills near Jeypore, Gupteswar, Jolapur, Balimela, Pedabial and the Arku Valley hills of Koraput district in Orissa (Patro and Kornel, 1987). Body colour is mostly ash grey and grey with copper color hairs. These buffaloes are reared only on grazing. They are semi-wild under forest grazing conditions but are docile when reared in villages. Only new born calves are provided some type of housing otherwise no housing facilities are provided. Buffaloes yield on an average 1.5 litres of milk per day. Bullocks are famous for longevity, hard work and length of working life.

2.1.2.6 *Nagpuri / Berari*

This breed is native of Vidarbha region of Maharashtra and commonly found in Nagpur and Wardha district. These buffaloes are used for heavy draught purpose. This breed is preferred by the farmers in the region mainly because of its low maintenance cost, efficiency of feed conversion, moderate production and disease resistance and adaptation to the hot climate.

2.1.2.7 *Pandharpuri*

The Pandharpuri buffaloes are native of Kolhapur, Solapur, Sangli and Satara districts of south Maharashtra. These buffaloes are found in adjoining Belgaum district of Karnataka. This breed is named after the name of the geographical area *i.e.* Pandharpur block of Solapur district. It is a hardy breed well suited to dry conditions.

2.1.2.8 *Surti*

The main home tract of Surti buffalo is south-western part of Gujarat (Anand, Nadiad, Khara and Baroda district). It is also spread in Udaipur, Rajsamand, Chittorgarh, Bhilwara and Sirohi district of Rajasthan. Surti is one the best suited buffalos of this region where erratic and uneven distribution of rainfall is very common (Tailor, 2004). Surti is a medum sized buffaloes with average milk production of 1200 kg in a lactation of 305 days. Although their average milk yield is comparatively less but they are very good for the areas where fodder is not as abundantly available. Surti buffaloes are preferred in coastal areas because of its compact size and better heat tolerance. Being hardy by nature the animals are easily adaptable to different climatic conditions. The breed is resistant to major infectious diseases that affect cattle in the region (The Hindu, 2006).

2.1.3 *Sheep Breeds*

Sheep is invariably considered as synonyms with drought conditions. This species has adapted to arid and semi arid ecosystem and now a great diversity in sheep breeds can be seen in these areas. Out of 42 breeds of sheep nearly 30 comes from hot arid and semi arid regions of the country. Sheep production is a traditional occupation of small and marginal farmers and landless laborers in the region of the country, where agriculture production is always a gamble due to unfavorable agro-climatic conditions. Sheep play a pivotal role in agrarian economy especially in dry regions of the country providing economic sustenance of about 5-6 million families with poor socio-economic background. Sheep are traditionally raised under extensive range management on public rangeland, fallow land, waste land and forest land. Sheep are able to collect their feed by close clipping and they follow cattle and buffaloes in rangelands with good pasture cover.

Most of the breeds of arid and semi-arid regions are well adapted to harsh climate, long migration, tropical diseases, poor nutrition and shortage of drinking water. The breeds like Chokla, Nali, Magra, Malpura, Pugal, Jaisalmeri, Marwari and Lohi are evolved in the arid and semi-arid regions of Rajasthan. Jaisalmeri, Marwari and Chokla are considered to be the best for migration. Epidemiological studies have shown that Chokla animals are more resistant for sheep pox than Nali (Sharma *et al.*, 1986). Most of these breeds can sustain with out water for long hours. More and Sahni (1973) reported that watering of Chokla sheep at intervals of 72 h did not pose any physical or managemental difficulty. Sonadi, Patanwadi, Muzzafarnagri, Jalauni, Munjal are some of the other breed found in the dry regions of the north western India. The breeds of western arid and semi-arid regions yield wool of 30-40 micron with 25 to 30 percent medullation and the wool is mostly used for carpet manufacturing.

However, sheep breeds of southern arid and semi-arid region produce mostly coarse hairy fibres, which are fully medullated. These sheep breeds have better efficiency of feed conversion for mutton production. Nilgiri breed found in this region is an exception as it produces fine wool. Moreover, there is no requirement of wool in this hot region and thus all the breeds of this region except Nilgiri have probably been selected against production of wool and in favor of high growth rate and body weight. Sheep breeds prevalent in the southern arid and semi arid region are Deccani, Bellari, Nelloer, Mandya, Hassan, Mecheri, Kilakarsal, Vembur, Coimbatore, Nilgiri, Ramnadwhite, Madaras Red, Trichiblack and Kenguri.

2.1.4 *Goat Breeds*

India possesses an enormous goat population numbering 124.6 million, which is the highest in south-east Asia. Goats are very important species of livestock in India, mainly on account of their short generation interval, high rate of prolificacy, and the ease with which the goats as also their products can be marketed. The goats can sustain on sparse forage and extreme climatic conditions where other species of animal may perish. Goats are the main meat animal of India, their meat is most preferred and hence the costliest of all. The distribution of the goats in the country is more homogenous than sheep, as the latter are mostly confined to relatively dryer and arid regions.

Arid and semi arid zone of northern region possess 11 goat breeds, that contribute about 33% of goat population of the country. Zalawadi, Gohilwadi, Kutchi, Marwari, Sirohi, Beetal, Jamunapari, Barbari, Surti, and Jakhrana are the important breeds prevalent in this region. Goat population is more intensified in western and southern plateau region. In major parts of Maharashtra, peninsular Karnataka and Andhra Pradesh, majority of the goat population is non-descript. Goat breeds prediominant in this region are Osmanabadi, Sangamneri, Kanni Adu and Malabari (Gokhle *et al.,* 2002; Kandasamy, 2001).

3.0 LIVESTOCK IN FLOOD PRONE AREAS

The states like Assam, west Bengal and north eastern sates, Bihar, Jharkhnad, and Orissa are characterized by high rainfall, poor drainage system, soil erosion and frequent flood. The flood is also common in the southern peninsular region that spreads along the east coast (Orissa, Andhra and Tamil Nadu) and west coasts (Maharashtra, Karnataka, Kerala and part of Tamil Nadu). Coastal region is blessed with heavy rainfall. Buffalo is the ideal animal species suitable for high rainfall areas. However, there are buffalo breeds which have evolved in arid and semi-arid regions of the country and are suitable for keeping in drought prone areas. Cattle normally prefer dry environment and can easily be maintained on low input in drought prone areas. Cattle breeds from the coastal region are adapted to thrive well in the flood prone areas. Small ruminants particularly goats are also found in the flood prone areas.

The water buffalo is potentially a most important tropical bovine species, especially in very hot areas where rivers and swamp abound (Webster and Wilson, 1980). The water buffaloes are used mainly for milk production and draught power in the paddy fields of India. The water buffalo is preferred highly in the tropics and subtropics since it thrives in hot conditions. The swamp buffalo is a semi-aquatic animal, spending the hottest part of the day (1.00 to 16.00 h) partly submerged in natural swamps or self made wallows (MacGregor, 1941). Because of its morphological and anatomical characteristics, buffalo is well suited to hot and humid climate and muddy terrain. Buffalo skin is covered with a thick epidermis, the basal cells of which contain many melanin particles that give the skin surface its characteristic black colour (Shafie, 1985). The melanin particles trap the ultraviolet rays and prevent them from penetrating through the dermis of the skin of the lower tissues. These rays are abundant in solar radiation in the tropics and subtropics, and excessive exposure of animal tissue could be detrimental to unprotected skin. Buffalo skin is almost devoid of hair, particularly in older animals, because of its exposure to water and mud. This beneficial characteristic is reinforced by its well developed sebaceous glands. With greater secretary activity than in cattle (Shafie and Abou El-Khair, 1970), these glands secrete a fatty substance called sebum, which emerges on the skin's surface and cover it with a lubricant, making it slippery for water and mud. This greasy sebum, along with thick layer of skin, prevents water and the solutes in it from being absorbed into the skin. In this way, the animal is protected from the harmful affects of the any deleterious chemical compounds in the water. Moreover, the sebum layer melts during hot weather and becomes glossier to reflect many of the heat rays, thus relieving the animal from excessive external heat load. Sebum also acts as fly replant. Compared to other animals, buffaloes have less number of sweat glands per unit of area of skin and this makes some difficulty for buffaloes to counter heat storm particularly during hot and humid environment. However, the behavior of wallowing helps the buffaloes to maintain thermal balance. With this behavioral and physical adaptation to the hot humid conditions buffaloes has become the ideal animal for high rainfall areas. Some cattle breeds have also acquired the characters required for sustenance in high rainfall areas (Ongole, Siri, Veture, Malenadu gidda and Dangi) prevalent in high rainfall areas, they also secrete sebum, which help their adaptation to the high rainfall zones.

3.1 Cattle Breeds

3.1.1 *Bachaur*

Animals are found in Sitamari and some parts of Madhubani, Darbhanga, Samastipur and Muzzafarpur districts of Bihar. Bachaur breed of Bihar is famous for its draught qualities and to thrive on poor quality fodder availability.

3.1.2 *Dangi*

Dangi breed of Maharashtra is famous for its excellent working qualities especially in rice fields and heavy rainfall areas. This breed has been selected against the heavy rainfall and have a thick skin which can with stand heavy rainfall (Ghotge, 2004). Skin of Dangi cattle exudes an oily secretion which protects the animals from heavy rains (Sadana, 2004).

3.1.3 *Hallikar*

Hallikar is found in Hassan, Mysore, Mandya, Bangalore, Kolar, Chitrdurga and Tumkur district of Karnataka. It is a draft breed used for road and field agricultural operation. They can also work in water logged fields.

3.1.4 *Kherigarh*

The main breeding tract of this breed is Kehri district in Uttar Pradesh. Animals of this breed are very active and can thrive on grazing alone. Bullocks are good for light draught and quick light transport but the cows are poor milker.

3.1.5 *Malenadu Gidda*

Malenadu gidda is a lessor known breed found in the Western Ghats of Karnataka. Selected cows of this breed can yield up to 3 litres milk per day. This breed is very well suited to the hilly tract and high rainfall area. The animals of this breed are reported to be resistant to foot and mouth diseases and mastitis.

3.1.6 *Ongole*

The animals of this breed are found in Nellore, Vizianagaram, Chittoor, Kurnool, Cuddapah, Ananatpur, Nalgonda, Mahabubnagar and Khammam district of Andhra Pradesh. It is an excellent dual purpose known for draught power and heat tolerance. Average milk yield ranges from 475 to 1000 kg in a lactation of 230 days (Nivasarkar *et al.*, 2000).This breed is well adapted to hot and humid climate, 85% of the heat from solar radiation is sent back to the environment by the Ongole animal through its white, reflective coat and rest 15% of heat immediately absorbed by the underlying black skin. Long legs of this breeds helps in minimizing absorption of solar radiated heat (www.ansi.okstate.edu/breeds/cattle.ongole/index.htm).

Ongole animals have more number of bigger, functional sweat glands per unit area of the skin, which secretes the zebum or sebum, a kind of substance which prevents the skin from high rain and make the animals resistant to tick borne diseases and make this breed suitable for high rainfall areas. Basal metabolic rate of Ongole cattle is low and these animals have the ability to reverse down metabolism during extreme of scarcity

3.1.7 *Ponwar*

This breed is found in Pilibhit and north-west part of Lakimpur Kheri district in Uttar Pradesh. They are primarily a darft purpose breed and the cows are poor milkers.

3.1.8 *Siri*

This breed can withstand the rugged terrains of mountains (high altitude breed) very well. This breed has evolved in areas with heavy rainfall and the animals have a thick skin which can withstand heavy rainfall (Ghotge, 2004).

3.1.9 *Umblacheri*

It is found in Tanjavur district of Tamil Nadu. Bullocks of this breed are excellent for wet ploughing.

3.1.10 *Vecher*

Venture is native breed of Kerala. It originated in a village called Veture in the Vaikom taluk of the Kottayam district of south Kerala. Heavy rainfall and hot humid climate of the area led to the natural selection of a small sized animal. It is the smallest cattle breed of the world and famous for production of maximum milk production proportionate to its body weight. The maximum height of Veture cow is 91 cm, weighing on an average 107 kg. It can give an average yield of 3 litres of milk per day (www.vechur.org). This breed is well adapted to hot, humid tropical climate and used for ploughing.

3.2 Buffalo Breeds

Buffaloes are well adapted to hot and hot humid climate and play a distinct role in the economy of farmers. Buffaloes found in our country are named as water or river buffalo and swamp buffalo. The morphology, purpose and the genetics of both these subspecies are different. River buffaloes are generally large in size, they prefer to enter in clean water and are reared primarily for milk and meat production besides also for draught purpose. On the other hand Swamp buffalo is reared mainly for draft purpose. Swamp buffalo is a semi-aquatic animal, spending the hottest part of the day (10.00 h to 16.00 h) partly submerged in natural swamps or self made wallows (MacGregor, 1941). Almost all the breeds of buffaloes can be reared under high rainfall areas. The breeds which have developed some unique characteristics with respect to the adaptation of the high rainfall zone, are described below.

3.2.1 *Murrah*

Murrah is one of the best breeds of buffaloes. It home tract lies in Rohtak, Jind, Hisar and Bhiwani district of Haryana, the union territory of Delhi and Patiala district of Punjab. However, animals of this breed can be found throughout the country. Buffaloes are good milkers and the males are good for draught purpose.

3.2.2 *Nili-Ravi*

Nili-Ravi buffaloes are found in Firozpur, Gurdaspur and Amritsar districts of Punjab. This breed is almost similar to Murrah in almost all characters except for white markings.

3.2.3 *Toda*

This breed is named after its herdsmen, the Toda tribes of the Nilgiris. This breed is found in Nilgiris district of Tamil Nadu. Geographical separation (hills separated from surrounding low lands) had helped in the evolution of this fine genetically isolated breed. High fat content, regularity in calving, longevity and adaptability to the extreme of climate of hilly tract are specific characters of this breed. These buffaloes are self-sustainable in low/zero input system (Kadirvel, 2004). Toda is one of the important breeds of buffaloes other than Kaziranga buffaloes of Assam which thrive well under the high rainfall and high humid areas (Nivasarkar *et al.*, 2000).

3.2.4 *Chilka*

Chilka buffaloes are found around Chilka lake in Orissa. These buffaloes possess special ability of entering deep into the salt lake of Chilka and feeding on the weeds, aquatic plants and other vegetations submerged in water. Chilka buffaloes are never provided with shed for shelter, they are let loose throughout the year. These buffaloes are raised under extensive system with almost zero input. They produce about 2.5 kg milk per day with high fat percentage (Das, 2004).

3.2.5 *Swamp*

Swamp buffaloes are found in Assam. These buffaloes are hardy and quite adapted to the local warm humid condition of the region. They survive mostly on naturally available feeds and fodder which comprise aquatic, semi-aquatic, hill and highland grasses. Animals are suitable for low lying water merged areas for ploughing paddy fields. Swamp buffaloes spend most of their day time by wallowing in muddy water or beneath the trees, and graze during morning time in winter and evening time in summer.

3.2.6 *Diara or Deshila Buffaloes*

South and north gangetic plains of Bihar are densily populated with clusters of medium sized buffalo population locally called Diara or Deshila. These buffaloes have full adaptation to the local agro-eco-socio-economic and climatic conditions prevailing in the gangatic plains particularily "Tal & Diara" areas of river Ganga, Sone, Gandak. These buffaloes are preferred by the farmers due to the zero management system utilizing locally coarse poor quality feed and fodder resources without any significant loss in morbidity and mortality (Singh *et al.*, 2006).

3.2.7 *South Kanara*

Buffalo of this are found around Mangalore region of Karnataka. Famous for agricultural operation particualrily in rice fields. The population of this breed is reported to be declining at a faster rate.

3.2.8 *Kuttanad*

Kuttanad buffaloes are found only in Kuttanad area of Kerala. Females are poor milkers. Males are exclusively used for working in paddy fields. The present population of these is reported to be around 500 only.

3.2.9 *Sikamese*

These buffaloes are the native of the Sikkim state. This is a small and hardy buffalo suitable for hilly terrain.

3.3 Sheep Breeds

Most of the sheep breeds are found in arid and semi-arid regions of the country. However, there are some breeds which are suitable for high rainfall areas of the states like Bihar West Bengal, Orissa, Assam and other eastern parts of the country characterized by hot and humid climates. Sheep breeds of hot humid climates are Sahabadi, Ganjam, Garole, Chhotanagapuri, Bolangir, Tibetan and Bonpala. The sheep breeds of this region are primarily reared for meat purpose. The quality of wool produced by the sheep of this region, in general, is small and extremely coarse, coloured and hairy. The fleece is rarely shorn. Most of the breeds of this region are small in size, high prolific, have short tail and can withstand with high rainfall. They generally give birth to twins, often triplets and occasionally quadruplets *e.g.* Garole sheep of Sunderban of West Bengal. The short tail not only eliminates the need for tail docking but also has advantage in the control of fly strike.

3.4 Goat Breeds

Goats in the region of high rainfall areas are dwarfer, having high prolificacy, early maturity, short kidding interval and superior quality meat and skin. The predominant breeds of this region are Black Bengal or its type and Assam hill, Ganjam or its type. Goats in this region are reared under extensive system (migration, sedentary type of rage grazing, transhumant range grazing) without supplementary feeding. The Bengal goat is well adapted to low input system, possibly by virtue of small body size, and high rate of nutrient utilization (Pan, 2006). Konkan Krishi Vidyapeth, Dapoli, district Ratnagiri has developed a tall breed of goat called Konkan Kanyal, which is suitable for heavy rainfall. Malabari and Attapaddy are the only recognized breeds of peninsular region. Andaman goats, feral goats, Kodi adu and Bidari are other less known strains of goats of high rainfall region.

4.0 BREEDING OF LIVESTOCK

Livestock breeding policy, particularly in the flood and drought prone areas, needs to be reviewed keeping in view the availability of feed and fodder, demand and supply of animal products like milk, meat, wool and draught power, farmers needs and perception, existing species/breed and their performance in different agro-climatic zones. Attention should be given to raise animal productivity from low to intermediate level rather providing genetic potential for productivity that can not be supported economically and might seriously hamper sustainability of natural resources.

4.1 Cattle

The Indian subcontinent is a treasure house of different native cow breeds. Before the introduction of exotic breeds, such as Holstein Frisian, Jersey, brown Swiss etc, from abroad, these native cattle were found in abundance in our country. Crossbreeding of these cattle with exotic bulls for increasing milk production has slowly and steadily wiped out a number of these native breeds. Though crossbreeding to an extent has increased the milk production, the true genetic characteristics of these native breeds have been lost. As a result a number of common breeds are gradually becoming extinct.

The overall annual growth rate in the cattle population from 1992 to 2003 was -1.14%. The annual growth rate for indigenous and crossbred cattle were -1.56 and +5.65%, respectively. As per the 17th Indian livestock census 2003, in 2003 there were only 156.85 million indigenous cattle as compared with 178.78 million and 189.37 million in 1997 and 1992, respectively. Conversely, the population of crossbred cattle has increased from 15.21 million in 1992 to 22.07 million in 2003 (Gandhi and Singh, 2006). The situation in different states is similar to the overall status in the country. In Karnataka, there has been a marked reduction in the number of indigenous breeds. Some of the known Indian breeds are on the verge of extinction. The number of Krishna Valley breed has come down by 97.96% and that of Deoni breed has come down by 72.22. The number of

Hallikar breed has come down by 39.97%, Khillari breed by 30.89% and Malenadu gidda breed by 24.96% (The Hindu, April 16, 2005). At the same time, the population of crossbred has increased. This situation is alarming. The diverse genetic base is becoming narrow, which is a threat to sustainable farming. If the population becomes uniform, its vulnerability to pests and disease will increase.

The experience on crossbreeding has shown that by and large crossbreeding programs have had no impact as per the expectations. The major problem is of survival due to nutrition and adaptability. Looking at the drawbacks of the crossbreeding program and the importance of indigenous breeds to all aspects of the life of the rural people, a proper breeding policy should be evolved. Indigenous breeds are known to have genes for disease resistance, draught quality and adaptability and heat tolerance. Genetic improvement and conservation of nationally important indigenous breeds of cattle through selective breeding in their home tracts to exploit their milk production potential and draught ability. Non- descript cattle of different regions may be upgraded using descript breeds of the region.

4.2 Buffalo

Since India has the best buffalo germplasm in the world therefore, to bring genetic improvement in this species selective breeding seems best fit tool for its improvement. Few important breeds like Surti and Bhadawari which are precious are dwindling in their number; therefore, there is an urgent need to look such type of situation. Different breeds of buffaloes which have adapted to their local environment need to be conserved and improved through selection. Simultaneously, the vast non-descript buffalo population need to be upgraded for increasing their productivity using descript breed of their region.

4.3 Goats

Indigenous goat breeds have an immense production potential and resistance against diseases and climatic fluctuations. Goat breeding strategies should be framed on the basis of production and economic efficiency in relation to local environment, resource utilization and sustainable production rather than absolute production. Goat breeds can be improved through selective breeding by supplying superior bucks. Attempts to improve the productivity of some breeds like Jamunapari, Beetal, barbari, Black Bengal and Malabari through selection have yielded encouraging results.

4.4 Sheep

Crossbreeding of native breeds with exotic sheep was undertaken to improve wool production and quality and meat production and it has been observed that crossbred developed have not performed well under farmer's conditions.

It has not been possible to produce fine wool with desirable staple length from higher crosses in arid and semi arid regions. Selective breeding is the best way to improve the productivity and to conserve the unique characteristics of the native breeds. All the native breeds may be improved through selective breeding by distributing good quality rams of the respective breeds. The program of selective breeding is under implementation in 9 districts of Rajasthan namely Barmer, Pali, Jalore, Jaisalmer, Jodhpur, Churu, Sikar and Nagaur. Interested farmers are supplied with selected rams of Marwari, Jaisalmeri, Magra and Nalil capable of producing more body weight and wool (Kaushish, 2006). Similar selective breeding programs may be initiated in other states also.

5.0 REFERENCES

Das, S.K. 2004. Buffaloes of Orissa and adjoining regions: Present status and perspectives. Buffalo production under different climatic regions (*Eds.* Kundu, S.S., Misra, A.K. and Pathak, P.S.), International Book Distribution Compny, Lucknow (India).

Frisck, J.E. and Vercoe, J.E. 1984. An analysis of growth of different cattle genotypes reared in different environments. *J. Agric. Sci. Camb.* **103**: 137-153.

Gandhi, R.S. and Singh, S. 2006. Impact assessment of crossbreeding of indigenous cattle with Europian Dairy breeds. *Indian Dairyman.* **58(2)**: 60-71.

Ghotge, N.S. 2004. Livestock and livelihoods: The Indian Context. Print Perfect Mayapuri. New Delhi.

Glass, E.J., Preston, P.M., Springbett, A., Craigmile, S., Kirvar, E., Wilkie, G. and Brown, C.G. 2005. Bos Taurus and Bos Indicus (Sahiwal) calves respond differently to infection with Theileria annulata and produce markedly different levels of acute phase protein. *Int. J Parasital.* **35(3)**: 337-47.

Gokhale, S.B., Gokhale, R.B., Phadke, N.L. and Desale, R.J. 2002. Status of village goat management practices in Maharashtra. *Indian Journal of Animal Science* **72**: 810-814.

Gunasekaran, M. and Reddy, A.O. 2006. Kangayam cattle: Dwindling population and strategic interventions. *In*: Compendium National Symposium on conservation and improvement of animal genetic resources under low input system: Challenges and strategies, February 9-10, 2006, Karnal.

Hunter, R.A. and Siebert, B.D. 1985. Utilization of low quality roughages by *Bos taurus* and *Bos indicus*. 2 the effect of rumen –degradable nitrogen and sulpher on voluntary food intake and rumen characteristics. *Br. J. Nutrition* **53**: 649-656.

The Hindu. 2004. Indigenous breed of cattle for dry regions. The Hindu (English daily), 29/07/2004.

The Hindu. 2006. Surti buffaloes yield milk with more fat content. www.thehindu.com/thehindu/seta/2006/04/20/stories/2006042004261700.htm

The Hindu. 2007. Banni bovine breed bequeaths big benefits. The Hindu (English daily), 08/02/2007.

The Hindu. 2005. Indigenous cattle breeds dwindling, scientists term it dangerous. The Hindu, (English daily) April 16, 2005.

Kadirvel, R. 2004. Toda buffalo rearing in southern plateau and hill regions of Tamil Nadu. Buffalo Production under different climatic regions (*Eds*. Kundu, S.S., Misra, A.K. and Pathak, P.S.), International Book Distribution Compny, Lucknow (India).

Kandasamy, N. 2001. Breeding technologies for augmenting chevon production in Southern India. Proceedings of IV National seminar of ISSGP&U at Bharathiar University, Coimbatiore, Tamil Nadu, November, 9-11.

Kaushish, S.K. 2006. Drought and livestock. *In*: Drought management strategies in India. ICAR, New Delhi.

MacGregor, R. 1941. The domestic buffalo. *Vet. Rec.*, **53**: 443-450.

Mitkari, V.R. and Thombre, B.M. 2006. Status of Deoni cattle from breeding tract in Maharastra. *In*: Compendium National Symposium on conservation and improvement of animal genetic resources under low input system: Challenges and strategies, February 9-10, 2006, Karnal.

More, T. and Sahni, K.L. 1973. Studies on water requirement of sheep. Annual Report of CSWRI, Avikanagar, Rajasthan.

Nivsarkar, A.E., Vij, P.K. and Tantia, M.S. 2000. Animal Genetic Resources of India: Cattle and buffalo. ICAR, New Delhi.

Prakash, O., Ahmed, M., Singh, P.K., Singh, O.P. and Singh, P.K. 2008. Gangatiri: A dual purpose cattle breed of Uttar Pradesh. *Livestock International*, **12(3)**: 12-13.

Pan, S. 2006. Small ruminant genetic resources of West Bengal and strategies for conservation under low input system. *In*: Compendium, National Symposium on conservation and improvement of animal genetic resources under low input system: Challenges and strategies, February 9-10, 2006, Karnal.

Patro, B.N. and Das, K. 1987. Paralakhemundi and Manda buffaloes of Orissa. Buffalo bulletin **6(3)**: 213-217.

Prabhu, B.H.M., Govindhaiah, M.G., Jayasankar, M.R., Nagaraja, C.S. and Kumarasamy, P. 2008. Evaluation of phenotypic confirmation and status of south Indian cattle breed Amritmahal. *In*: *Proceedings of National Symposium on redefining role of indigenous animal genetic resources in rural development,* February 15-16, 2008, Bangalore.

Sadana, D.K. 2004. Indigenous cow breeds of India. *The Indian Cow* **13**: 6-25.

Banerjee Sandeep. 2006. Gaini, a dwarf breed of cattle in Rajasthan, India. *In*: Compendium, National Symposium on conservation and improvement of animal genetic resources under low input system: Challenges and strategies, February 9-10, 2006, Karnal.

Shafie, M.M. 1985. Physiological responses and adaptation of water buffalo. Yousef, M.K. (ed.). *In*: *Stress Physiology in Livestock,* vol. 2: Ungulates. Florida, USA, CRC.

Shafie, M.M. and Abou El-Khair, M.M. 1970. Activity of the sebaceous glands of bovine in hot climate (Egypt). *J. Anim. Prod.* UAR, **10**: 81-98.

Sharma, M.M., Uppal, P.K., Lonakr, P.S. and Mathur, P.B. 1986. Epidemiology of sheep pox outbreak in mutton and fine wool type sheep at organized far. *Indian Journal of Animal Sciences*, **56(12)**: 1183-1186.

Sharma, R.J. and Kumar Sanjay. 2008. Effective exploitation of unique characteristics of indigenous livestock. *In*: *Proceedings of National Symposium on Redefining Role of Indigenous Animal Genetic Resources in Rural Development,* February 15-16, 2008, Bangalore.

Silanikove, N. 2000. The physiology of adaptation in goats to harsh environments. *Small Ruminants Research*, **35(3)**: 181-193.

Singh, S.R., Mandal, K.G., Singh, P.K., Ranjeet Kumar and Singh, R.N. 2006. Phenotypic parameters of a true breeding buffalo population in Tal and Diara areas in Bihar. *In*: Compendium National Symposium on conservation and improvement of animal genetic resources under low input system: Challenges and strategies, February 9-10, 2006, Karnal.

Tailor, S.P. 2004. Breeding and feeding of management of Surti breed of buffaloes in southern Rajasthan. Buffalo Production under different climatic regions *Ed.* Pathak....

Technical Bulletin/Folder No. 07/04. Bannu buffalo- incredible germplasm of Kachchh. S.D. Agriculture University, Sardarkrushinagar - 385 506, Gujarat.

Webster, C.C. and Wilson, P.N. 1980. Agriculture in the tropics, p. 390-400. London, UK, Longman. (ELBS, 2nd edn.).

World Watch list for domestic animal diversity, 2nd edition (*Eds.* Beate D. Sherf), F.A.O. Rome, September 1995.

20

STRATEGIES FOR DISASTER MITIGATION AND MANAGEMENT

N. Das

ABSTRACT

Disaster Management is a collective term encompassing all aspects of planning for and responding to disasters, including both pre- and post-disaster activities. It is important to make a plan of action for dealing with natural calamities in advance and it needs to be updated periodically. The National Disaster Management Act envisaged the creation of the National Disaster Management Authority (NDMA), headed by the Prime Minister, and State Disaster Management Authorities (SDMAs) headed by respective chief ministers, to spearhead and implement a holistic and integrated approach to disaster management in India. The primary objectives of NDMA are human resource development, coordination of various organizations involved in disaster management, establishment of national level information base, promotion of international and regional partnerships and assisting various states in strengthening their disaster management systems. The Disaster Management Act has also mandated the constitution of a specialist response force National Disaster Response Force (NDRF) which will be a multi-disciplinary, multi-skilled, high-tech force for all types of disasters capable of insertion by air, sea and land. The four phases of disaster management are mitigation, preparedness, response and recovery. Disaster management plan for animals include -i) epidemiological study in disaster prone areas, ii) action plan for rescuing and treatment of injured animals, iii) planning of resources which interalia includes man power, logistical needs, store, equipments, ambulance, medicine and facility for grazing and watering, iv) training of professionals and administrators and v) allied planning (organization) like augmenting political and administrative support and involving N.G.O.'s, media, goshalas, animal

Indian Grassland and Fodder Research Institute, Jhansi (UP), India

welfare organizations and other volunteer groups. The subject of natural disaster education traditionally fell in the domain of different departments / disciplines each having their own priorities. Disaster education should be balanced and without undue bias to any particular discipline.

1.0 INTRODUCTION

For a countries like India which has substantial animal population and where the livelihood of majority of rural poor are dependent on the income from livestock rearing operations, disaster preparedness, management and recovery plans that include a livestock component are important and essential. A view that disaster plan especially for livestock is unaffordable in developing countries like India is unjustified and can be harmful in the long run not only to animals, but also to humans. Natural and man-made disasters caused severe damages and loss of lives for centuries. How people handle with them depends primarily on i) the periodicity of such disasters, ii) the financial resources and physical infrastructures available, and iii) the level of preparedness. In places, where disasters are repeated at regular intervals, the local community/society has some inbuilt socio-cultural systems to deal with them. For example, Somali pastoralists have developed elaborate and effective survival mechanisms that enable them to mitigate the impact of droughts and increase their survival and that of their livestock. The pastoralists usually keep two or more species as a buffer against dramatic losses during drought, and clan members help out less fortunate members by collecting a voluntary donation of an animal, if not more, from all members who can afford (Dirie and Mohammed, 1999). In India, the sheep owners in the desert migrate with their animals (which may include animals from other community members) from one state to another in every summer. In some parts of rural India, when one poor farmer lost his animal, he could take one or two animal for rearing on loan basis from an affluent farmer; he took a portion of the income arising from sale of products and also retained few of the new born young.

Disaster Management as "a collective term encompassing all aspects of planning for and responding to disasters, including both pre and post disaster activities. It may refer to the management of both the risks and consequences of disasters". Clearly the term management has emerged as an umbrella term that encompasses the entire disaster cycle, including mitigation. This needs careful noting and wide spread awareness because traditionally the term management was restrictively used to address only post disaster operations. Although it is not always possible to predict and prevent natural disasters, but a state of preparedness and ability to respond quickly to natural disasters will definitely mitigate the suffering and loss of life to a great extent. It is important to make a plan of action for dealing with natural calamities in advance and it needs to be updated periodically

(EMI, 1998). The plan must be prepared in advance of any possible natural disaster and must be tested with simulation model in order to validate the plans. The National Disaster Management Act of 2005 brings about a paradigm shift in India's approach to disaster management. The centre of gravity stands visibly shifted from calamity relief to preparedness, prevention and planning simultaneously as the national disaster response.

2.0 HISTORY OF DISASTER MANAGEMENT

The history of disaster management (DM) in India can be divided into two eras.

2.1 Pre-National Disaster Management Act (2005) Era in India

In the British time, a scarcity relief division was setup in the then agriculture ministry in 1878, in order to address famine. Eventually, in 1883, a Famine Commission was constituted, and it was then that the first famine code was promulgated by the government. In the post independent India, the subject of disaster management did not find any special mention in the Indian constitution. Generally financial assistance was provided to the state governments for meeting expenditure on six identified natural calamities for calamity relief. The approach to disaster management was fragmented with controls resting with individual ministries as noted below.

Disaster	Nodal Ministry
Air accidents	Ministry of Civil Aviation
Civil strife	Ministry of Home Affairs
Railway accidents	Ministry of Railways
Chemical disasters	Ministry of Environment & Forest
Biological disasters	Ministry of Health
Nuclear accident inside or outside the country	Dept of Atomic Energy
Natural disasters	Ministry of Agriculture

Government of India set-up a High Powered Committee (HPC) on disaster management in August 1999. The HPC in its October 2001 report suggested to create a separate ministry of disaster management. A national institute of disaster management and a national centre for calamity management were also recommended to assist in the operational of the plans and procedures of the higher level bodies. A cabinet committee was envisioned to effectively manage cataclysmic events in the hour of a national crisis. The disaster management chain extended to panchayats with the objective to ensure the involvement of local government, who are usually the first responders, at the grass roots level. A chapter on disaster management also entered in to the Tenth Five Year Plan prepared by the Planning Commission of India. Realization increased so much that the terms of reference of the Twelfth Finance Commission were changed to include concerns of disaster management.

2.2 Post-National Disaster Management Act Era

A committee was constituted on 11 January 2005 by the Government of India to draft the disaster management bill. Rajya Sabha passed the bill with amendments on 28 November 2005 and Lok Sabha did so on the 12/13 December 2005. President of India signed the bill on the 23 December 2005 and the bill became the National Disaster Management Act. Government of India has also constituted cabinet committee on management of natural calamities and cabinet committee on security. Besides above there are high level cabinet committee and inter-ministerial committee and inter ministerial group in place.

2.2.1 *Disaster Management Act*

The new act provides that

- There shall be a National Disaster Management Authority of which the prime minister of India will be the chairperson, helped by a vice chairperson.
- There shall be creation of State Disaster Management Authorities, expected to be chaired by the chief minister of the concerned state.
- There shall be a District Disaster Management Authorities co-chaired by district collector and president of the elected body of the district.
- There shall be an institution to study and do research in science and develop technologies to manage the disasters.
- There shall be disaster management funds available to the union, State and district authorities to meet the immediate needs of providing rescue and relief to the victims of disasters.

According to the Act, disaster management means a continuous and integrated process of planning, organizing, coordinating and implementing measures which are necessary or expedient for

(1) Prevention of danger or threat of any disaster
(2) Mitigation or reduction of risk of any disaster or its severity or consequences
(3) Capacity building
(4) Preparedness to deal with any disaster
(5) Prompt response to any threatening disaster situation or disaster
(6) Assessing severity or magnitude of effects of any disaster

(7) Evacuation, rescue and relief, and
(8) Rehabilitation and reconstruction.

2.2.2 *National Disaster Management Authority (NDMA)*

In recognition of the importance of disaster management as a national priority, Government of India (GOI) set up a High-Powered Committee (HPC) in August 1999 for making recommendations on the preparation of disaster management plans and suggestion for effective mitigation mechanisms. On 23 December 2005, the Government of India enacted the Disaster Management Act, which envisaged the creation of the National Disaster Management Authority (NDMA), headed by the prime minister, and State Disaster Management Authorities (SDMAs) headed by respective chief ministers, to spearhead and implement a holistic and integrated approach to disaster management in India. The primary objectives of NDMA are

- Human resource development covering multiple aspects of disaster management and to play a lead role in national level policy formulation.
- To coordinate various role players within the field of disaster management: government, non-governmental organisations, public and private sector and international organisations.
- To establish an exhaustive national level information base on disaster policies, prevention mechanisms, mitigation measures, and region wise preparedness and response plans as well as resource spent on mitigation and response for various types of disasters.
- To forge, promote and sustain international and regional partnerships for launching joint, synergistic projects and programmes.
- To assist various states in strengthening their disaster management systems and capacities, and in preparation of their plans and strategies for hazard mitigation and disaster response.
- To set up linkages with other international institutions in the region for mutual benefits and sharing of experience.

Besides prime minister as chairman, NDMA has a vice chairman with the status of cabinet minister, and eight members with the status of ministers of state. Each of the members has a well defined functional domain covering various states as also disaster specific areas of focus and concern. To carry out the mandated functions, a functional and operational infrastructure has been built which is appropriate for disaster management involving uncertainties coupled with desired plans of action. The concept of the organization is based on a disaster divisions-cum-secretariat system. Each member of the NDMA heads disaster-specific divisions for specific disaster and functional domains. Each member has also been given the responsibility of specified states and UTs for close interaction and coordination. The NDMA secretariat, headed by a secretary is responsible to provide secretarial support and continuity. It is proposed to have two Disaster Management (DM) wings under the secretariat. They are:-

- DM I wing dealing with mitigation, preparedness, plans, reconstruction, community awareness and dealing with financial/ administrative aspects.
- DM II wing is composed of the National Disaster Management Operations Centre with the state-of-the-art multi-redundant communication systems, to carry out the tasks of capacity development, training and knowledge management.

Table 1 Ministries Responsible for Various Categories of Disasters

Disasters	Ministries
Earthquakes and Tsunami	Ministry of Home Affairs/Ministry of Earth Sciences
Floods	Ministry of Home Affairs/Ministry of Water Resources
Cyclones	Ministry of Home Affairs/Ministry of Earth Sciences
Drought	Ministry of Agriculture
Biological disasters	Ministry of Health and Family Welfare
Chemical disasters	Ministry of Environment & Forests
Nuclear disasters	Ministry of Atomic Energy
Air accidents	Ministry of Civil Aviation
Railway accidents	Ministry of Railways

The NDMA prepared national policy framework which is in conformity with the International Strategy for Disaster Reduction, the Rio Declaration, the Millennium Development Goals and the Hyogo Framework 2005-2015. The central ministries, states and other stakeholders have been involved in the participatory and consultative process of evolving policies and guidelines. The themes underpinning this policy are:-

- Community-based disaster management, including last mile integration of the policy, plans and execution.
- Capacity development in all related areas.
- Consolidation of past initiatives and best practices.
- Cooperation with agencies at national, regional and international levels.
- Compliance and coordination to generate a multi-sectoral synergy.
- The objectives of the national policy include:
  - √ Promoting a culture of prevention and preparedness – by centre-staging DM as an overriding priority at all levels and at all times.
  - √ Encouraging mitigation measures based on state-of-the-art technology and environmental sustainability.
  - √ Mainstreaming DM concerns into the development planning process.

- √ Putting in place a streamlined institutional techno-legal framework in order to create and preserve the integrity of an enabling regulatory environment and a compliance regime.
- √ Developing contemporary forecasting and early warning systems backed by responsive and fail-safe communications and Information technology (IT) support.
- √ Promoting a productive partnership with the media, NGOs and the corporate sector in the areas of awareness generation and capacity development.
- √ Ensuring efficient response and relief with a caring humane approach towards the vulnerable sections of the society.
- √ Making reconstruction an opportunity to build back better and construct disaster-resilient structures and habitats.

2.2.2.1 *Role and responsibilities of NDMA*

NDMA as the apex body is mandated to lay down the policies, plans and guidelines for disaster management to ensure timely and effective response to disasters. Towards this, it has the following responsibilities:-

- Lay down policies on disaster management;
- Approve the national plan;
- Approve plans prepared by the ministries or departments of the government of India in accordance with the national plan;
- Lay down guidelines to be followed by the state authorities in drawing up the state plan;
- Lay down guidelines to be followed by the different ministries or departments of the Government of India for the purpose of integrating the measures for prevention of disaster or the mitigation of its effects in their development plans and projects;
- Coordinate the enforcement and implementation of the policy and plan for disaster management;
- Recommend provision of funds for the purpose of mitigation;
- Provide such support to other countries affected by major disasters as may be determined by the central government;
- Take such other measures for the prevention of disaster, or the mitigation, or preparedness and capacity building for dealing with the threatening disaster situation or disaster as it may consider necessary;
- Lay down broad policies and guidelines for the functioning of the National Institute of Disaster Management.

2.2.3 *National Disaster Response Force (NDRF)*

The Disaster Management Act has mandated the constitution of a specialist response force to a threatening disaster situation or a disaster. This force will function under the NDMA which has been vested with its control, direction and general superintendence. This will be a multi-disciplinary, multi-skilled, high-tech force for all types of disasters capable of insertion by air, sea and land. All the eight battalions are to be equipped and trained for all natural disasters including four battalions in combating nuclear, biological and chemical disasters. Presently this force is constituted of eight battalions, two each from the BSF, CRPF, CISF and ITBP. Each battalion will provide 18 self-contained specialist search and rescue teams of 45 personnel each including engineers, technicians, electricians, dog squads and medical/ paramedics. These NDRF battalions are located at nine different locations in the country based on the vulnerability profile to cut down the response time for their deployment. During the preparedness period/in a threatening disaster situation, proactive deployment of these forces will be carried out by the NDMA in consultation with state authorities.

3.0 COPING WITH DISASTER

A common man in a disaster prone area needs the answer of the following questions:

(1) What might possibly happen?
(2) How likely it is?
(3) What damage, injury and loss may result?
(4) What will be the impact of the event?
(5) What can be done about it?

These answers to these vital questions can not be properly answered unless we follow a systematic approach both in letters and spirit. Fragmentary approach may prove to be deceptive, expensive and even wasteful. **There are four phases of disaster management *viz.* mitigation, preparedness, response and recovery.** Professionals are involved in all the four phases. These phases bring together a variety of professionals from different spheres: meteorologists, earth scientists, seismologists, emergency medical teams, social scientists, social welfare workers, psychologist, land use planners, architects, construction engineers and workers, insurers, risk assessment experts, hospital staff, policy makers, administrator, support staffs and others. A large number of people are involved to solve problems created before, during or after a disaster and they are forced to work within temporal and budgetary constraints to solve the problems. An 'ALL HAZARDS' approach would make better integration among professionals in dealing disaster management efforts. This approach is based on the concept that regardless of the impact of different types of disasters, the socio-economic consequences of disaster on a country, including on animal husbandry are similar. Biological threats could arise in disasters even if the originating event is not biological in nature. For example, earthquakes disrupt water supply leading to unhygienic conditions and diseases; flood which is a geophysical event, lead to stagnation of water creating conditions conducive for decomposition of dead animals and other organic materials (Dipak, 2002).

Veterinarians have a role to play in all the four phases from mitigation to recovery; however it is during the relief works when they play an intensive and significant role to safeguard the life of the animals as well as their future production potential. Veterinarians also contribute significantly in 'search and rescue' team which generally employs dogs to detect the injured and dead persons/animals. Veterinarians provide the necessary care (like first aid and treatment) to the dogs injured in the rescue operation. In rural areas the number of animals could be greater than the number of humans and in urban areas; pets could form a sizeable population. In the USA, special Veterinary Medical Assistance Team (VMAT), comprising veterinarians, veterinary pathologists, epidemiologists, toxicologist, technicians etc. actively involved in emergency situations to assess veterinary needs of the animals. The VMAT teams also include private citizens who are activated in times of disaster and they are paid a salary and protected under the Federal Toirt Claims Act against personal liability.

3.1 Mitigation

The basic aim is to prevent or minimize the effect of future emergencies. This phase includes any activities that prevent an emergency (like construction of earthquake proof building, preparation of fodder bank etc.), reduce the likelihood of occurrence (construction of dams, water conservation measures) or reduce the damaging effect of unavoidable hazards (buying flood and fire insurance for farmhouse). According to the Indian Act 2005, "mitigation means measures aimed at reducing the risk, impact or effects of a disaster or a threatening disaster situation". There is the urgent need to ensure that disaster mitigation strategies get enmeshed and integrated with the very development process. Actions to be taken by veterinary institutions (Dipak, 2002) are mentioned below:

- Ensure that bio-security system and high risk establishments have adequate facilities for waste treatments methods, control of water and air pollution. The buildings are also resistant to flood and earthquake.
- Isolation and quarantine station should be available during disaster.
- Extension work to propagate the knowledge to the animal owner about the animal health, bio-security of the farm, preventive actions that can check the pollution of environment.

3.2 Preparedness

This is the planning about 'What to do?', 'Where to go?' and 'Where to call for help?' before an event occurs. Such actions will help improve the chances of successfully dealing with an emergency. For example quick access of emergency telephone numbers, installing cyclone and tsunami warning system, keeping emergency stock of feed and fodder, holding disaster drills at regular interval are all preparedness measures. Dipak (2002) has suggested the following measures to be adopted by the veterinary/animal husbandry institutions.

- Keeping in view all the available facilities/infrastructure available and the probable risk, a disaster plan is to be created and team identified to handle situations during disaster. Copy of the disaster plan should be available to each of the team member and a copy should be stored on-site and off-site. A material stockpile needs to be created and renewed regularly.
- Policies and measures in respect of animals made for the different phases of disasters need to be justified and explained to the public through the appropriate media.
- Introduce regular safety drill involving all the staff and local community. Regular monitoring of broadcasts and news about impending disaster and communication about them to the community.
- Networking should be done with other nearby institutions about logistical assistance (like diagnostic facilities, fodder, vaccine etc.) during disaster.
- Member of the disaster management teams, including veterinarians, should be trained for coordinated work and compensated for their work during disaster and provided health insurance and legal cover for their actions during disasters.
- Special high-risk facilities in which works on pathogenic organisms are conducted (like vaccine production unit, microbiological laboratory etc) should take adequate precautions so that biological materials cannot be stolen or released into the environment inadvertently or intentionally. Building plans with key features like emergency exits, main switch panels, fire extinguishers, hydrants, emergency equipment storage areas, telephone and safe areas should be clearly displayed in public area. Names of contact persons and their telephone numbers should be displayed.
- Fire equipments, hydrants, open area, safety exit etc. in the farm/ animal building should be in working condition and checked at regular interval.
- Inventories of all immovable and movable items are to be listed and a back-up list to be maintained off-site. Important records (not in

use) should be documented and stored in safe place. A fool-proof storage and back-up system for computer networks need to be created.

- Veterinary institutions should review the library collection and categorized into rare, valuable and easily available or other relevant categories. Rare and valuable publications should be scanned and stored on microfilm or appropriate medium. A copy of such publications could be stored off-site in a secure location.

3.3 Response

It is the reaction to emergency. It includes actions taken to save lives and prevent further damage in an emergency situation. Response is executing the preparedness plans into action. Seeking shelter from a cyclone or leaving out of the buildings at the time of earthquake are response activities. In the short term, efforts should be made to mobilize the necessary basic resources (like feed and fodder, water, shelter, treatment and medication to the livestock). Emphasis should be on saving human lives, animal lives and then on rescuing and preserving material resources. In the event of disease outbreaks, information should pass on immediately to the agencies responsible for diagnosis of samples, surveillance and monitoring, vaccine production, agencies dealing with enforcement of quarantine. Medical assistance including vaccination of both humans and animals must be implemented. Efforts should be made to make an inventory of materials, damaged and destroyed and pass on this information to the appropriate agency for future use. The long term plan includes resettlement programmes, economic and legal needs (*e.g.* documentation for loan/grant, construction of sheds, shifting of animals etc.) People should be informed about the diseases and social behaviors (*e.g.* theft) associated with disasters and that may be exacerbated by deteriorated living conditions. Although rescue and relief measures are emotional issues and volunteers with different background and the army are the backbones of such measure in India, it must be recognized that certain tasks (like disposal of high risk dead bodies, mercy killing of animals, quarantine of animals suspected of disease etc.) should be performed by qualified professionals teams (which include veterinarians) who have received specialized training for disaster situations.

3.4 Recovery

It includes action taken to return to a normal or an even safer situation following an emergency. Getting financial and physical assistance from aid agencies or government for repair or reconstruction of shed is a recovery activity. Veterinarians or those specialized in epidemiology should observe the community and their animals closely for any signs of new or re-emerging diseases or problems. Replacement of damaged materials and livestock should be made after obtaining special funds for such emergencies. Veterinarians should also help the farmers with appropriate interventions so that they can again reach to the pre-disaster level. Interventions should be culturally and socio-economically similar to the pre-disaster situations.

3.4.1 *Aim at Swift Response to Disaster*

Swift response will not come only by capacity development and training. It will require capacity to anticipate problems and judicious decision making and people's participation in real time. One-way would be to revisit some of the well studied disasters to project future possible disaster scenario based on current state of knowledge about the area. A FUTURE EARTHQUAKE scenario projected for the Kawasaki City of Japan is a good example to emulate.

- "Kawasaki is a highly industrial coastal city in Japan with a population about 1.2 million. It is located south-west of Tokyo across the Tamagawa River, bordered on the south-west side by Yokohama city. The city covers a 30 km long stretch of land, 5 km wide, from the south-east to the north-west with its south-east edge forming a coastal area along Tokyo bay.
- It is assumed that on a winter weekday evening at 5 p.m. an earthquake of magnitude of 7.9 strikes Sagami bay.
- The Urgent Earthquake Detection and Alarm System (UrEDAS) of the Railway Technical Research Institute (RTRI) detects the arrival of P-wave from its own observation network and determines the magnitude and location of the earthquake instantly. Using the information it stops the bullet trains (shinkansen) which travel at around 250 km or above on the Tokyo, Osaka line.
- Simultaneously, RTRI's Hazard Estimation and Restoration Aid System (HERAS) go into operation to collect damage estimations relating to railway system. The HERAS system is capable of providing damage information on railways and associated facilities in about 5 min after an event.
- The above information will be used in the recovery and restoration operations to the railway systems.
- At the same time, the Seismic Information Gathering and Network Alert system (SIGNAL) of Tokyo Gas company goes in operation shutting off gas supply automatically in areas where ground shaking exceeded a prescribed threshold, in order to prevent fires from secondary damage such as gas leaks.
- On the government side, the damage assessment and support system of Kawasaki City, which became operational in 1994, is activated. The system gathers real time earthquake information form a dense

observation network, calculates ground motion and estimates probable damage by combining the ground motion estimates with the information from a Geographic Information System (GIS) consisting of soil, land cover, infrastructure, population distribution, housing conditions and lifeline information.

- The estimated damage due to this earthquake will be on the order of 1.3 trillion yen, with about 91% of the 33 thousand wooden buildings predicted to be destroyed by fire at the prevailing wind speed of 6 m/s from NNW direction. Fatalities will be up to 3120, road damages would be up to about 250 locations.
- There would also be widespread destruction to lifeline systems such as water supply and gas supply, which would require more than a month for complete recovery.

4.0 PEOPLE CENTRIC EARLY WARNING

Although it is not possible to predict earthquakes at the present level of scientific knowledge but early prediction of disasters like cyclones, storms, floods, and landslides is possible. National and state governments should create adequate facilities for installation of sophisticated instrumentations, facilities for real time data processing and timely early warnings. Continuous modernization of early warning systems and of the communication network should be undertaken at regular intervals. However timely early warnings will not always save lives if public response to the warnings is not swift and matured. Greater emphasis should be given to developing capacities that are relevant and responsive to the needs of the local communities. There are examples where commonsense approach succeeded where official warnings failed. For example seismological instruments were successful in recording the massive earthquake off the coast of Sumatra that triggered the recent Tsunami (26 December 2004) but there was no way to warn people of catastrophe in the making. The result was well over 2,850,00 people killed or missing. However, in the same tsunami, 3630 lives were saved in Nallavadu on the eastern coast of India by a simply phone call. According to World Disasters Report 2005 published by International Federation of Red Cross and Red Crescent Societies, the phone call based on radio news was made by Vijay Kumar Gunasekaran from Singapore to his family in Nallavadu. His family in turn alerted the villagers. The early-warning process can be split into three related phases.

4.1 Forecast

Scientific and technical observations and their analyses can most of the time indicate that a natural disaster is imminent. The aim of the forecast is to log the imminent event according to magnitude, localization, and duration.

4.2. Warning

Forecast has to be transformed into adequate warning and/or recommendations for further action, which then has to be communicated to the people quickly in appropriate form. The institutional and political involvement in the early-warning process is needed. Make people aware that their area/region may also suffer from natural disaster; nobody can claim a particular area/region is free from any type of natural disaster. Ensure that institutional mechanism and infrastructural facilities for early warning are in place and in operation. Prediction and warning about the disaster periodically are essential component of preparedness. Technical advances like satellite imagery can help to make early warning. During two cyclones of similar intensity that struck the coast of Andhra Pradesh in 1977 and 1990, the number of human deaths in 1977 was over 10,000 whereas it was less than 1000 in 1990. The reduction in death is due to the technological advances in forecasting and warning systems, effective administration in evacuating people, construction of shelters and public awareness.

4.3 Reaction

The reaction of the warned persons is crucial for successful early-warning. The risk perception and decision-making on the part of people are crucial. A well established response mechanism at the local level and high level of community awareness can avert disasters, if the warnings are timely.

5.0 DISASTER EDUCATION

In our country, the subject of natural disaster education traditionally fall in to the domain of agriculture scientists, earth scientists, seismologists, hydrologists and civil engineers, each discipline having their own priorities and is restricted to fulfill their own objectives. Their course curriculums are being drawn heavily upon the text books developed on traditional lines. Such subject specific education fails to provide the students the holistic view of natural disasters and devoid of the realization of local problems. Disaster education should be balanced and without undue bias to any particular discipline. A good foundation course needs to be first introduced before anything more serious or higher in education is attempted. Disaster education needs to be for all kinds of people, men or women, young or old, literate or illiterate and it must address their specific needs. For meeting the challenge, both formal and informal education are to be introduced.

5.1 Short Duration Training Programme (1 to 3 months)

Teaching through training programmes is a popular way of educating people. The persons who actually experience the natural disasters and of those who pass through the traumatic times because of natural disasters, can teach better besides providing very useful clues to reconstruct the whole disastrous event. Different types of short training programmes are:

- Science and Technology Orientation Programmes (STOP) for all Public
- Awareness and Training (PAT)
- Tailor-made Training for Trainers (TTT)
- Professional Education and Training (PEAT)

5.2 Higher Education and Training (Heat)

This would include Diploma, Graduate, Post Graduate, Doctoral and Post Doctoral Programmes in the area of disaster mitigation and management. However by mere opening of new institutions, or re-orienting existing institutions and courses we may not meet the challenge of disaster education. Since the scope of disaster education spreads even beyond the entire disaster cycle, design of education programmes should be done very carefully with extra-ordinary vision, imagination and effort. Distance learning, e-learning and learning-while-you-earn can also be used.

6.0 DISASTER MANAGEMENT PLAN FOR ANIMALS

Considering the economic, political, social and environmental importance of animals and life system of the small holder who depend on them in the developing countries, one may have to consider animal disaster management in that light. A disaster management plan for animals shall essentially include: -

6.1 Epidemiological Study in Disaster Prone Areas

It includes:

- Data collected, interpreted and analyzed on the basis of which some prediction can be made.
- *Herd profile*: the total animal population (herd number), vulnerable animal population as per their species, breed, age, sex etc.
- Community profile, total population, animal owning population, their age, sex, socio-economic status, education, cultural distribution etc.
- Animals at risk; nature of hazard, intensity of impact and mortality rate (immediate or delayed).

- *Risk factor analyses*: it is the analysis of type of risk (identification & analyses)

6.2 Action Plan

Since following disaster, animals are to be rescued and collected in relief camps, the immediate priority would be controlling and combating disease. The animal health component of disaster mitigation will include:

- Promotional herd health care such as nutrition, pregnant animal care, care of new-born and young animal etc.
- Prevention of risk through vaccination, vector control, sanitation etc.
- Specific therapy by way of early diagnosis and treatment.
- *Rehabilitation*: help animals to recover from any trauma or fear.
- *Disposal of dead animals*: Carcass utilization is one method. Many animals in which treatment is unlikely to be beneficial may have to be put to sleep *i.e.* euthanasia ("mercy killing").

6.3 Resources Planning

- Assessment of available man power *i.e.* veterinary doctors, para-veterinary staff, ancillary staff.
- Store and equipments include the medicine, surgical and medical appliances, diagnostics, life saving equipment etc.
- *Logistical needs*: that is the need for fuels, lighting equipment, tents, sheds, grass bedding, trolleys, material for sanitation, storage of feed and fodder and water.
- *Ambulance and out reach facility*: - transporting animals is more cumbersome when they are sick, injured and non-ambulatory.
- Veterinary medical facilities as veterinary hospital, mobile veterinary units etc.
- Grazing and watering facility.

6.4 Training Plan

- Training veterinary personnel, para-veterinarians, attendants etc.
- Training administrators like R.D.O./B.D.O, telephones and fire service personnel, civil defense personnel, sarpanch/village headmen, administrators.
- Animal health awareness for trainees such as social workers, volunteers.
- ICAR Institutes, KVKs, Regional Centres may organize training programmes on the above.

6.5 Allied Planning (Organization)

- Augmenting political and administrative support
- Involving N.G.O.'s, media, goshalas, animal welfare organisations and other volunteer groups.
- Eliciting commitment and allocation of funds
- Formation of veterinary service groups at state and central government level.
- Organisation of district area level bodies, assigning specific tasks and responsibilities.
- Establishing communication channels, alternate channels like ham radios.
- Establishing alternate source of power, energy etc.
- Plan for monitoring and supervision.
- Publicity and public relation activities (Vet-PR)
- Plan for mitigation and rehabilitation of animal owners along with animals

6.6 Anticipation of Sequence of Events

The sudden increase in deaths and causality would increase the demand for veterinary personnel. Early relief period would therefore be critical. Here it is important to understand animal behavior, in each disaster situation. Importance of animal resource information is equally important in rescue, relief and damage assessment. As has been discussed herd profile depicting the total animal population (herd number), vulnerable animal population as per their species, breed, age, sex etc. would immensely help mitigation including search, rescue, relief and rehabilitation. The nodal agency of the country in charge of disaster management (In India it is the Ministry of Agriculture) should specifically draw its attention to disaster management in animals. A separate cell has to be established for the purpose in the department dealing with veterinary service. Disaster prevention measures and preparedness are to be organized by this department with the cooperation of their counterparts. Animal population profile and distribution profile should be clearly drawn and plans for salvaging and rescuing the animals in major disasters like cyclones and flood to be undertaken on the basis of this data.

Agencies like goshalas, dairy and NGO's are to be organized in such a manner that their services can also be utilized at short notice for disaster management. Cyclone shelters are to be specifically devised for animals in cyclone prone areas of the country. Drought relief programmes like arrangement of scarcity fodder, drinking water and subsidized food be arranged as the stockpiles along with drugs and vaccines required to combat with allied disasters. Each state should organize a disaster management group in the department dealing with animal husbandry and veterinary service with specially trained staff, epidemiological data and communication facilities. The required field staff should be kept in constant readiness throughout the vulnerable months of the year. During lean period, the team should undertake preparedness and relief exercise to test their efficacy and preparedness. This will help develop a well oiled working system. There has to be a separate control cell attached to the Central Control Room dealing with National Animal Disaster Management. The epidemiological and statistical information collected before, during and after disaster should be compiled and discussed among the states of each zone on a scheduled timetable. Preventive actions taken or preparedness should be constantly monitored and lacunae thereof are reported and corrected for future use.

6.7 Provision of Compensation

Disaster being an emergency situation need be attended with the urgency it deserves. It is seen that accountability may have to be compromised to an extent to make room for faster service delivery. Relevant information/ data should be made use of in loss assessment, for efficient disposal of dead animals and animal wastes and should be used for estimating and monitoring of compensation; this process would therefore ensure transparency, impact and social justice.

7.0 SAFETY CODES FOR BUILDINGS AND CONSTRUCTIONS

Bureau of Indian Standards (BIS) has published Standards for all important types of disaster. These standards need to be followed for having safer buildings and constructions.

7.1 Eartbhquake

(1) IS 1893:1984 - Criteria for Earthquake Resistant Design of Structures (initially published in 1962 and revised in 1966, 1970, 1975 and 1984– divided into five parts to deal with different structure types

(2) IS 4326:1993 - Earthquake resistant design and construction of buildings

(3) IS 13827:1993 - Improving earthquake resistance of earthen buildings

(4) IS 13828:1993 - Improving earthquake resistance of low strength masonry buildings

(5) IS 13920:1993 - Ductile detailing of reinforced concrete standards subjected to seismic forces

(6) IS 13935:1993 - Repair and seismic strengthening of buildings

(7) IS4967:1968 - Recommendations for seismic Instrumentation for River Valley Projects.

7.2 Cyclones

(1) IS 15498:2004 - Guidelines for improving the cyclone resistance of low rise houses and other buildings/structures.

(2) IS 15499:2004 - Guidelines for survey of housing and building typology in cyclone prone areas for assessment of vulnerability of regions and post cyclone damage estimation.

7.3 Floods

(1) IS 13739:1993 - Guidelines for estimation of flood damages. It also mentions scientific procedure for collection of flood damages along with methods to quantify losses.

7.4 Landslides

(1) IS 14496 (Part2):1998 - Guidelines for preparation of landslide hazard maps in mountainous terrain,

(2) IS 14458 Guidelines for retaining walls for hilly areas (3 Parts),

(3) IS 14680:1999 - Guidelines for landslide control,

(4) IS 14804:2000 - Guidelines for design and selection of materials for residential buildings in hilly areas.

7.5 Fire Hazards

(1) National Building Code, Part 4 provides comprehensive recommendations for minimum standards of fire protection. It specifies

- The demarcation of fire zones.
- Restrictions on construction of buildings in each fire zone.
- Classification of buildings based on occupancy.
- Types of building construction according to fire resistance of the structural and non structural components.
- Other restrictions and requirements necessary to minimize danger to life from fire, smoke, fumes or panic before the building can be evacuated.

8.0 REFERENCES

Dipak, N. 2002. Coping with new age disaster: Need for veterinary institutional preparedness and special disaster management teams in India. National Seminar on Biological Threats and Disaster Management – Role of Veterinarians. 9–11 August, 2002. Punjab Agriculture University, Ludhiana.

Dirie, M.F. and Mohammed, H.A. 1999. Emergency management of disasters involving livestock in developing countries, *In*: *Management of Animal Health Emergencies. Rev. Sci. Tech. Off. Int. Epi.* **19(1)**: 256-271.

Bhandari R.K. http://www.vitcdmm.org/journel_on_disasters.pdf

VMAT Website: http://www.vmat.org/

NDMA Website HTTP://NDMA.GOV.IN/NDMA/INDEX.HTM

SUBJECT INDEX

A

B

C